NON-WOOD FOREST PRODUCTS

14

RATTAN
Current research issues and prospects for conservation and sustainable development

Edited by
**John Dransfield, Florentino O. Tesoro
and N. Manokaran**

Based on the outcome of the
FAO Expert Consultation
on Rattan Development held at FAO, Rome, Italy
5-7 December 2000

Jointly organized by the International Network on
Bamboo and Rattan (INBAR) and FAO
and co-funded by the Swedish International
Development Cooperation Agency (SIDA)

FOOD AND AGRICULTURE ORGANIZATION OF THE UNITED NATIONS
Rome, 2002

The designations employed and the presentation of material in this information product do not imply the expression of any opinion whatsoever on the part of the Food and Agriculture Organization of the United Nations concerning the legal status of any country, territory, city or area or of its authorities, or concerning the delimitation of its frontiers or boundaries.

ISBN 92-5-104691-3

All rights reserved. Reproduction and dissemination of material in this information product for educational or other non-commercial purposes are authorized without any prior written permission from the copyright holders provided the source is fully acknowledged. Reproduction of material in this information product for resale or other commercial purposes is prohibited without written permission of the copyright holders. Applications for such permission should be addressed to the Chief, Publishing and Multimedia Service, Information Division, FAO, Viale delle Terme di Caracalla, 00100 Rome, Italy or by e-mail to copyright@fao.org

© **FAO 2002**

FOREWORD

According to an estimate made by the International Network for Bamboo and Rattan (INBAR), the global local usage of rattan is worth US$ 2.5 billion and external trade of rattan is estimated to generate US$ 4 billion. Seven hundred million people worldwide use rattan. Most of the raw material for local processing and for supplying the rattan industry is still obtained by harvesting of unmanaged, wild rattan resources in natural tropical forests. Only a very small share is obtained from rattan plantations. The huge economic and social importance of the rattan sector is based on a fast dwindling stock of wild rattan resources available in the forests, particularly of tropical Asia, and therefore compromising its future outlook.

In light of the above, the Food and Agriculture Organization of the United Nations (FAO) and the International Network for Bamboo and Rattan, with support from the Swedish International Development Cooperation Agency (Sida), organized an Expert Consultation at FAO, Rome, from 5 to 7 December 2000. The Expert Consultation provided a unique forum where representatives of key stakeholders in the rattan sector and experts in the field made a critical assessment of the present situation of the rattan sector and set the stage for the elaboration of proposals at the global level towards a more sustainable development of the rattan sector. In addition, the Consultation fostered further collaboration among key agencies and has helped to identify short and medium-term activities that will enhance the sustainable development of the rattan sector.

These proceedings include the papers presented at the Consultation, as well as a number of extra papers prepared during the Consultation by the participants or submitted after, that were thought to be of sufficient interest to merit inclusion. The papers included in this publication are a synthesis of the current state of the knowledge on rattan resources and its utilization in general and related issues surrounding their exploitation, processing and trade and provide a unique overview of updated and checked information on the rattan sector throughout Asia and Africa and on its outlook.

Wulf Killmann
Director
Forest Products Division, FAO
Rome
Italy

Ian R. Hunter
Director General
International Network for Bamboo
and Rattan
Beijing
People's Republic of China

ACKNOWLEDGEMENTS

The Consultation and these proceedings would not have been possible without the hard work of a number of institutions, individuals and the participants themselves. Full acknowledgement should be given to the organizing team of the Forest Products Division of FAO's Forestry Department, without whose facilities and considerable logistical support the Consultation would not have run as smoothly as it did. Special mention must also be made to INBAR and Sida who facilitated the preparation and the Consultation in a highly professional manner. Finally, these proceedings would never have happened without the commitment, dedication, and many long hours of hard work from the authors of the contributions and the editors. Overall guidance and coordination for the preparation, review and publication of the final document was provided by Paul Vantomme. Thanks and appreciation to them, and to the others who participated in the process.

LIST OF ACRONYMS

AAC	annual allowable cut (Philippines)
APAFRI	Asia-Pacific Associations of Forestry Research Institutions
APKINDO	Indonesia Wood Panel Association
ASMINDO	Association of Furniture and Handicraft Industry (Indonesia)
ADB	Asian Development Bank
C&I	Criteria and Indicators
CARPE	Central African Regional Programme for the Environment
CBFM	Community-based Forest Management
CDC	Cameroon Development Corporation
CDC	Commonwealth Development Corporation
CENRO	Community Environment and Natural Resources Officer (Philippines)
CGIAR	Consultative Group on International Agricultural Research
CIFOR	Center for International Forestry Research
DENR	Philippine-German forest resources inventory project
DFID	Department for International Development (United Kingdom)
DWLC	Department of Wildlife Conservation (Sri Lanka)
EU	European Union
FAO	Food and Agriculture Organization of the United Nations
FDPM	Forestry Department Peninsular Malaysia
FELDA	Federal Land Development Authority
FORSPA	Forestry Research Support Programme for Asia and the Pacific
FPP	Forest Products and People programme
FPRDI	Forest Products Research and Development Institute (Philippines)
FRIM	Forest Research Institute Malaysia
GEF	Global Environment Facility
GTZ	German Agency for Technical Cooperation
HTI	Pulpwood Plantations (Indonesia)
ICSB	Innoprise Corporation Sendirian Berhad (Indonesia)
IDRC	International Development Research Center (Canada)
IFAD	International Fund for Agricultural Development
IMP	Malaysian Industrial Master Plan
INBAR	International Network for Bamboo and Rattan
IPAS	Integrated Protected Area System (Philippines)
IPGRI	International Plant Genetic Resources Institute
ITTA	International Tropical Timber Agreement
ITTC	International Tropical Timber Council
ITTO	International Tropical Timber Organization
IUCN	World Conservation Union
IUFRO	International Union of Forestry Research Organizations
JBIC	Japan Bank of International Cooperation
KFRI	Kerala Forest Research Institute (India)
LLNP	Lore Lindu National Park
MARA	Trustee Council for Indigenous Proples
MIFF	Malaysia International Furniture Fair
MITI	Ministry of Trade and Industry (Malaysia)
mlm	million linear meters
MNRD	Ministry of National Rural Development (Malaysia)
MPOB	Malaysian Palm Oil Board
MTIB	Malaysian Timber Industry Board
MYS	Ministry of Youth and Sports (Malaysia)
NARS	National Agricultural Research Systems

NBCA	National Biodiversity Conservation Area (Laos)
NDC	National Development Corporation (Philippines)
NFA	National Forestry Act (Malaysia)
NFI	National Forest Inventory (Malaysia)
NGO	Non-Governmental Organization
NRAP	National Resources Accounts Programme
NSO	National Statistics Office (Philippines)
NTFP	Non-timber Forest Product
NWFP	Non-wood Forest Product
ODA	Overseas Development Administration (United Kingdom)
PCS	production-to-consumption systems
PFE	Permanent Forest Estate
PICOP	Paper Industries Corporation of the Philippines
PPE	personal protective equipment (Malaysia)
PRF	Permanent Reserve Forests (Malaysia)
R&D	Research and Development
RAPD	random amplified polymorphic DNA
RFD	Royal Forest Development (Thailand)
RIC	Rattan Information Centre (Malaysia)
RISDA	Rubber Industry Smallholders Development Authority
RITF	Research Institute of Tropical Forestry (China)
RLI	relative light intensity
SAFODA	Sabah Forest Development Authority
SFM	Sustainable Forest Management
Sida	Swedish International Development Cooperation Agency
SIRIM	Standards and Industrial Research Institute of Malaysia
SPA	Seed production area
SSEDV	Small Scale Entrepreneurs Development Unit
TOTEM	Transfer of Technology Model (INBAR)
UNDP	United Nations Development Programme
UNIDO	United Nations Industrial Development Organization
YNTBI	Yunnan Tropical Botanical Institute (China)

CONTENTS

INTRODUCTION AND SUMMARY
OF THE RESULTS
OF THE EXPERT CONSULTATION

Introduction

An Expert Consultation on Rattan Development was jointly organized by the Food and Agriculture Organization of the United Nations (FAO) and the International Network for Bamboo and Rattan (INBAR), and co-funded by the Swedish International Development Cooperation Agency (Sida). The meeting was held at FAO Headquarters in Rome, Italy, from 5 to 7 December 2000, and attended by 23 experts from 16 countries, selected on the basis of their specialized knowledge and their role in the management and development of rattan resources in their respective countries; in addition to representatives of the International Fund for Agricultural Development (IFAD), the International Tropical Timber Organization (ITTO), the International Plant Genetic Resources Institute (IPGRI); the Centre for International Forestry Research (CIFOR), INBAR; Sida, Kew Gardens, the German Agency for Technical Cooperation (GTZ), Tropenbos, the private sector, and universities. A number of colleagues from different units within FAO (Forestry, Agriculture and Technical Cooperation Departments) also attended the meeting.

The focus of the meeting was on the sustainable development of the rattan sector worldwide, but with particular emphasis on Asia and Africa. Some attention was given to Latin America in view of its potential for rattan introduction.

The objectives of the Expert Consultation were to review and analyse:

(a) essential baseline information on the rattan sector in producing countries, the critical global supply situation and key requirements to guarantee a sustainable future supply of rattan;
(b) the needs and methods for better cooperation and coordination among key agencies and stakeholders in relation to their ongoing activities on rattan development; and
(c) the desirability of developing an international programme aimed at promoting and undertaking rattan development activities with partner institutions in the various regions and strengthening global networking in rattan research and development.

Key observations

Based on the papers presented and their discussion, the Expert Consultation emphasized the economic, socio-cultural and ecological importance of rattan to a large number of people in the world and noted that rattan resources in their natural range of tropical forests in Asia and Africa were being depleted through overexploitation, inadequate replenishment, poor forest management and loss of forest habitats. There was a need to ensure a sustainable supply of rattan through improved and equitable management.

The meeting recalled that:

- There were approximately 600 species of rattans, of which some 10 percent were commercial species. Many species, including some of commercial importance, had very restricted natural ranges. The majority of the world rattan resources (by volumes and by number of species) were in one country – Indonesia.
- Rattan was an important commodity in international trade. At the local level, it was of critical relevance for rural livelihood strategies as a primary, supplementary and/or emergency source of income. Rattan collection complemented agriculture in terms of seasonal labour and as a source of capital for agricultural inputs.
- The rattan sector was characterized by a variety of stakeholders with different needs and interests, such as rattan growers, raw material collectors, manufacturers and traders, and it functioned within a complex and dynamic socio-economic, political and ecological context. Rattan was gathered by unorganized or organized collectors, the latter either under contract or in debt relationships with traders and farmers/cultivators. In addition, there was a loss of traditional rattan management practices, while at the same time increasing competition for resources. Linkages between industry, traders, collectors/cultivators and research and development efforts were weak. Rattan manufacturing and trade were fragmented and diverse in size and markets, with a focus on export.

The meeting highlighted that taxonomic knowledge on species was patchy and available information conflicting. Likewise patchy was the knowledge of biological aspects, e.g. pollination and gene flow. In spite of the Red List of the World Conservation Union (IUCN) review of 1998, the conservation status of rattans was not well known and it was difficult to assess and monitor. In addition, rattan species were assumed not to be "safe" in protected areas or in national parks, as harvesting in such areas was usually permitted or tolerated. It was also assumed that the genetic basis of rattan species was narrowing. Some species were likely to be at risk of extinction.

The meeting underlined that there could be no sustainable supply of rattan, if the forests in which they grew were not managed sustainably. In its natural habitat, rattan was not as yet managed, and rattan received low priority in national forest and conservation policies. There was no dedicated rattan development institution in any country as rattan was usually subsumed within the forestry services, and the few existing national rattan programmes were weak and with limited research and development capacity. With a few exceptions, national forest inventories did not include rattan, and information on the resource base was scarce. However, in large tracts of degraded and logged-over forests, (re-)introduction and management of rattan had the potential to complement significantly the value of these forests.

The meeting was informed that significant advances had been made in the understanding of growing rattan as a plantation crop. Although community-based or smallholder rattan gardens could be profitable in some situations, the profitability of industrial-scale plantations in Asia was currently uncertain, as other land uses were more lucrative. As a result of this, private-sector investment in industrial-scale rattan plantations had declined. The meeting took note that existing rattan plantations had been converted into more profitable crops like oil palm.

The meeting was further informed that rattan production was also affected by the low return to gatherers, resulting in weak incentives for sustainable rattan harvesting and management. A number of factors contributed to the low returns. Foremost among these were uncertain property rights, the dispersed nature of production and inconsistent cane quality. In several countries, prices and competition had been affected by the remoteness of collecting areas and poor transportation; "illegal" harvesting; poor market information; lack of organization among collectors; large post-harvest losses due to insect and fungal infestation; prohibitive tax policies; export barriers; and informal taxes that depressed raw material prices.

The meeting noted that international agencies such as INBAR, CIFOR, IPGRI, FAO and ITTO addressed rattan management, either directly or indirectly, within their programmes. National focal points for member countries of INBAR on rattan information had been established with the primary function to identify key stakeholders and their increasing involvement, to collect statistical data and exchange information in the local languages.

Conclusions

In the light of the above, the meeting concluded that there was a wide variety of potential interventions that could assist the different stakeholder groups. Raw material producers and smallholders could be encouraged to, and assisted in, managing local resources on a more sustainable and productive basis, through the establishment of community forest management practices, long-term concessions, local land-use planning and the provision of resource and/or land tenure rights, in conjunction with approved management plans.

At the processing level, needs were particularly great at the artisanal level. Potential interventions that might assist industry include improving entrepreneurship and competitiveness; training of advisers; improving post-harvest treatment and quality control; market deregulation and improved market information; establishment of design centres; and trade fairs. Also, given the nature of the resource users and the industry being generally cottage and small scale, employing socially disadvantaged groups, rattan products could become ideal commodities for promotion as rainforest conservation products.

The meeting identified the following key actions to be initiated immediately for enhancing a more sustainable supply of rattan:

- Resources:
  - intensifying *ex-situ* and *in-situ* conservation efforts in a more coordinated and organized manner among countries in the regions;
  - developing suitable methods for resource assessments, including studies on growth, yield, basic biology and taxonomy of rattan species;
  - improving techniques of enrichment planting and management of rattan in degraded forests, and a wide dissemination of the available guidelines for rattan planting.

- Products:
  - increasing the knowledge of the properties of commercial species and of the potential of underutilized/lesser known species;
  - improving and disseminating technologies for reducing post-harvesting losses, biological deterioration, improved storage and processing;
  - introduction of quality grading.

- Policies and institutional support:
  - awareness raising on the importance of the rattan sector to decision-makers at all levels;
  - institutional strengthening and coordination regarding rattan conservation, management and processing issues, including the promotion of more government and private sector cooperation/coordination to enhance the contribution of rattan for poverty alleviation and economic prosperity;
  - providing tenure security to rattan gatherers and planters by incorporating them into community-based forest management schemes;
  - introducing incentive schemes for rattan cultivation to increase the economic benefits for rural households and smallholder plantations in Asia, such as providing credit and technical assistance for small-scale plantation development and favourable harvesting and marketing arrangements;
  - introducing market deregulation to benefit rattan collectors and traders (i.e. removing transport barriers; support for improved collection and dissemination of market information; extension in processing techniques);
  - providing comprehensive training and support to local specialists in rattan-producing countries in taxonomy, management and processing, complemented with "twinning arrangements" among relevant institutions in the regions.

Recommendations

The Expert Consultation recommended for immediate follow-up

- to FAO to:
  - develop consistent terms and definitions on rattan and its products;
  - harmonize existing measurement concepts and methodologies for rattan resources inventories and for accurate collection of statistics on rattan products.

- to INBAR to:
  - establish a list server on rattan;
  - conduct, with the assistance of CIFOR a study on the economics of large- and small-scale rattan plantations;
  - conduct, with the assistance of FAO, a study on improving forest policies and relevant regulations with regard to rattan;
  - commission a study on potential alternative market mechanisms to provide greater transparency and competition in rattan trade (e.g. auction mechanisms);
  - update and publish the existing rattan bibliography on its web site.

The Expert Consultation recommended that the <u>governments</u> of countries with rattan resources be encouraged:

- At the national level to:
  - develop and implement a national rattan strategy involving all stakeholders in a participatory process;
  - include rattan as an integral component of national forest and conservation policies, as well as forest management plans and, where appropriate, by giving due attention to rattan in the national and relevant regional processes on Criteria & Indicators for Sustainable Forest Management;
  - establish specific pilot projects focused on critical issues such as property rights and management institutions, opportunities and constraints to community-based resource management, and post-harvest treatment;
  - strengthen national research programmes/activities through enhancing the network of rattan research and development activities, including the establishment of "rattan scholarships".

- In support to actions at the international level to:
  - commission the development of a five-year international rattan development programme with the primary objective of promoting and undertaking rattan development activities with partner institutions in the various regions, and in order to strengthen global networking in rattan research and development. This international programme would enhance national institutional capabilities and examine the possibilities and merits of INBAR establishing/strengthening nodal point(s) in national institutions as permanent focal point(s) to continue long-term programmes on rattan research and development;
  - revive the Rattan Information Centre (RIC) established in 1982 in Malaysia;
  - support awareness-raising campaigns on conservation, management and processing of rattan (e.g. impacts of insufficient taxonomic/biological knowledge of rattan conservation issues) with senior policy and decision makers at international development, conservation, research and funding agencies, as well as senior government officials in rattan producing/consuming countries.

The Expert Consultation emphasized the potential of enhancing regional cooperation through information exchange; collaborative research and development; training; and material exchange to promote rattan as a vehicle for achieving social, economic and environmental sustainability in rattan-producing countries. To this end, the expert consultation called for a concerted effort of governments, the private sector, non-governmental organizations (NGOs) and relevant international agencies.

THE RESOURCE, ITS USES AND PRESENT ACTION PROGRAMMES

SPECIES PROFILES RATTANS
(*Palmae: Calamoideae*)

Terry C.H. Sunderland and John Dransfield

1. Introduction

Rattans are climbing palms exploited for their flexible stems that form the basis of a significant market for cane and cane products. The thriving international and domestic trade in rattan and rattan products has led to substantial over-exploitation of the wild rattan resource. This exploitation, coupled with the loss of forest cover through logging and subsequent agricultural activities, is threatening the long-term survival of the rattan industry, particularly in Southeast Asia (Dransfield, 1988). The detrimental impact of the decline of wild rattan resources on local rattan collectors who harvest the majority of the traded cane, is often over-shadowed by the more publicised concerns of the rattan industry itself. In many areas, the sustainable exploitation of the rattan resource is hindered by the lack of a sound taxonomic base in order that meaningful inventories and studies of population dynamics can be carried out. In addition, lack of resource tenure also precludes any attempts at long-term and sustainable harvesting; the fact that rattan is considered an "open-access" resource throughout much of its range hinders any effective attempts at long-term management of rattan in the wild.

2. What are rattans?

The word rattan is derived from the Malay "rotan", the local name for climbing palms. Rattans are spiny palms found in the old world tropics and subtropics and are the source of cane for the well-developed rattan industry, currently worth some US$6.5 billion per annum (ITTO, 1997). Most of the cane entering the world trade originates from Southeast Asia and is collected, in the main, from wild populations, although considerable efforts have recently been focused on the future provision of raw cane from cultivated sources (Dransfield and Manokaran, 1993).

As well as forming the basis of a thriving export market, rattans contribute significantly to the subsistence economies of forest-based communities throughout their range. Many of these people utilise the cane resource as a means of direct cash income, especially during periods when other products are seasonally unavailable. Rattan harvesting is often a secondary activity for many farmers, who rely on the cash sale of rattan canes to invest in developing their agricultural base or during times when immediate cash is needed for household support, such as the payment of school fees or medical expenses.

3. Rattan Taxonomy and Distribution

Rattans are climbing palms belonging to the Palm family (*Palmae or Arecaceae*). There are around 600 different species of rattan belonging to 13 genera and these are concentrated solely in the old world tropics; there are no true rattans in the new world[1]. Rattans belong to the *Calamoideae*, a large sub-family of palms. All of the species within the *Calamoideae* are characterized by overlapping reflexed scales on the fruit and all of these climbing palms are spiny, a necessary pre-adaptation to the climbing habit (Dransfield, 1992b).

---

[1] In the new world two other groups of palms have climbing representatives and are often mistaken as being closely-related to true rattans. *Chamaedorea* (sub-family Ceroxyloxideae; tribe Hyophorbeae) and *Desmoncus* (sub-family Arecoideae; tribe Cocoeae) climb through the means of reflexed terminal leaflets. Indeed, *Desmoncus* is often exploited for its cane-like qualities and is used in the same way as the true rattans (Henderson and Chávez, 1993). A climbing palm (*Dypsis scandens*) (subfamily Arecoideae; tribe Areceae) has also been discovered in Madagascar (Dransfield & Beentje 1995); it too is not related to the true rattans.

Of the 13 genera of rattan, three are endemic to Africa: Laccosperma (syn. *Ancistrophyllum*), *Eremospatha* and *Oncocalamus*. Although some species within these genera are utilised locally and form the base of a thriving cottage industry, they have not, until recently, attracted much attention from commercial concerns (Dransfield, 1992b; Sunderland, 1999).

The largest rattan genus is *Calamus*, with ca. 370 species; it is represented in Africa by one, very variable species, *C. deërratus*. *Calamus* is predominantly an Asian genus and ranges from the Indian subcontinent and south China southwards and east through the Malaysian region to Fiji, Vanuatu and tropical and sub-tropical parts of eastern Australia. Most of the best commercial species of rattan are members of this genus. The remaining rattan genera, *Daemonorops, Ceratolobus, Korthalsia, Plectocomia, Plectocomiopsis, Myrialepis, Calospatha, Pogonotium* and *Retispatha*, are centred in Southeast Asia and have outliers further eastwards and northwards (Uhl and Dransfield, 1987; Dransfield, 1992a).

Table 1. The rattan genera: number of species and their distribution
(Modified from Uhl and Dransfield, 1987)

Genus	Number of species	Distribution
Calamus L.	ca. 370-400	Tropical Africa, India and Sri Lanka, China, south and east to Fiji, Vanuatu and eastern Australia
Calospatha Becc.	1	Endemic to Peninsular Malaysia
Ceratolobus Bl.	6	Malay Peninsula, Sumatra, Borneo, Java
Daemonorops Bl.	ca. 115	India and China to westernmost New Guinea
Eremospatha (Mann & Wendl.) Wendl.	10	Humid tropical Africa
Korthalsia Bl.	ca. 26	Indo-China and Burma to New Guinea
Laccosperma *(Mann & Wendl.) Drude*	5	Humid tropical Africa
Myrialepis Becc.	1	Indo-China, Thailand, Burma, Peninsular Malaysia and Sumatra
Oncocalamus (Wendl.) Wendl.	4	Humid tropical Africa
Plectocomia Mart.	ca. 16	Himalayas and south China to western Malaysia
Plectocomiopsis Becc.	ca. 5	Laos, Thailand, Peninsular Malaysia, Borneo, Sumatra
Pogonotium J. Dransf.	3	Two species endemic to Borneo, one species in both Peninsular Malaysia and Borneo
Retispatha J. Dransf.	1	Endemic to Borneo

Despite the commercial importance of rattan, basic knowledge of the resource is somewhat limited and the rattan flora of Africa and much of Southeast Asia and Malaysia remains poorly known. Taxonomic work of this sort is not purely an academic exercise; it is an essential basis for the development of the rattan resource and under-pins the conservation and sustainable development objectives much advocated for rattans. It is essential that species delimitation be clearly understood; we need to know which species are of commercial importance and how they may be distinguished from other species. This knowledge is essential in order to undertake meaningful inventories of commercially important taxa and to be able to assess the silvicultural potential of each species, based on sound ecological knowledge. Reference to a structured systematic framework also ensures that any experimental work undertaken is replicable.

Table 2. Available rattan floras by region to date

Region	Reference
Peninsular Malaysia	Dransfield, 1979
Sabah	Dransfield, 1984
Sarawak	Dransfield, 1992a
Brunei	Dransfield, 1998
Sri Lanka	de Zoysa & Vivekanandan, 1994
India (general)	Basu, 1992
India (Western Ghats)	Renuka, 1992
India (south)	Lakshmana, 1993
Andaman and Nicobar Islands	Renuka, 1995
Bangladesh	Alam, 1990
Papua New Guinea	Johns & Taurereko, 1989a, 1989b (preliminary notes only)
Irian Jaya	Currently under study at Kew (Baker & Dransfield)
Indonesia	Dransfield and Mogea [in prep.]; more field work needed
Laos	Currently in prep. (Evans)
Thailand	Hodel, 1998
Africa	Currently in prep. (Sunderland)

4. The Uses of Rattan

The most important product of rattan palms is cane; this is the rattan stem stripped of its leaf sheaths. This stem is solid, strong and uniform, yet is highly flexible. The canes are used either in whole or round form, especially for furniture frames, or split, peeled or cored for matting and basketry.

Other plant parts of some species of rattan are also utilised and contribute to the indigenous survival strategies of many forest-based communities. A summary of these uses is listed in Table 3 below.

However, it is for their cane that rattans are most utilised and rattan canes are used extensively across their range by local communities and play an important role in subsistence strategies for many rural populations (Burkill, 1935; Corner, 1966; Dransfield, 1992d; Sunderland, 1998). The range of indigenous uses of rattan canes is vast; from bridges to baskets; from fish traps to furniture; from crossbow strings to yam ties. Despite these many uses, there is a common misconception that all rattans are useful, and therefore all have potential commercial applications. However, whilst there may indeed be substantial spontaneous use for many species (Dransfield and Manokaran, 1993; Johnson, 1997; Sunderland, 1998) it is estimated that only 20% of the known rattan species are of any commercial value (Dransfield and Manokaran, 1993) with the remaining species not being utilised due to inflexibility and being prone to breakage or possessing other poor mechanical properties, or due to biological rarity.

Table 3. Some traditional uses of rattans, excluding cane

Product / Use	Species
Fruit eaten	*Calamus conirostris;; C. longisetus; C. manillensis; C. merrillii; C. ornatus; C. paspalanthus;; C. subinermis; C. viminalis; Calospatha scortechinii; Daemonoropsingens; D. periacantha; D. ruptilis*
Palm heart eaten	*Calamus deerratus; C. egregius; C. javensis; C. muricatus; C. paspalanthus;; C. siamensis; C. simplicifolius; C. subinermis; C. tenuis; C. viminalis; Daemonorops fissa; D. longispatha; D. margaritae; D. melanochaetes; D. periacantha; D. scapigera; D. schmidtiana; D. sparsiflora; Laccosperma secundiflorum; Plectocompiopsis geminiflora*
Fruit used in traditional medicine	*Calamus castaneus; C. longispathus; Daemonorops didymophylla*
Palm heart in traditional medicine	*Calamus exilis; C. javensis; C. ornatus; Daemonorops grandis; Korthalsia rigida*
Fruit as source of red resin exuded between scales, used medicinally and as a dye (one source of "dragon's blood")	*Daemonorops didymophylla; D. draco; D. maculata; D. micrantha; D. propinqua; D. rubra*
Leaves for thatching	*Calamus andamanicus; C. castaneus;; C. longisetus; Daemonorops calicarpa; D. elongata; D. grandis; D. ingens; D. manii*
Leaflet as cigarette paper	*Calamus longispathus; Daemonorops leptopus*
Leaves chewed as vermifuge	*Laccosperma secundiflorum*
Roots used as treatment for syphilis	*Eremospatha macrocarpa*
Leaf sheath used as toothbrush	*Eremospatha wendlandiana; Oncocalamus sp.*
Leaf sheath/petiole as grater	*Calamus sp (undescribed, from Bali);*
Rachis for fishing pole	*Daemonorops grandis; Laccosperma secundiflorum*

Table 4. The major commercial species of rattan as identified for Asia by Dransfield and Manokaran (1993) and for Africa, by Tuley (1995) and Sunderland (1999)

Species	Distribution	Conservation status
Calamus caesius *Bl.*	Peninsular Malaysia, Sumatra, Borneo, Philippines and Thailand. Also introduced to China and south Pacific for planting	Unknown
Calamus egregius Burr.	Endemic to Hainan island, China, but introduced to southern China for cultivation	Unknown
Calamus exilis Griffith	Peninsular Malaysia and Sumatra	Not threatened
Calamus javensis Bl.	Widespread in Southeast Asia	Not threatened
Calamus manan Miq.	Peninsular Malaysia and Sumatra	Threatened
Calamus merrillii Becc.	Philippines	Threatened
Calamus mindorensis Becc.	Philippines	Unknown
Calamus optimus Becc.	Borneo and Sumatra. Cultivated in Kalimantan	Unknown
Calamus ornatus Bl.	Thailand, Sumatra, Java, Borneo, Sulawesi, to the Philippines	Unknown
Calamus ovoideus Thwaites ex Trimen	Western Sri Lanka	Threatened
Calamus palustris Griffith	Burma, southern China, to Malaysia and the Andaman Islands	Unknown
Calamus pogonacanthus Becc. ex Winkler	Borneo	Unknown
Calamus scipionum	Burma, Thailand, Peninsular Malaysia, Sumatra,	Unknown

Loureiro	Borneo to Palawan	
Calamus simplicifolius Wei	Endemic to Hainan island, China, but introduced to southern China for cultivation	Unknown
Calamus subinermis (eddl. ex Becc.	Sabah, Sarawak, East Kalimantan and Palawan	Unknown
Calamus tetradactylus Hance	Southern China. Introduced to Malaysia	Unknown
Calamus trachycoleus Becc.	South and Central Kalimantan. Introduced into Malaysia for cultivation	Not threatened
Calamus tumidus Furtado	Peninsular Malaysia and Sumatra	Unknown
Calamus wailong Pei & Chen	Southern China	Unknown
Calamus zollingeri Becc.	Sulawesi and the Moluccas	Unknown
Daemonorops jenkinsiana (Hance) Becc.	Southern China	Unknown
Daemonorops robusta Warb.	Indonesia, Sulawesi and the Moluccas	Unknown
Daemonorops sabut Becc.	Peninsula Malaysia and Borneo	Unknown
Eremospatha macrocarpa (Mann & Wendl.) Mann & Wendl.	Tropical Africa from Sierra Leone to Angola	Not threatened
Eremospatha haullevilleana de Wild.	Congo Basin to East Africa	
Laccosperma robustum (Burr.) J. Dransf.	Cameroon to Congo Basin	
Laccosperma secundiflorum (P. Beauv.) Mann & Wendl.	Tropical Africa from Sierra Leone to Angola	Not threatened

5. A Word about Local Classification

The development of extensive indigenous classification systems for rattans often reflects the social significance of rattan, and these taxonomies have developed to reflect rattan as it grows in the forest, as well as how it is used. For example, a widespread species may be referred to by many names, as its range encompasses a number of dialect groups. Often, one species can be given many names reflecting the different uses of the plant or the various stages of development (from juvenile to adult with very distinct morphological differences between the two). Commonly, blanket names for "cane" are given to a wide range of species.

Some species that have no use are often classified according to their "relationship" to those that are utilised. These are often along kinship lines and species may be referred to as "*uncle of...*" or "*small brother of...*" reflecting their perceived relationship and similarity to species that are widely utilised. In the past, serious confusion has arisen from the uncritical use of vernacular names and has contributed to the misconception that all species are of commercial potential. Local names should be used in conjunction with classical Linnean taxonomic methods and not on a mutually exclusive basis.

6. The Commercial Rattan Trade

6.1 Introduction

The international trade in rattan dates from the mid-19th century (Corner, 1996) and this trade is now currently worth some US$6.5 billion a year (ITTO, 1997). A conservative estimate of the domestic markets of Southeast Asia by Manokaran (1990) suggested a net worth of US$2.5 billion. This latter market includes the value of goods in urban markets and rural trade, as well as the value of the rural usage of the material and products. Dransfield and Manokaran (1993) estimate that 0.7 billion of the world's population use, or are involved in the trade of rattan and rattan products.

6.2 Southeast Asia

The majority of the international rattan trade is dominated by countries of Southeast Asia. By the early 20th century, Singapore, despite having a very small rattan resource itself, became the clearing-house for nearly the whole of Southeast Asia and the western Pacific. Between 1922 and 1927, up to 27,500 tonnes of cane and cane products were exported mainly to Hong Kong, the United States and France from there. During this same period, exports from Kalimantan and Sulawesi increased dramatically: from 9,400 to 19,300 tonnes and from 10,300 to 21,800 tonnes respectively, with much of this also being re-exported through Singapore (Dransfield and Manokaran, 1993).

By the 1970s, Indonesia had become the supplier of about 90% of the world's cane, with the majority of this going to Singapore for processing and conversion (from which Singapore earned more than US$21 million per annum). Thus, in 1977, Hong Kong imported some US$26 million of rattan and rattan products which, after conversion, was worth over US$68 million in export value. By comparison, Indonesia's share of the trade, mainly of unprocessed canes, was a mere US$15 million (*ibid.*).

In the last twenty years, the international trade in rattan and rattan products has undergone rapid expansion. The increases in the value of exports from the major producing countries are indeed staggering 250-fold for Indonesia over 17 years; 75-fold in the Philippines over 15 years; 23-fold over 9 years in Thailand and 12-fold over 8 years in Malaysia (Manokaran, 1990). By the late 1980s, the combined value of exports of these four countries alone had risen to an annual figure of almost US$400 million, with Indonesia accounting for 50% of this trade. The net revenues derived from the sale of rattan goods by Taiwan and Hong Kong, where raw and partially finished products were imported and then processed, together totalled around US$200 million per annum by the late 1980s.

During this same period, Thailand, the Philippines, Indonesia and Malaysia banned the export of rattan, except as finished products. These bans were imposed to stimulate the development of rattan-based industries in each country and ensuring that the value of the raw product was increased, and (theoretically) to protect the wild resource. Recently, however, given the recession that has hit many countries in Southeast Asia, Indonesia has lifted the ban on the export of raw cane and is currently flooding the market with relatively cheap supplies of cane, negatively impacting the cultivation industry of Malaysia in particular (Sunderland and Nkefor, 1999a).

6.3 Africa

The restrictions in the trade of raw cane by some of the larger supply countries outlined above has encouraged some rattan dealers and gross users to investigate non-traditional sources of rattans, predominantly Indo-China, Papua New Guinea and more recently, Africa. Some raw cane has recently been exported from Ghana and Nigeria to Southeast Asia and there is a flourishing export trade of finished rattan products from Nigeria to Korea (Morakinyo, *pers. comm.*). In addition, trade within and between countries is reported to be growing significantly across West and Central Africa (Falconer, 1994; Morakinyo, 1995; Sunderland, 1999).

Historically, there has been a significant trade in African rattans. Cameroon and Gabon supplied France and its colonies (Hedin, 1929), and Ghana (formerly the Gold Coast) supplied a significant proportion of the large United Kingdom market during the inter-war period (Anon, 1934). The export industry was not restricted to raw cane and in 1928 alone over 25,000 FF worth of finished cane furniture was exported from Cameroon to Senegal for the expatriate community there (Hedin, 1929). More recently, an initiative promoted by the United Nations Industrial Development Organization (UNIDO) in Senegal was exploiting wild cane for a large-scale production and export (Douglas, 1974), although this enterprise folded not long after its establishment due to problems securing a regular supply of raw material.

7. Rattan Ecology and Natural History

7.1 Introduction

The large number of rattan species and their wide geographic range is matched by great ecological diversity. The majority of admittedly crude ecological preferences for rattan species have been generally made during taxonomic inventory work, yet these broad ecological summaries are invaluable as a basis for establishing cultivation procedures. A major gap in the knowledge of rattans, even of the commercial species, is an understanding of population dynamics and demography. The knowledge of the population structure, distribution, rate of regeneration, the number of harvestable stems per hectare etc. of each species is essential and forms the basis of an understanding of potential sustainability.

7.2 Forest Type and Light Requirements

Throughout their natural range, rattan species are found in a wide range of forest and soil types. Some species are common components of the forest understorey, whilst some rely on good light penetration for their development; hence several species are found in gap vegetation and respond very well to canopy manipulation, particularly that caused by selective logging. Other species grow in swamps and seasonally inundated forest whilst others are more common on dry ridge tops.

Despite this wide range of ecological conditions, the majority of rattans need adequate light for their development. Cultivation trials on many of the Southeast Asia species, as well as recent germination trials of the African taxa, have indicated that seeds will germinate under a wide range of light conditions. The resultant seedlings will remain for long periods on the forest floor awaiting sufficient light for them to develop, such as a tree fall. This seedling bank is a common feature of the regeneration of most species and is a well-recognised component of forests where rattans occur.

7.3 Life Form

Rattans can be clustering (clump-forming) or solitary; some species, such as *Calamus subinermis*, can be both. Other species are acaulescent, having no discernible stem at all. Clustering species sometimes possess up to 50 stems of varying ages in each clump and produce suckers that continually replace those stems lost through natural senescence, or through harvesting. Some clumps can be harvested many times on a defined cycle if the light conditions are conducive to the remaining suckers being able to develop and elongate. Ensuring that stem removal through harvesting does not exceed that of stem replacement is the crux of rattan sustainability.

An even more crucial component of sustainability is the monitoring of the exploitation of solitary species. *Calamus manan*, one of the most commonly exploited rattan species, is single-stemmed; thus the impacts of harvesting of this species are much greater than harvesting from clustering rattans. Sustainability of such species relies on recruitment through sexual means, rather than through vegetative means.

7.4 Flowering

Another ecological feature of palms that is extremely important in terms of management is that rattans display two main modes of flowering: hapaxanthy and pleonanthy. In hapaxanthy, a period of vegetative growth is followed by the simultaneous production of inflorescence units from the axils of the uppermost leaves. Although often described as a terminal inflorescence, which it undoubtedly resembles, morphologically this is not the case. Flowering and fruiting is followed by the death of the stem itself. In single-stemmed palms species, the whole organism dies after the reproductive event. However, in clustering species of rattan the organism continues to regenerate from the base and it is only the individual stem that dies. In pleonanthic species, axillary inflorescences are produced continually and flowering and fruiting does not result in the death of the stem. All the species of *Korthalsia, Laccosperma, Plectocomia, Plectocomiopsis* and *Myrialepis* and a few species of *Daemonorops* are hapaxanthic. All other rattan species are pleonanthic. In terms of silviculture, the mode of flowering will affect the cutting regime and stem selection for harvest, particularly if the cultivated resource is to supply seed for further trials. Furthermore, in many hapaxanthic species, stems tend to be of low quality due to the presence of a soft pith which results in poor bending properties. These stems are also more prone to subsequent insect attack due to increased starch deposition.

7.5 Fruit and Seed

Rattan fruits are often brightly coloured (white, yellow, orange or red) and the sarcotesta is also attractive to birds and mammals. Birds (e.g. hornbills) and primates are the main dispersers of rattan seeds in both Southeast Asia and Africa with primates and elephants also sharing a preference for the ripe fruit. Fruits are often ingested whole, where they pass through the intestinal tract of the agent concerned, or are sucked and spat out, with the seed intact.

In the Asian taxa, the seed is often covered with a sarcotesta (a fleshy seed coat). Incomplete removal of this sarcotesta often results in delayed germination suggesting that it contains some chemical germination inhibitors. However, once this outer layer is fully removed, the germination of commercial species such as *Calamus manan* and *C. caesius*, is both rapid and uniform. In contrast, propagation trials on the commercially-important species in Africa have noted a physically-induced dormancy resulting from the presence of a relatively robust seed coat surrounding the endosperm and experiments have shown that, because of the barrier to imbibition, the germination times of some African rattan species can be rather prolonged and it may take between 9-12 months before germination commences. This physical dormancy has provided some difficulties in the cultivation of some species and numerous trials have been undertaken to reduce these germination times (Sunderland and Nkefor, 1999b). Physical seed treatments, such as scarification or chitting, somewhat effective means of reducing the germination times for some of the species, however, soaking the seeds in water for at least 24 hours prior to sowing is probably the most effective means of inducing early germination (*ibid.*).

7.6 Rattan Relationships

Several species of rattan (*Laccosperma, Eremospatha, Korthalsia, Calamus* and *Daemonorops*) have developed morphological adaptations that provide nesting sites for ants. These adaptations include the hollowing out of acanthophylls, interlocking spines that form galleries or recurved proximal leaflets that tightly clasp the stem, or inflated leaf sheath extensions (ocreas). This relationship is extremely complex and has yet to be fully investigated. The "farming" of scale insects by ants is also a common relationship. The scale insects feed on the rattan phloem cells, secreting a sweet honeydew that the ants then feed on and the ants, in turn, protect the rattan from other predators (unfortunately also including rattan harvesters and unwitting botanists).

8. Harvest and Management

8.1 Growth Rates

Rattans are vigorous climbers with relatively high growth rates, and are thus able to be harvested on a short cycle. For the majority of rattans, stem production from the rosette stage (and the seedling bank) is initiated by exposure to adequate light. Stem elongation is also affected by light and, whilst continuous, varies, usually on a seasonal basis. Whilst no data on the growth rates of rattans in the wild exists, long-term studies have been undertaken in cultivation.

Table 5. The growth rates of some commercial rattans in cultivation
(Modifed from Dransfield and Manokaran, 1993)

Species	Growth rates (m / year)
Calamus caesius	2.9-5.6
Calamus egregius	0.8
Calamus hainanensis	3.5
Calamus manan	1.2
Calamus scipionum	1.0
Calamus tetradactylus	2.3
Calamus trachycoleus	(5.0)
Daemonorops jenkinsiana	(2.0-2.5)

(Figures in brackets are estimates)

8.2 Management and Harvesting

Examples of long-term *in situ* management of rattans in the wild are rare (Belcher, 1999). However, based on experimental work in Southeast Asia, four production systems of rattan exploitation can be identified:

- *Natural regeneration in high forest.* This level of management requires the development and implementation of management plans based on sound inventory data and an understanding of the population dynamics of the species concerned. This is particularly appropriate for forest reserves, community forests and other low-level protected areas. These "extractive reserve" models are highly appropriate for rattan: a high value, high yielding product that relies on the forest milieu for its survival.

- *Enhanced natural regeneration*, through enrichment planting and canopy manipulation, in natural forest. This is especially appropriate where forest has been selectively logged. Management inputs are fairly high, with the clearance of competing undergrowth vegetation and subsequent selective felling to create "artificial" gaps has been practised in India, with some success for the rattan resource. Rattan planting in forest in East Kalimantan has also proved successful. The Forest Research Institute of Malaysia (FRIM) suggests that enrichment planting is perhaps the most beneficial form of cultivation, both in terms of productivity and the maintenance of ecological integrity (Manokaran, 1985; Supardi, *pers. comm.*).

- *Rattan cultivation as part of shifting cultivation or in formal agroforestry systems.* The incorporation of rattan into traditional swidden fallow systems in some areas of Southeast Asia is well known (Connelly, 1985; Siebert and Belsky, 1985; Peluso, 1992; Weinstock, 1983; Kiew, 1991). The general principle is that, on harvesting ephemeral or annual crops, rattan is planted and the land is then left fallow. When the rotation is repeated, usually on a 7-15 year cycle, the farmer first harvests the rattan and then clears the plot again to plant food crops. The income generated from the harvesting of rattan in this way is significant.

- *Silvicultural trials* have concentrated on the incorporation of rattan into tree-based plantation-type systems. The need for a framework for the rattan to grow on is imperative and the planting of rattan in association with tree cash crops was begun in the 1980s. In particular, planting under rubber (*Hevea brasiliensis*) and other fast-growing tree crops has proven relatively successful and both silvicultural trials and commercial operations are commonly encountered throughout Southeast Asia.

The harvesting techniques employed in the extraction of rattan have an impact on potential sustainability, particularly for clustering species. The mature stems selected for harvesting are those without lower leaves (i.e. where the leaf sheaths have sloughed off) and usually only the basal 10-20m is harvested; the upper "green" part of the cane is too soft and inflexible for transformation and is often left in the canopy. In many instances, all the stems in a clustering species may be cut in order to obtain access to the mature stems, even those that are not yet mature enough for exploitation and sale. This is particularly an issue where resource tenure is weak.

In general, two simple interventions can be implemented to improve upon rattan harvesting practices:

For clustering species:

- Younger stems, often indiscriminately cut during harvesting should be left to regenerate and provide future sources of cane. Rotational harvesting systems could be increased if this was the case. However, better "stool management" relies on adequate resource tenure.

For all species:

- Harvest intensity and rotation should be based on long-term assessments of growth rates and recruitment.

8.3 Inventory

Rattan inventory has proved to be a somewhat imperfect science. Initial attempts to determine stocking and yield have often been thwarted by a poor taxonomic base from which to begin; it is essential to know which species are being enumerated. Furthermore, lack of sampling the correct parameters also led to much inventory information being simply discarded. When planning a rattan inventory, it is essential to:

- Know the species concerned (collect herbarium specimens when in doubt).
- Measure the correct parameters. These include:
    - Number of stems per clump (for clustering species)
    - Number of clumps per hectare, or for solitary canes, stems per hectare
    - Total stem length
    - Harvestable stem length (the lower stem portion with the leaves sloughed off.
- Establish a protocol for measuring over time to determine growth rates and recruitment; this will determine the potential harvest and hence sustainable extraction rates.

8.4 Land Tenure and Socio-Economic Issues

Rattan management of whatever kind will only be a success if those involved have clear access to the forest, and/or long, and easily renewable resource rights on it (de Fretas, 1992). Currently, rattan collectors rationally maximise their income by harvesting the best and most accessible canes, because they are paid on a per item basis. Larger canes bring the best prices and it is also important to minimise the opportunity costs of collection (i.e. the rattans closer to the community will be harvested first, and probably more intensively).

Traditionally, many communities in Southeast Asia and Africa have benefited from the harvesting of their rattan resource. Many of these groups are dependent on high-value forest products, such as rattan, for access to the cash economy. However, local scarcity caused by uncontrolled harvesting is denying many local people access to this traditional means of income, let alone access to the resource for their own subsistence needs.

8.5 Cultivation

Many examples exist of rattans being cultivated in agroforestry systems in forest lands controlled by local communities (Weinstock, 1983; Connelly, 1985; Siebert and Belsky, 1985; Peluso, 1992) and a small proportion of cane from such systems supplies the formal markets. When it first became clear that rattans were becoming scarce, the associated market implications (scarcity = price increases), during the post-war period in particular, considerable attempts were made to include rattans into commercial-scale cultivation and silvicultural systems. Across Southeast Asia, many plantations and trials were established. However, many of these plantations and cultivated sources of cane are owned and managed by sovereign forestry departments or private companies. Hence, the revenues accrued do not often find their way to local communities as it would if they were harvesting directly from their own agroforestry systems. In many ways, commercial cultivation leads to the removal of a resource from the informal forest economy and into the formal forestry sector; a system renowned for its inequity in terms of ensuring that benefits accrue to local people (Belcher, 1999). However, as discussed above, the attractiveness of these intensive cultivation systems has now been questioned and many are being converted to oil palm plantations.

Table 6. Commercial-scale rattan trials and plantations
(Modified from Dransfield and Manokaran, 1993)

Country	Cultivation
Bangladesh	Trials of *Daemonorops jenkinsiana* established in early 1980s
Cameroon	1 ha trial plot of *Laccosperma secundiflorum* under obsolete rubber near Limbe.
China	1970s - 30,000 ha of enrichment planting of forest on Hainan Island with *Calamus tetradactylus and Daemonorops jenkinsiana (as D. margaritae)*; Plantations of *C. egregius* & *C. simplicifolius* in Guangdong Province; Cultivation trials of many species have been initiated since 1985.
Indonesia	Trials of *C. manan* begun in the 1980s in Java; 1988-1993 several thousand hectares of *C. caesius* planted by forestry department, and to a lesser extent *C. trachycoleus*, in Java and East Kalimantan.
Kenya	Trial plot *C. latifolius* under *Gmelina arborea* near Lake Victoria
Malaysia	1960 – *C. manan* planted in Ulu Langat Forest Reserve; 1972 – Cultivation trial of *C. manan* initiated in Pehang 1975 – FRIM cultivation trials of *C. scipionum* and *C. caesius* planted under rubber, 1,100 ha in total; 1980-81, Sandakan area – 4,000 ha plantation in logged forest planted with *C. caesius* and *C. trachycoleus* and 2,000 ha of abandoned rubber, *Acacia mangium* and logged forest planted with *C. manan, C. caesius* and *C. merrillii*; 1982-1983 – Two trial plots of *C. optimus* established in Sarawak; 1990 – large scale planting in Sarawak with *C. manan, C. caesius, C. optimus* and *C. trachycoleus*.
The Philippines	Cultivation trials of *C. merrillii* and *C. ornatus* var. *philippinensis* established in Quezon in 1977; 5,000 ha plantation of *C. merrillii* established in Mindanao; Early 1990, 500 ha of *C. merrillii* and *C. ornatus* var. *philippinensis* planted under *Endospermum peltatum* (matchwood tree plantation) in Mindanao
Sri Lanka	*C. ovoideus* and *C. thwaitesii* trials established in recent years
Thailand	213 ha of *C. caesius* in Narathiwat Province, established by 1978; *C. caesius* trials established in 1979 in Ranong, Surathani and Chuporn Provinces; 1980-1987, *C. caesius* and *C. manan* trials established – 930 ha in Narathiwat Province

9. Rattans – Potential for Certification

The ecology and nature of rattans make them one of the few products that can be sustainably harvested, given the availability of crucial baseline information, the implementation of an appropriate management regime and adequate land and resource tenure. Rattan exploitation, be it in natural forest or in agroforestry systems, relies on an intricate and multi-layered ecological balance between the rattan resource and the trees that it needs to support it. Rattan cannot be grown outside of this system and therefore its management lends itself to the ecological and management criteria for sustainability set by certifiers (Viana *et al.*, 1996).

Rattan is relatively fast growing and high yielding and can be harvested on relatively short rotations. This allows for relatively short to medium scale returns for those involved in its management, be they local communities or companies. However, limited ecological data on growth rates and estimates of recruitment necessary to set harvesting levels and quotas will continue to hinder field assessments of sustainability, and more basic ecological and applied research is urgently needed before management regimes for species of commercial interest can be drawn up and implemented. In this respect it should be noted that the ecological sustainability of managed high forest and/or agroforestry systems are more easily monitored than intensively managed natural forest systems.

Chain of custody (or "production-to consumption") issues are complex for rattans harvested from the wild (Belcher, 1999), and in many areas collectors are small-scale and widely dispersed, meaning that the trade is dominated by a series of middlemen. In areas where rattan is more intensively managed by communities, such as in east Kalimantan, however, the chain of custody of material should be more straightforward as there are very clear, long-established links between these communities and the markets they supply.

Managed high forest and/or agroforestry systems also offer potentially great social rewards. By re-enforcing traditional tenure and management systems, certification can promote rights over resources and lands not commonly held by communities managing rattan. However, technical issues remain important and there are currently few documented traditional cultivation models for rattan and little research has been undertaken on small-scale cultivation of rattans.

Commercial rattan plantations, albeit under a parent crop, do not usually reflect the diversity and complexity of traditionally managed crops. In addition, the fact that the majority of commercial plantations are government or privately-owned means that such systems offer only limited benefits to communities (de Fretas, 1992). However, when designed and maintained efficiently, commercial plantations can be both productive and highly profitable and should necessarily be excluded from potential certification.

On a final note, and most importantly, the vigorous international and domestic markets for cane and cane products suggests that the additional costs incurred from certification, including better management practices and the equitable distribution of benefits, can be easily absorbed by consumers paying a "green premium" for certified products.

REFERENCES

Alam, M.K., 1990. *The rattans of Bangladesh.* Bangladesh Forest Research Institute. Dhaka.

Anon, 1934. *Shipments of rattans to the UK and USA from the Gold Coast.* Gold Coast Forestry Department Annual Report.

Basu, S.K., 1992. *Rattan (canes) in India: a monographic revision.* Rattan Information Centre. Kuala Lumpur.

Belcher, B., 1999. A production to consumption systems approach: lessons from the bamboo and rattan sectors in Asia. In: E. Wollenberg and A. Ingles (eds). Incomes from the forest: methods for the development and conservation of forest products for local communities. CIFOR/IUCN.

Burkill, I.H., 1935. *A dictionary of the economic products of the Malay Peninsula.* Crown Agents. London. 2 vols.

Conelly, W.T., 1985. Copal and Rattan Collecting in the Philippines. *Economic Botany,* 39(1): pp 39-46.

Corner, E.J.H., 1966. *The Natural History of Palms.* Wiedenfield and Nicolson. London.

De Fretas, Y., 1992. Community versus company-based rattan industry in Indonesia. In: S. Counsell & T. Rice (eds). *The rainforest harvest: sustainable strategies for saving the tropical forest.* Friends of the Earth. London.

De Zoysa, N. & K. Vivekenandan, 1994. *Rattans of Sri Lanka.* Sri Lanka Forest Department. Batteramulla.

Defo, L., 1999. Rattan or porcupine? Benefits and limitations of a high-value non-wood forest products for conservation in the Yaounde region of Cameroon. In: T.C.H. Sunderland, L.E. Clark & P. Vantomme (eds). The non-wood forest products of Central Africa: current research issues and prospects for conservation and development. Food and Agriculture Organization. Rome.

Douglas, J.S., 1974. *Utilisation and Industrial Treatment of Rattan Cane in Casamance, Senegal (Return Mission).* United Nations Industrial Development Organization.

Dransfield, J., 1979. *A Manual of the Rattans of the Malay Peninsula.* Malayan Forest Records No. 29. Forestry Department. Malaysia.

Dransfield, J., 1984. *The rattans of Sabah.* Sabah Forest Record No. 13. Forestry Department, Malaysia.

Dransfield, J., 1988. Prospects for rattan cultivation. In: M.J. Balick. The Palm-Tree of Life: Biology, Utilization and Conservation. *Advances in Economic Botany* 6: 190- 200.

Dransfield, J., 1992a. *The Rattans of Sarawak.* Royal Botanic Gardens, Kew and Sarawak Forest Department.

Dransfield, J., 1992b. The taxonomy of rattans. In: Razali, W.M., J. Dransfield & N. Manokaran (eds). *A Guide to the Cultivation of Rattan.* Forest Research Institute: Forest Record No. 35. Kuala Lumpur. Malaysia.

Dransfield, J., 1992c. The ecology and natural history of rattans. In: Wan, R.W.M., J. Dransfield & N. Manokaran (eds). *A Guide to the Cultivation of Rattan.* Forest Research Institute: Forest Record No. 35. Kuala Lumpur. Malaysia.

Dransfield, J., 1992d. Traditional uses of rattan. In: Wan, R.W.M., J. Dransfield & N. Manokaran (eds). *A Guide to the Cultivation of Rattan.* Forest Research Institute: Forest Record No. 35. Kuala Lumpur. Malaysia.

Dransfield, J., 1998. *The rattans of Brunei Darussalam.* Forestry Department, Brunei Darussalam and the Royal Botanic Gardens, Kew, UK.

Dransfield, J. & H. Beentje, 1995. *The Palms of Madagascar.* Royal Botanic Gardens Kew and the International Palm Society, HMSO, Norwich.

Dransfield, J. & N. Manokaran (eds.), 1993 . *Plant resources of Southeast Asia – Rattans.* PROSEA. Indonesia.

Falconer, J., 1994. *Non-timber Forest Products in Southern Ghana - Main report.* Natural Resources Institute.

Hedin, L., 1929. Les rotins au Cameroun. *Rev. Bot. Appl.* Vol.9: 502-507.

Henderson A. & F. Chávez, 1993. *Desmoncus* as a useful palm in the western Amazon basin. *Principes.* 37:184-186.

Hodel, D., 1998. *The palms and cycads of Thailand.* Allen Press. Kansas. USA.

International Tropical Timber Organisation, 1997. Bamboo & Rattan: Resources for the 21st Century? *Tropical Forest Update.* Vol.7. No.4.

Johns, R. & R. Taurereko, 1989a. *A preliminary checklist of the collections of Calamus and Daemonorops from the Papuan region.* Rattan Research Report 1989/2.

Johns, R. & R. Taurereko, 1989b. *A guide to the collection and description of Calamus (Palmae) from Papuasia.* Rattan Research Report 1989/3

Johnson, D.V., 1997. *Non-wood forest products: tropical palms.* Food and Agriculture Organization. Rome.

Lakshmana, A.C., 1993. *The rattans of South India.* Evergreen Publishers. Bangalore. India.

Manokaran, N., 1985. Biological and ecological considerations pertinent to the silviculture of rattans. In: Wong, K.M. and Manokaran, N. (Editors): Proceedings of the rattan seminar, Kuala Lumpur, 2-4 October 1984. The Rattan Information Centre, Forest Research Institute, Kepong. pp. 95-105.

Manokaran, N., 1990. The state of the rattan and bamboo trade. Rattan Information Centre Occasional Paper No. 7. Rattan Information Centre, Forest Research Institute Malaysia, Kepong. 39 pp.

Morakinyo, A.B., 1995. Profiles and Pan-African Distributions of the Rattan Species (Calmoideae) Recorded in Nigeria. *Principes,* 39(4), pp 197-209.

Peluso, N.L., 1992. The Rattan Trade in East Kalimantan, Indonesia. *Advances in Economic Botany.* 9:115-127.

Renuka, C., 1992. *Rattans of the Western Ghats: A Taxonomic Manual.* Kerala Forest Research Institute, India.

Renuka, C., 1995. *A manual of the rattans of the Andaman and Nicobar islands.* Kerala Forest Research Insititute, India.

Siebert, S. & J.M. Belsky, 1985. Forest-product Trade in a Lowland Filipino Village. *Economic Botany,* 39(4): pp 522-533.

Sunderland, T.C.H., 1998: The rattans of Rio Muni, Equatorial Guinea: utilisation, biology and distribution. Report to the Proyecto Conservación y Utilización Regional de los Ecosistemas Forestales (CUREF) – Fondo Europeo de Desarrollo – Proyecto No.6 – ACP-EG 020.

Sunderland, T.C.H., 1999. New research on African rattans: an important non-wood forest products from the forests of Central Africa. In: T.C.H. Sunderland, L.E. Clark & P. Vantomme (eds). The non-wood forest products of Central Africa: current research issues and prospects for conservation and development. Food and Agriculture Organization. Rome.

Sunderland, T.C.H. & J.P. Nkefor, 1999a. Technology transfer between Asia and Africa: rattan cultivation and processing. African Rattan Research Programme Technical Note No. 5.

Sunderland, T.C.H. & J.P. Nkefor, 1999b. The propagation and cultivation of African rattans. Paper presented at the international workshop: *African rattans: a state of the knowledge,* held at the Limbe Botanic Garden, Cameroon 1-3 February 2000. [Proceedings in prep.]

Tuley, P., 1995. *The Palms of Africa.* Trendrine Press. UK

Uhl, N.W. & Dransfield, J., 1987. *Genera palmarum: a classification of palms based on the work of H.E.Moore Jr.,* pp.610. The International Palm Society & the Bailey Hortorium, Kansas.

Viana, V.M., A.R. Pierce & R.Z. Donovan, 1999. Certification of non-timber forest products. In: V.M. Viana, J. Erwin, R.Z. Donovan, C. Elliot & H. Gholz (eds). *Certification of forest products: issues and perspectives.* Island Press, Washington DC.

Weinstock, J.A., 1983. Rattan: Ecological Balance in Borneo Rainforest Swidden. *Econ. Bot.* 37(1): 58-68.

GENERAL INTRODUCTION TO RATTAN – THE BIOLOGICAL BACKGROUND TO EXPLOITATION AND THE HISTORY OF RATTAN RESEARCH

John Dransfield

1. Introduction

Rattans are old world climbing palms belonging to subfamily *Calamoideae* that also includes tree palms such as *Raphia* (Raffia) and *Metroxylon* (Sago palm) and shrub palms such as *Salacca* (Salak) (Uhl & Dransfield, 1987). There are 13 different genera of rattans that include in all some 600 species. Some of these species in fact do not climb, being shrubby palms of the forest undergrowth; nevertheless reproductive features link them with other species that are climbers and they are hence included in the rattan genera. The climbing habit in palms is not restricted to *Calamoideae*, but has also evolved in three other evolutionary lines – Tribes Cocoeae (*Desmoncus* with c. 7–10 species in the New World tropics, all climbers) and Areceae (one species of the large genus *Dypsis* in Madagascar) in subfamily Arecoideae, and Tribe Hyophorbeae (one species of the large genus *Chamaedorea* in Central America) in Subfamily Ceroxyloideae. Of these only *Desmoncus* spp furnish stems of sufficiently good quality to be used as rattan substitutes. Rattan genera, their size and distribution have been discussed in detail elsewhere (Uhl & Dransfield, 1987, Dransfield & Manokaran, 1993) where a great deal of basic introductory information is also available.

In the absence of rattans, local communities throughout the tropics use a variety of plant resources for purposes similar to those for which rattan is used – basketry, matting, binding, etc., but few products other than true rattan have the strength and flexibility to be utilised for furniture production. Where a variety of plant resources including rattan are available, rattan is almost always regarded as the preferred resource for basketry and binding. In terms of furniture production, very few resources can be substituted for rattan – where resources other than rattan are used, then this normally involves significant changes to the design and construction of the furniture. Bamboo, *Raphia* (petioles) and buri (*Corypha utan* – petioles) are used locally for some sorts of furniture but in the international market they do not command prices comparable to those of rattan furniture. While rattan substitutes may be locally significant and socially important, there is no doubt that rattan is pre-eminent.

Rattans are predominantly plants of primary rain and monsoon forest (Dransfield & Manokaran, 1993). Some species may be adapted to growing in secondary habitats, but these are the exception. Furthermore, rattan entering world trade is, overwhelmingly, collected in the wild, with only a very small proportion coming from cultivated sources. Not all rattans are equally useful. Stem diameter varies enormously from 2–3 mm diameter among the smallest species to 10 cm in exceptionally large species. Species of different diameters are used for different purposes. Furthermore, within a size class, not all species are of equal quality, some being brittle, others of poor external appearance. Ranging from sea level to over 3000 m elevation, from equatorial rain forests to monsoon savannahs and the foothills of the Himalayas, there is a huge range of ecological adaptation. In discussing rattan in broad terms, it is essential that we know which species we are dealing with. For this an accurate taxonomic framework is crucial.

2. History of Rattan Research

The history of research into rattans and their utilisation in the southeast Asian region over the last thirty years is convoluted. Research in the region has not always delivered products of value to the rattan industry; indeed, some research may never have been intended to do so.

There is a long history of rattan exploitation in Asia and Malaysia. Rattans were pulled from the forest and traded down rivers to coastal entrepreneurs who marketed them on to traders in Singapore and Hong Kong who in turn exported raw cane to Europe and North America for manufacture into furniture. Almost all this rattan was from wild stocks. Independently in several places in the region, local villagers planted rattan to provide a local supply of high quality cane for their domestic needs of basketry and matting. A good example is the First Division of Sarawak, where up to twelve different species of rattan have been cultivated, supplying a variety of cane colours and texture that are incorporated into the elegant baskets of the Bidayuh people (Dransfield, 1992). In one area of Central Kalimantan rattan was planted since the latter part of the nineteenth century on a rather extensive scale and was harvested as a cash crop (Dransfield & Manokaran, 1993). Periodically in the twentieth century there were perceived shortages of rattan that stimulated forest departments to pay some attention to the wild stocks of rattan. By the late 1950s forest departments in Malaysia had established small trials in various places, but these without exception were soon forgotten as the stocks of wild rattan recovered and the need for planting seemed to disappear. Then in the early 1970s it became apparent that in parts of the region shortages were genuine and were beginning to affect the market of rattan. The Forest Research Centre at Gunung Batu in Bogor, Indonesia, mounted an expedition in 1973, which I joined, to investigate the long-standing cultivation of rattan (*Calamus caesius* and *C. trachycoleus*) in Barito Selatan. Although this excursion was fraught with problems and little time was spent in the rattan gardens, it was nevertheless extremely important for me, as I learned at first hand that cultivation of small diameter rattan could be successful, at least on a smallholding basis, and it provided me with basic information on the silviculture of such species (Dransfield, 1977a). Shortly after this, the Forest Research Institute Malaysia (FRIM) realised the importance of an integrated approach to rattan shortages in Peninsular Malaysia and started a rattan research programme with assistance from the United Kingdom government via the Colombo Plan. I joined this project as a rattan taxonomist and ecologist, with the remit to survey the wild rattans of Peninsular Malaysia, to provide a manual for their identification, to select good quality species of silvicultural potential and to establish small scale trials in various forest types, both natural and man-made. We have to remember that the rattan flora of Peninsular Malaysia consists of over 100 species, so a guide to the identification of the wild resource was seen as an essential basis for rattan research (Dransfield, 1979). As a plant taxonomist I still feel of course that this is the case with any floristic region.

During the period that we conducted the research in Peninsular Malaysia two important events occurred that would have significant effects on the development of rattan research. Both of them involved visits to the Forest Research Institute to my counterpart, Dr. Manokaran, and myself to discuss rattans. The first was a visit by a representative from the International Development Research Center (IDRC) in Canada to FRIM to discuss the potential role of rattan in the alleviation of rural poverty in Southeast Asia. This visit, by Giles Lessard, was the start of a major involvement by IDRC in rattan in the region and led, ultimately, to the founding of the International Network for Bamboo and Rattan (INBAR). The other event was the visit of a group of Malaysian Chinese economists from a small company called Markiras Corporation to FRIM to discuss the potential of rattan as a cash crop to alleviate the poverty of a group of Murut tribes-people in Sabah, people who had been moved from their traditional homelands on the border with Indonesia to an administratively easier-to-control area nearer to the local administrative centre. These two visits, with such similar aims, resulted in two major strands of research – the former became the politicised research programme involving many nations in the region, the latter became the tightly-managed economically-driven research that resulted in the establishment of the first commercial-sized rattan plantations in the region and the reorganisation of the rattan trade in Sabah.

As I have mentioned above, Murut people had been moved away from a sensitive border area into an area in the centre of Sabah, an area of poor soils, where the people had, not surprisingly, not prospered. The Sabah government, concerned with the need to improve the lot of these villagers, hired Markiras Corporation to come up with suggestions for poverty alleviation and Markiras suggested rattan cultivation. With expertise from FRIM and Kew, the Sabah Forest Development Authority (SAFODA) and the Sabah Forest Department then started a rattan project, once again cataloguing the known rattan diversity (resulting in the publication of a rattan manual (Dransfield, 1984) and, incidentally, the identification and selection of one of the most promising of the large diameter canes – *Calamus subinermis*) and the selection of species suitable for cultivation). A major event was the importation by SAFODA of seeds of *Calamus trachycoleus* from Indonesia. At the time this was not strictly banned – now such a seed export would be prohibited. *Calamus trachycoleus*, the elite small diameter species for alluvial soils prone to flooding, freed SAFODA to develop a rattan plantation on an area of land with reasonably good soils not used for any major perennial agricultural activity.

This aggressive, rapidly growing species (the stems of which will grow as fast as 7 m a year) produces a good quality cane and rapidly establishes itself under the right conditions. Although the initial impetus for this research was the need to find a source of income for rural poor people, the project developed as a commercial forestry rather than a social forestry project. Now, over 20 years on, the first estate has been ploughed up and planted with oil palm, the rattan having succumbed to poor management, inefficient harvesting techniques and the ravages of the wild fires that affected Sabah in the 1980s and 1990s. Undoubtedly, the estate was also affected by the considerable effects of the fluctuations in rattan prices in Indonesia, the country that dominates world supply of rattan. Perhaps the whole scale of the enterprise was too large – had the same effort been put into agricultural extension for rattan planting to the very people for whom the project was originally conceived – smallholders – the result may have been more lasting.

However, in the early days of the SAFODA estates, it seemed that rattan was the answer for everyone, for growing in forest in marginal areas. New estates, both government and private, sprang up all over Malaysia and, increasingly, elsewhere in the region. Eventually the Commonwealth Development Corporation (CDC) together with a Malaysian private company, established an estate in Sarawak in the mid 1980s. This, the most tightly managed estate in the region, is still extant, but its days are probably numbered. If any commercial estate was to have been economically successful it was this one – but there are many problems, such as how to harvest rattan successfully and not leave significant amounts of cane in the forest canopy. This problem has yet to be solved. The most serious problem, however, is the recent economic instability of the whole rattan growing region, that has resulted in a major fall in rattan prices to the extent that rattan in the Keresa Estate is now no longer economically viable.

With so much interest in rattans in the 1980s, rattan research started to burgeon in the region, rattan workshops and international rattan seminars were held. The research seemed to explode in all directions, from taxonomy, ecology, tissue culture, anatomy to mechanical properties. A major rattan bibliography was published (Kong-Ong & Manokaran, 1986) and a newsletter begun by the Rattan Information Centre in Kepong, with funding from the IDRC.

In what follows I have tried to provide broad summaries of the more important results of this research.

3. State of Taxonomic Knowledge of Rattan

Every time I speak on rattan, I emphasise the importance of taxonomy to the development of rattan. With 600 different species to choose from, it is essential that we know what we are dealing with. A good taxonomy provides the means for reliable transfer of information and for predicting the properties of rattan.

Lest it be thought that taxonomic surveys are of mere academic interest, it is worth remembering that until 1979, *Calamus subinermis* was known as a single herbarium collection in the Kew Herbarium, collected by Hugh Low in the early 1900s. Yet, once this species was taxonomically properly circumscribed and understood, it has proven to be one of the most important large-diameter canes with excellent silvicultural potential. Another example is provided by *Calamus poilanei*. It was known for some time that the most important high quality large diameter cane in Laos was a species called locally *Wai toon*. So little was known about this species, the mainstay of the cane furniture industry in Laos, and the target of many Vietnamese rattan collectors when they came over the border from Vietnam, that when herbarium material became available, it was not at first identifiable and we wondered whether it might be a new species. Eventually it was matched up with *Calamus poilanei*, also known from Vietnam.

Most taxonomic studies have been country based. This is not surprising as with 600 species to deal with the taxonomists have to devise achievable short-term goals, usually provided by funding agencies that specify particular countries or parts of countries. Such an approach allows the building up of an intimate knowledge of local variation and ecology of the rattan species but in concentrating on a defined political area, there is a tendency for unidentified species to be described as new local endemics when they may well be species described and well known in neighbouring areas. Problems occur, for example in Indochina, where species distributions transgress political boundaries, particularly where diverse early European influence occurred. This is an area of complex inter-digitation of vegetation types dependent on different climatic influences and superimposed on this are the political boundaries of Myanmar, Bangladesh, India, China, Thailand, Lao PDR, Vietnam and Cambodia. Ex colonial powers Britain and France are involved along with two countries that have remained independent throughout (Thailand and China). At the end of last century and the early part of this century, palms from the entire region were described mostly by the great Italian botanist Beccari, who had material from throughout the region at his disposal. Not only was he able to resolve political over-description, but also tended not to describe the same thing twice. After his death, no one continued this overall approach and Burmese, Bangladeshi and Indian plants tended to be described by British botanists, whereas those of Vietnam, Cambodia and Lao PDR were described by French botanists. The account of the rattans in the Flora of China (Pei & Chen, 1991) illustrates the problem further.

There has been a tendency for the description of new taxa from China without sufficient comparison with material from over the borders in neighbouring Asian countries to see if the species have not already been described. This has been coupled with what I regard as a rather narrow species concept, perhaps influenced also by chauvinism, so there has been a proliferation of new names. Resolving some of the taxonomic and nomenclatural problems related to this has been difficult, given language barriers and the difficulty of exchanging material. In particular there are several rattans where new Chinese varieties of extra-Chinese taxa have been described. *Calamus giganteus* Becc.(Malayan) is now regarded as a synonym of *C. manan* Miq. (Dransfield, 1977b) that is not known further north than the Isthmus of Kra in Thailand. What then is *Calamus giganteus* var. *robustus* Pei & Chen, recently described (Pei & Chen, 1991) Tom Evans, taking a regional approach, has been able to resolve this and several other problems of over-description, in his joint project on the rattans of Laos. So little was known of Lao rattans at the beginning of the Darwin Initiative funded rattan project that he had perforce to look at the rattans of the neighbouring countries. Painstaking work in regional and European herbaria has allowed him to resolve almost all the major taxonomic problems of Indochinese rattans.

Another example of a regional approach is provided by Terry Sunderland's African project. The rattans of Africa consist of four genera, one of which, *Calamus*, is represented by a single variable species, while the other three genera are endemic to Africa and represent a separate evolutionary line within the *Calamoideae*. Rattan taxonomies developed in anglophone West Africa seemed different from those in Francophone West Africa and differed further from those in Central Africa. Could we be certain that the rattan referred to as *Laccosperma secundiflorum* in Ghana was the same as the rattan with the same name in Cameroon? The only way to solve this was to take a regional approach – and this is exactly what Terry Sunderland did in his study of African rattan taxonomy. In fact, what he found was that three large rattans with different ecology and different cane quality had at some time all been referred to as *Laccosperma secundiflorum*. Sorting out this basic taxonomy is essential if research results are to be successfully applied.

The state of rattan taxonomic knowledge is summarized in Table 1.

That there are major gaps in our knowledge is obvious from the table. Priorities for future survey work are Myanmar, Sulawesi, Maluku and Papouasia. For Sumatra and Thailand, accounts could easily be prepared based on available herbarium material.

Underlying the global taxonomic effort on rattans is a nomenclatural database held at Kew compiled in the ALICE database program (Dransfield, 1999).

Table 1. State of rattan taxonomic knowledge

Country	Genera/approx. no. of species	Representation in herbaria	State of taxonomic knowledge	Identification Aids
Africa	4/21	Moderate	Good	Manual (in press) CD-ROM (in preparation)
India – Subcontinent	4/c.50	Good	Good	Manual
India – Andamans & Nicobars	3/17	Good	Good – There is a possibility of further new records	Manual
Sri Lanka	1/10	Good	Very good	Manual
Bangladesh	2/7 (likely to be several more)	Poor	Poor – There are likely to be many more taxa than the 7 recorded	Manual
Myanmar	5/25	Very poor	Very poor	No recent account
China	3/45 (but probably fewer)	Moderate	Good, but there will be many name changes as the taxonomies of the different parts of Indochina and China are integrated	Flora
Vietnam	3/21 (probably more)	Moderate	Quite good, but there will be name changes and new records as the taxonomies of the different parts of Indochina and China are integrated	Field guide and interactive CD-ROM
Laos	6/32	Good	Very good	Field guide and interactive CD-ROM
Cambodia	5/11	Poor	Poor	Field guide and interactive CD-ROM
Thailand	6/62	Good	Very good	Popular palm book (but several major flaws and missing species)
Malay Peninsula	9/105	Very good	Very good	Manual
Borneo (whole)	8/150	Good	Good	Interactive CD-ROM in preparation
Brunei Darussalam	8/80	Very good	Very good	Manual, interactive CD-ROM
Sabah	7/82	Very good	Very good	Manual
Sarawak	8/107	Good	Good	Manual
Kalimantan	8/c.90	Moderate	Quite good	None available
Sumatra	7/90	Good	Quite good	None available
Java and Bali	5/27	Very good	Very good	Flora (but several major flaws and missing species)
Philippines	4/80	Good	Good	Checklist
Sulawesi	3/33 (but likely to be more)	Moderate	Poor	None available
Maluku	3/18	Poor	Poor	None available
New Guinea	3/55 (but likely to be more)	Moderate	Poor, but currently under intensive study	Field guide and full monograph in preparation
Western Pacific	1/3	Moderate	Moderate	Palm field guide
Australia	1/8	Very good	Very good	Popular local floras

4. Phylogeny

Studies on the phylogeny of rattans and other members of the *Calamoideae* using molecular and morphological datasets have been carried out at the Royal Botanic Gardens Kew by Bill Baker and J. Dransfield and produced results of particular significance in understanding the evolution of rattans (Baker *et al.*, 2000a, 2000b). The results seem to indicate that the large genus *Calamus* is probably unnatural and includes elements of several evolutionary lines. This diversity could either be expressed in the taxonomy by including all members of Calaminae in a single genus, *Calamus*, with over 500 species, or the further division of *Calamus* into several further genera. Currently we do not feel that we have sufficient data to produce a stable taxonomy but whichever approach is eventually taken, there will be name changes that cause problems for the users of rattan taxonomies. Another interesting result of the study has been to suggest that the climbing habit – the rattan habit – has evolved at least three different times.

5. Anatomy

Cane anatomy was exhaustively studied by Weiner and Liese (e.g., 1989) who were able to show significant differences between genera, groups of species within genera, and even in some cases, differences between closely related species. These studies also showed the anatomical basis for differing qualities of cane. Work on the anatomical basis for cane maturation is not yet complete. Going beyond the characterisation of anatomical detail, Jack Fisher and Barry Tomlinson have recently investigated the vasculature of rattans from a functional perspective. As also shown in the molecular phylogeny work, rattans display diversity of construction – in fact there are at least two different vascular architectures, again suggesting that the rattan habit has evolved more than once (Tomlinson, *pers. comm.*). Incidental to this, Fisher and Tomlinson have discovered the longest recorded xylem vessels in any vascular plant – in a species of *Calamus*, with vessel length over 7 m, an astonishing record (Tomlinson, *pers. comm.*). Work is currently underway at Harvard University and in the field in Malaysia, by Alex Cobb to investigate the physiology and natural history of water movement in rattan stems.

6. Pollination

Pollination studies were pioneered by Lee Ying Fah in Sabah (Lee *et al.*, 1995) who established that *Calamus subinermis* was primarily pollinated by nocturnal microlepidoptera. Anders Bøgh (1996a), working in southern Thailand, failed to find microlepidoptera and concluded that pollination in the four species of *Calamus* he studied was primarily by diurnal hymenoptera. Both studies have indicated that there is almost certainly a diversity of pollination syndromes in the rattans and that individual genera, and perhaps even individual species, may be pollinated by different organisms in parts of their geographical range.

7. Demography

Demographic studies of wild rattan populations are essential to the development of models for sustainable use. Far too few data are available on this vital subject. Examples of recent demographic work are as follows. Anders Bøgh (1996b) studied the demography of four different sympatric species of rattan in southern Thailand. His results show surprisingly low growth rates and recruitment in natural forest, rates that provide a very pessimistic outlook for the sustainable harvesting of rattan from natural forest. Nur Supardi studied palms as part of the large project sponsored by the Malaysian Forest Department and the Overseas Development Administration (ODA) in London conducted at Pasoh forest reserve into the effects of logging on biodiversity. Supardi showed that rattans represent a major component of palm diversity and within rattan species most individuals are seedlings. Supardi's study also showed that even after 40 years the palm flora has not recovered to what it was likely to have been before the logging took place and that overall there is a significant loss in rattan diversity. Perhaps surprisingly, there was little increase in the species, which appear to be adapted to seral habitats in the Malaysian lowland forest (Supardi *et al.*, 1999).

8. Genetic Diversity

The intraspecific variability of rattans remains a poorly known area of research, much in need of further study. Provenance and progeny trials have been carried out in Sabah on *Calamus subinermis*, *C. manan* and some other species (Lee, 1999). As part of the EU-funded rattan project, Bon and associates (1997) investigated genetic variability in *Calamus subinermis*. She demonstrated considerable variability in wild populations of this species in Sabah and also showed that the variability had some correlation with geographical distribution, with populations occurring near to each other being more closely related genetically than those occurring further afield.

9. Seed Physiology

Several attempts have been made to develop methods for long-term storage of rattan seed. Pritchard and Davies (1999) in screening a range of rattan and other calamoid palm seed, confirmed that the seed is recalcitrant but also demonstrated the possibility of short term hydrated storage of rattan seed for periods of up to 6 months.

All these and other studies not mentioned have contributed significantly to our understanding of the natural history of rattans and provide a wealth of data that can be applied in the management of wild rattan stands and in the development of rattan cultivation.

10. State of Knowledge of Cultivation

Although trial plots of rattans have been established by forestry departments from time to time since the 1920s, research on rattan cultivation did not begin in earnest until the mid 1970s when local shortages of cane and growing awareness of the social significance of rattan alerted forest departments in Southeast Asia to the need for action to safeguard the future of the resource. Since then a wide range of studies have been conducted on various aspects of the growing of rattan. Studies have covered growth of rattan in primary forest, in logged forest, in secondary habitats, and intercropping with plantation crops. Both large and small-scale cultivation have been investigated. The heyday of rattan research was the decade of the 1980s when the local investment climate and general optimism concerning the future of rattan led to the establishment of several large estates, estates established before rattan cultivation trials had matured and could provide recommendations. Nevertheless, these large scale plantations themselves provided a huge amount of valuable data. The results, both final and interim, of numerous trials were summarised in the Guide to the Cultivation of Rattan (Wan Razali *et al.*, 1992) that I worked on under the sponsorship of FAO between 1988 and 1989. This publication aimed to draw together the great wealth of data that had accumulated from trials and to make concrete recommendations for the cultivation of rattan. The studies on which this book was based were largely carried out in the perhumid areas of Southeast Asia rather than in monsoon areas, so there are limitations to their applicability. Shortly after the publication of this book, a new project was started, based in Sabah and funded by the European Union (EU), aimed at increasing productivity of rattan in plantation by improving the genetic resource of rattan available for planting and by improving plantation practices. Seriously blighted by increased protectionism governing the movement of rattan seed within and outside the region, the study nevertheless was able to carry out provenance trials within Malaysia, investigating genetic variability, investigating seed physiology for improved seed storage, and carrying out carefully based cultivation trials. As adjuncts there were also studies of pollination, and biodiversity of rattan in Papuasia.

Now in year 2000 we have a firm if incomplete base of knowledge for the cultivation of about 10 species of rattan, both within the perhumid area of Southeast Asia and further to the north in monsoon areas, with detailed knowledge of the cultivation of three species – *Calamus manan*, *C. trachycoleus* and *C. caesius* – sufficient for the establishment of plantations under various systems. There are of course many aspects of rattan growth that require further study, but nevertheless a great deal is now known. We also know that under optimal conditions in plantations rattan can grow at astonishingly fast rates, rates that provide forecasts of excellent productivity. We should therefore ask, given the wealth of research on rattan and the breadth of the research results, why no new plantations have been established in the last decade and why the remaining large scale commercial plantation in Sarawak is in the final throes of being converted into an oil palm plantation?

Major problems that have been faced by all estates are maintenance of the optimal light regime within plantations to give maximum growth rates, control of pests and diseases and problems of harvesting. This last is a particularly serious problem and can have disastrous results on the final productivity of the plantations – if significant amounts of cane are not harvestable because they cannot be pulled and if pulling damages remaining canes and support canopy, then the overall production is reduced. However, even more important has been the effects of major economic crises and political change within the region. Indonesia, as supplier of 85–90 % of all rattan entering world trade plays a proportional role in affecting rattan prices. When the 'rupiah' lost much of its value, cheap rattan flooded the international market, undermining the investment climate for rattan cultivation elsewhere. Furthermore, controls over access to the resource appear to have weakened, further adding to the flood of cheap rattan on the market. Another factor has been the strength in value of palm oil and refinement of cultivation practices that have resulted in oil palm cultivation being feasible and more attractive than rattans on the very marginal lands that might have been suitable for large-scale rattan cultivation. All this has created an environment inimical to long-term investment in rattan plantation. Yet when we look at growth rates, estimates of returns, etc., all looks excellent for both small and large diameter canes in large-scale cultivation. One has only to look at the huge stands of *Calamus manan* growing excellently in Inhutani plantations of rattan in Sukabumi Selatan and elsewhere in West Java to realise how spectacular growth can be. The problem is, of course, that the economics of the rattan trade mean that at present such excellent "green" large diameter rattan just cannot compete with wild-harvested canes from other areas.

Figure 1: *Calamus pilosellus* in natural forests of Kalimantan (van Valkenburg)

Yet we know that cultivation is possible – the rattan gardens of Central Kalimantan provide sufficient proof, as do the trials conducted at Kepong over the 1970s and 1980s. If the rattan resource is to be safeguarded for the future of the industry then an important part of its future must, therefore, surely be in cultivation on a small scale, by smallholders, rather than in large estates. Sustainable management of the wild resource presents great challenges. All studies of wild rattan populations so far suggest much lower growth rates than those achieved for the same species in cultivation – rattan harvesting on a sustainable basis form natural forest would require great restraint, long gaps between harvests and will be to all intents and purposes very difficult to achieve.

11. Conservation of the Resource

The threats to rattan come from several sources: (1) Decreasing natural forest cover leading to loss of habitat (although rattans are frequently regarded as weedy, few of them seem capable of rapid colonisation of secondary habitats – for the most part they are plants of primary forest habitats). (2) Selective exploitation of stems for the furniture industry. (3) More encompassing exploitation for handicrafts. (4) Exploitation for palm hearts (one of the most damaging of all threats, but currently largely restricted to Laos). (5) Biotic factors such as the increase of wild pig populations (due to clearing of land for agriculture, and loss of predators) in certain parts of Peninsular Malaysia that result in churning of forest soils and removal of seedlings (Supardi *et al.,* 1998).

Of the approximately 600 species of rattan, 117 are recorded as being threatened to some degree (Walter & Gillet, 1998); of these, 21 are Endangered, 38 are regarded as Vulnerable, 28 as being Rare and 30 as Indeterminate (IUCN Red List Categories). While this listing may give some indication of the global threats to rattan species, very few of the listed species known in any detail. The assigning of species names to these categories is based on crude estimates of distribution and threat rather than on the sort of detailed studies that allow us to say that, for example, the Giant Panda is an endangered species. At one level we know a little about the precise distribution and conservation status of a very few species such as *Ceratolobus glaucescens* – species that are restricted geographically and edaphically and because remaining forest is so restricted we can actually estimate the degree of threat. At the other end of the scale there are highly exploited species such as *Calamus poilanei* that are widespread in their natural range, but over-exploitation throughout their range means that we cannot use factors such as remaining forest cover and distribution of soils to estimate the remaining population and the degree of threat. We can monitor the amount of cane emerging from the forest and, from on the ground surveys in the forest we can suggest that the population is severely threatened but we can do little more than this. Conserving highly sought after top quality canes will require properly controlled and policed reserves, where rattan harvesting is strictly forbidden and this enforced – yet this has in all reserves in Southeast Asia proved virtually impossible.

Reviews of rattan utilisation and conservation status have been published in Palms for Human Needs in Asia (Johnson, 1991) and the IUCN Status Survey and Conservation Action Plan for Palms (Johnson, 1996).

12. Rattan Research – A Personal Perspective

In the 1980s, rattan research started to burgeon in the Southeast Asian region and rattan workshops and international rattan seminars were held. The research seemed to explode in all directions, from taxonomy, ecology, tissue culture, anatomy to mechanical properties. All too often, I believe, this research ended up being research for its own sake and contributed little to our understanding of how to provide a sustainable source of rattan for the future, surely the real aim of focused research.

As an example I cite the effort that was expended on research into the tissue culture of rattan. All the really significant commercially important canes in the region are species of *Calamus*, a genus we all know to be dioecious; that is, there are separate male and female plants and the species are thus obligatory out-crossing. In Peninsular Malaysia in the early 1980s there was a perception that there was an extreme shortage of seeds of the elite species *Calamus manan*, this despite the fact that whenever the financial incentive was sufficient and the season right, the Malay Aborigine plant collectors at FRIM were always able to find plenty of seeds wild in the forest. Research was started to produce planting stock from tissue culture. Palms are usually difficult to propagate in tissue culture and a few years elapsed before a successful technique was developed. Although successful, the time required to produce the plantlets and then acclimatise them to growth, first in the nursery and then in the plantation, exceeded the normal time for seeds to germinate and grow to the point they can be planted out – and, of course, such seedlings do not require the same sort of acclimatisation that a propagule raised in a test tube requires. Added to this, by the time the research had reached the point where large-scale production of plantlets could be achieved, rattans in many of the trials and even commercial plantings of *Calamus manan* had reached sexual maturity and were producing so many fruits that there was a plentiful supply of planting material. Currently, I can envisage only one major potential use for the tissue culture techniques, that is in the propagation of selected provenances of elite rattans, material that cannot be achieved by conventional techniques of seed production, because of out-breeding, and at the moment there seems not be a demand for major expansion of rattan planting.

Some of the research on rattans in the region that has at least been focussed on issues important for an understanding of sustainable management of rattan, has failed because it was conducted in way that was simply unscientific and cannot be reproduced. The reliance on vernacular names has had a particularly blighting effect on certain aspects of rattan research in Southeast Asia – if rattan research is to be scientific then it is essential that we know which species we are using for the research. For that we need to establish the scientific identity of the species and voucher our research with specimens that can be checked against the type specimens that provide the essential reference points for plant taxonomy.

Finally, I come back to the problem of the scale of research, and its application. I believe we have seen in southeast Asia that the large scale cultivation of rattan – the big rattan estates – are fraught with problems and that many of these problems do not affect the small scale planting of rattan by small-holders. Given ownership of land and the rattans growing on it, villagers have a vested interest in how the rattans perform, they make every effort to harvest canes carefully and in a mixed village economy where rattan represents only one potential source of income, they do not need to harvest canes if the price is too low – the canes can be left until the price is more favourable.

REFERENCES

Alam, M.K., 1990. *Rattans of Bangladesh*. Bulletin 7. Bangladesh Forest Research Institute, Chittagong.

Backer, C.A. & Bakhuizen van den Brink, R.C., 1968. *Flora of Java Vol. 3.* 761 pp. Wolters-Noordhoff, Groningen.

Baker, W.J., Dransfield, J. & Hedderson, T., 2000. Phylogeny, character evolution and a new classification of Calamoid palms. *Sys. Bot.* 25: 297–322.

Baker, W.J., Dransfield, J. & Hedderson, T., 2000. Molecular phylogenetics of subfamily Calamoideae (Palmae) based on nrDNA ITS and cpDNA *rps16* intron sequence data. *Molecular phylogenetics and Evolution* 14: 195–217.

Basu, S.K., 1992. *Rattans (canes) in India; a monographic revision*. 141 pp. RIC, Kepong, Kuala Lumpur.

Bøgh, A., 1996a. The reproductive phenology and pollination biology if four *Calamus* (Arecaceae) species in Thailand. *Principes* 40: 5–15.

Bøgh, A., 1996b. Abundance and growth of rattans in Khao Chong National Park, Thailand. *Forest Ecol. and Management* 84: 71–80.

Bon, M.C., 1997. *Ex situ* conservation and evaluation of rattan resources. pp. 165 – 172. In Rao, A.N. & Rao, V.M. (eds.). *Rattan – Taxonomy, Ecology, Silviculture, Conservation, Genetic Improvement and Biotechnology*. IPGRI, INBAR.

De Zoysa & Vivekanandan, K., 1995. *Rattans of Sri Lanka: an illustrated guide*. 82 pp. Sri Lanka Forest Department.

Dowe, J., 1989. *Palms of the Southwest Pacific*. Publication Fund, PACSOA. Milton, Australia. 198 pp.

Dransfield, J., 1977a. *Calamus caesius* and *Calamus trachycoleus* compared. *Gdns' Bull. Singapore* 30: 75–78.

Dransfield, J., 1977b. The identity of "rotan manau" in the Malay Peninsula. *Malay. For.* 40(4): 197–199.

Dransfield, J., 1979. *A manual of the rattans of the Malay Peninsula.* Mal. For. Records 29. 270pp. Forest Dept. Malaysia.

Dransfield, J., 1984. *The rattans of Sabah*. Sabah Forest Records 13. pp182. Sabah Forest Department.

Dransfield, J., 1992. *The rattans of Sarawak*. Royal Botanic Gardens Kew & Sarawak Forest Department. pp 213.

Dransfield, J. & Manokaran, N. (eds), 1993. *Rattans*. PROSEA volume 6. Pudoc, Wageningen. pp137.

Dransfield, J., 1997. *The rattans of Brunei Darussalam*. 217 pp. Forestry Dept.Brunei Darussalam and Royal Botanic Gardens Kew.

Dransfield, J., 1999. Rattan biodiversity issues. Pp 2–12. In Bacilieri, R. & Appanah, S. (eds) *Rattan cultivation: achievements, problems and prospects*. CIRAD, Malaysia.

Fernando, E.S., 1990. A preliminary analysis of the Palm flora of the Philippine Islands. *Principes* 34: 28–45.

Johnson, D. (ed.), 1991. *Palms for human needs in Asia*. Balkema, Rotterdam. 258 pp.

Johnson, D., 1996. *Palms: their conservation and sustained utilization.* IUCN/SSC Palm Specialist Group, IUCN, Gland and Cambridge. 116 pp.

Jones, D.A., 1984. *Palms in Australia*. 279 pp. Reed Books, NSW.

Kirkup, D., Dransfield, J. & Sanderson, H., 1999. *The rattans of Brunei Darussalam interactive key on CD-ROM*. Forestry Department, Brunei Darussalam and Royal Botanic Gardens, Kew.

Kong-Ong, H.K. & Manokaran, N., 1986. *Rattan: a bibliography*. RIC, Kepong, Malaysia. 109 pp.

Lee, Y.F., Jong, K., Swaine, V.K. Chey, V.K. & Chung, A.Y.C., 1995. Pollination in the rattans *Calamus subinermis* and *Calamus caesius. Sandakania* 6: 15–39.

Lee, Y.F., 1999. Morphology and genetics of the rattan *Calamus subinermis* in a provenance cum progeny trial. Pp 38–50. In Bacilieri, R. & Appanah, S. (eds) *Rattan cultivation: achievements, problems and prospects*. CIRAD, Malaysia.

Pei, S.J. & Chen, S.Y., 1991. *Flora Reipublicae Popularis Sinicae*. Tomus 13(1). Science Press, Beijiung. 172 pp.

Pritchard, H.W. & Davies, R.I. ,1999. Biodiversity and conservation of rattan seeds. pp. 51–60. In Bacilieri, R. & Appanah, S. (eds) *Rattan cultivation: achievements, problems and prospects*. CIRAD, Malaysia.

Renuka, C., 1995. *A manual of the rattans of Andaman and Nicobar Islands.* 72 pp. Kerala Forest Research Institute, Peechi.

Supardi, N., Dransfield, J. & Pickersgill, B., 1998. Preliminary observations on the species diversity of palms in Pasoh Forest Reserve, Negri Sembilan. pp. 105–114. In Lee, S.S. *et al.* (eds.). *Conservation, Management and Development of Forest Resources*, Forest Deparrtment, Malaysia.

Supardi, N., Dransfield, J. & Pickersgill, B., 1999. The species diversity of rattans and other palms in the unlogged lowland forest of Pasoh Forest Reserve, Negeri Sembilan, Malaysia. pp 22–37. In Bacilieri, R. & Appanah, S. (eds) *Rattan cultivation: achievements, problems and prospects*. CIRAD, Malaysia.

Uhl, N.W. & Dransfield, J., 1987. *Genera palmarum: a classification of palms based on the work of H.E.Moore Jr.* pp 610. The International Palm Society & the Bailey Hortorium, Kansas.

Walter, K.S. & Gillett, H.J., (eds). 1998. *1997 IUCN Red List of Threatened Plants*. IUCN, Gland and Cambridge.

Wan Razali, Dransfield, J. & Manokaran, N. (eds), 1992. *A guide to rattan planting*. Forest Research Institute Malaysia. pp 293.

Weiner, G. & Liese, W., 1989. Anatomical structures and differences of rattan genera from Southeast Asia. *J. Trop. Forest Science* 1: 122–132.

CHALLENGES AND CONSTRAINTS IN RATTAN
PROCESSING AND UTILIZATION

Walter Liese

1. Introduction

Since the first Workshop on Rattan in Singapore in 1979, a large number of national, regional and international conferences, seminars and training courses have been held with many presentations and well-formulated recommendations, the last one just three weeks ago at FRIM. In addition, an impressive series of case studies have been undertaken, just to mention the numerous comprehensive INBAR Working Papers on " Production-to-Consumption Systems" (PCS), which have resulted in many valuable recommendations (Belcher, 1999). In all the various contributions the challenges and constraints for further improvement of rattan processing and utilization were elaborated, and it is hardly possible for anyone to read all of them. It has always been emphasized that processing is a key issue for any further development and utilization. A task of this consultation will be to review critically the progress obtained during the last 20 years since the first conference in Singapore and to highlight the gaps still existing.

Since the audience includes really knowledgeable persons from the producing countries, I can only summarize well-known facts and conclusions. They are supplemented by some observations and results from laboratory work on structural characteristics of rattan as a material, although it is realized that this review is perforce limited.

2. Major Characteristics of Rattan Processing

A comprehensive overview of the major and minor rattans was compiled by Dransfield and Manokaran (1993) with detailed descriptions of their properties and utilization.

The various stages in the preparation of rattan stems have recently been outlined by Gnanaharan and Mosteiro (1997). They include primary processing of deglazing to remove the silicified epidermis, fumigation, bleaching, oil curing and drying. A major problem faced by rattan processors is wastage during harvesting, cutting to size and infections by staining fungi and beetle attack. The loss due to wastage is estimated to be at least 30 percent. The cane quality and the product value on the local and international markets determine the choice of processing methods. Post-harvest treatments are necessary to avoid defects and to increase processing possibilities and market value.

3. Defects by Staining and Insect Attack

Defects occur as discoloration by fungi, pinholes or worm holes by beetles and also as scars or bruises. Rattan is much liable to be infested by fungi and insects due to its high starch content. Fungal invasion can occur within one day after cutting. The required prophylactic treatment within 24 hours or drying is seldom possible or applied due to the harvesting procedure, storage and transport. Canes arriving at the processing site for air seasoning are often pre-infested. It is estimated that about 20 percent of the harvested canes become stained. The most common cause is blue stain fungi that penetrate with their hyphae deep inside the stem, utilizing starch and sugar. Stained canes are often coloured during processing to hide the defect. Due to intensive marketing, furniture in various colours has become fashionable. Heavily fungi-stained material cannot be used for furniture because of reduction in bending strength; it is often utilized for baskets and other perishable products, or even as mere fuel material. Stain control can be achieved by spraying or soaking in preservative solution, mainly done rather late at the depot due to the transportation time. Preservation is often a neglected area of cane utilization. The danger of environmental pollution and regulations against the use of chemicals (if existing) restrict its application. Under the limit of 20 percent moisture content poles can be discoloured by surface moulds, if stored under humid conditions, but also during transport in containers. Contrary to blue stain, this discoloration is only superficial and can be wiped off. Nevertheless, the surface shine is reduced (Kumar, 1993; Mohanan, 1993).

At a higher moisture level also decay fungi can attack the stem, which is often noticed only at a later stage, when the fruiting bodies appear. The resulting structural degradation of rattan in service is most serious.

Seasoned poles with moisture content of 50–100 percent are liable to insect attack, mostly by the powder-post beetle. The beetles deposit their eggs into the big pores at the ends of the poles and the larvae are nourished from the starch content. The occurrence of light-yellowish powder beneath the poles is an indication of an ongoing infestation, so that the material has to be sterilized or burnt. For protection an insecticide has to be applied very early, mostly by dipping or soaking, considering again the consequences of pollution. The availability and legal acceptance of suitable preservatives are different among countries. Sterilization of goods for export can be done in containers at the harbour side by an approved agency. Slightly infested materials may be fumigated or disposed of by burning at the discretion of the authorities and the buyer.

4. Seasoning

Harvesting and drying should preferably be done during the dry months to reduce the initial moisture and to speed up air seasoning. The moisture content of fresh stems varies between 160 and 130 percent with an increase from the base to the top. Seasoning has to start in the forest. Traditionally bundles of 20-30 cane pieces are kept in erect position against a tree for about a week to drain off the sap/water. They are then spread out on the ground in an open yard before delivery to the processing site. If curing is not foreseen the poles are placed in a wigwam fashion for about two–three weeks, so that the moisture content is reduced below 20 percent before further processing. During the drying process curved parts can be straightened by placing weights over horizontally stacked poles.

A high demand for furniture may lead to improperly dried canes, which results in lower quality products.

5. Curing

Curing means the immersion of canes in a hot oil bath to prevent deterioration. It is often an integral part of the processing line. Many reports have dealt with the methods and their effects, e.g. Bhat and Dhamodran (1993) and Silitonga (1989).

Fresh stems cut to desired length and bundled are soaked for a given time in an oil bath within one–two days after harvesting. The oil penetrates the cane axially, while the radial penetration through the skin is almost nil due to its refractory anatomical structure. This process reduces the moisture content so that bio-deterioration is prevented. The rattan skin attains an ivory white colour, which is much desired.

There are many investigations on the best methods for curing. Different combinations of diesel oil, kerosene, palm oil, coconut oil are used, depending on availability and cost, and are applied at varying temperatures between 80°C and 150°C for a duration of 10 to 60 minutes. Whereas the cane diameter determines the latter parameter, the influence of species has apparently not been established. In general, a treatment with kerosene oil at a temperature of 100–105°C for 20-45 minutes, depending on the stem diameter, appears to be best for enhancing the skin colour. After curing the stems are drained of excess oil and rubbed with sawdust, core or waste cloth to remove the waxy substances and silica deposits on the skin. During the subsequent sun drying, often in a wigwam-like fashion, the colour changes from green into ivory white. After one–three weeks, depending on species and weather conditions, the canes are stored under cover.

6. Grading

The grading of rattan stems is a most important step in processing, albeit still quite controversially applied. It is a crucial step in trading and will affect the producers, processors, exporters, importers and also the end user. Grading stages and rules are discussed in many papers and were explicitly outlined by Bhat (1996). Although the situation is quite different in the various countries, a certain amount of simplification and unification is necessary for internal and external trade.

At the harvesting site hardly any grading is done, exceptionally according to the species collected. The bundled canes are brought in small quantities either to the villagers for local processing or in larger quantities to the middlemen or trade centres.

The first stage in grading criteria is dimensions, such as thickness, length of cane and internodes, hardness and defects. According to stem thickness, the categories large- and small-diameter canes are widely applied, with 18 mm being the most widely used criterion and a corresponding length classification.

Hardness is tested by bending the stem by hand. Hard stems regain their original form either quickly or slowly, soft ones break. These differences are related to anatomical characteristics, differences among species or due to thinner fibre walls in younger stem parts.

A further grading is done after processing on the basis of surface colour, mainly to improve the aesthetic value. Since surface appearance is the main criterion for marketing, several measures can be applied to improve quality.

Grey-brown canes can be bleached with hydrogen peroxide or other chemicals for a better finish. Artificial colouring is often applied to discoloured canes with a large range of colours. The uptake of the colouring liquid by the outer stem layer is quite good. Also melamine coating is used for a smooth finish. Fumigation with sulphur dioxide does not only sterilize the canes, but it also improves surface quality.

In most producing countries the rattan grading rules are not precisely formulated. Often they present a confusing terminology, non-standard grading practices and, consequently, a production of sub-standard rattan goods. On the basis of his extended survey, Bhat (1996) proposed model rattan grading rules with a standardized terminology (containing 20 terms), defects (9 terms), methods for clarification, nomenclature of commercial rattan species, as well as grading rules for large-diameter and small-diameter canes and split rattan.

The application of general grading rules would lead to advantages in the trade, in market standardization and less material wastage.

7. Secondary Processing

Secondary processing involves steaming, bending, splitting, dyeing, sanding and finishing, details of which are not to be outlined here.

Peeling and splitting are often done by hand with simple traditional tools or by simple machines. To obtain the inner core, the outer layers, called peel, are removed. Especially some *Calamus* species possess a highly silicified epidermis, which is taken off in different ways (deglazing), whereby the silica layer snaps off in flakes. The shaped components are sanded, scraping is required to remove the burn marks of blowtorch, which is used by the villagers to soften the respective portion of a stem for bending. Bending at processing plants can also cause damage, when the necessary facilities are missing, such as steam chambers to soften the stems.

Rattan processing is done at various levels of competence and intensity: as a cottage-type, at small and medium-sized factories and at larger companies.

8. Some Technological Constraints

8.1 Structural properties

The processing and utilization of the rattans are determined to a great extent by the structural composition of the stem, which exhibits considerable differences along the stem length and between species.

A major constraint consists in the limited utilization of the many rattan species. Out of a total of about 600 species only about 50 are said to be utilized commercially. In the Philippines, 12 out of 68 belong to this category (Tesoro, 1988). IPGRI and INBAR (1998) list 21 *Calamus* species in order of their priority using as criteria cane size, commercial potential and use for rural industries. This selection relates partly to the quantities available, partly to unsuitable dimensions, and partly to inferior properties for processing and utilization.

Anatomical investigations have explored the structural details and changes within the stem from base to top, as well as the differences between species and genera. In addition, the characteristics of commercial species versus non-commercial ones have been analysed and considered for quality improvement (Bhat, 1992a and 1992b; Bhat *et al.*, 1990 and 1993; Abd. Latif and Siti Noralakmam, 1993; Liese, 1994; Renuka *et al.*, 1987; Weiner, 1992; Weiner and Liese, 1988, 1990, 1991, 1992 and 1993; Weiner *et al.*, 1996).

From our previous work some examples are presented of factors which influence processing and utilization.

A cross-sectional view of a rattan stem reveals its principal architecture, consisting of two zones – the periphery with epidermis and cortex; and the central cylinder with vascular bundles with fibrous sheaths embedded in thin-walled parenchymatous ground tissue. The outer part, the epidermis, is often heavily encrusted with amorphous silica, which hinders processing and is therefore removed. The silica content varies considerably among species (0.9–2.7%). For specific products, such as ropes, binds and splits, species with low silica content should be selected. Skin colour is an important criterion, thus *Calamus caesius* is desirable for high-valued products because of its yellowish-cream colour with good lustre, but the anatomical base for such appearance is still unknown. The fibre sheath surrounding the phloem and metaxylem vessel(s) consists of thick-walled fibres with a polylamellate wall structure. Fibre length varies between 0.9 and 2.9 mm with an average of 1.3-1.8 mm. Density varies between. 0.3 and 0.6 g/ cm^3.

There is a definite structural pattern within the stem and between species. The general structures change not only along the stem from the periphery to the centre of a given internode but also from the basal to the top internodes: fibre percentage and fibre cell wall thickness decrease, whereas vessel diameter increases. This increase with age is caused by a fibre wall thickening within the individual lamellae by secondary processes, as well as by an additional deposition of further wall lamellae. Contrary to softwoods and hardwoods, the fibres in rattans are still alive. The increase of wall thickness with age is more pronounced in fibres than in parenchyma cells.

These structural changes result in a higher density at the lower stem parts, whereas the upper part exhibits a higher moisture content and higher volumetric shrinkage. This appears to impart stiffness and determines the breaking behaviour of rattan both within a stem and among the species. The early harvesting of immature stems or the use of the upper portion leads to high shrinkage and warping. If the top portion of a stem is integrated as furniture component, it may break more easily due to its smaller fibre walls, whereas the base is stronger than the middle positions. On a species level *Calamus metzianus* may be mentioned; it breaks easily due to its exceptional low fibre content, thinner-walled fibres and relatively wide xylem vessels.

Figure 2: A solitary cane growing in a degraded forest patch (Sastry)

In the general stem structure, there are distinct differences in the composition of the central part among the 13 genera and even among some species of one genus. Of special significance is the number of vessels and phloem fields per vascular bundles, the type of ground parenchyma, fibre rows in the cortex and special features, such as fibre sclereids ("yellow caps"), ducts, raphids (calcium oxalate) and SiO_2 particles in small parenchyma cells, called stegmata. The differences in the anatomical make-up are so typical that a dichotomic identification key has been developed for all 13 genera on the basis of 284 species investigated. This diversification between genera is of practical value especially for the identification of processed material and can be helpful in trade disputes.

The basic information on rattan anatomy obtained in recent years has been used in the study of 433 rattan species to clarify if certain structural features are characteristic for a "commercial" cane since, of the total of around 600 species, only a small portion is utilized. It became clearly evident that a commercial cane is characterized by definite criteria such as:
- even distribution of vascular bundles over the stem, cross-sectional composition of about 20-25 percent fibres, 45 percent conducting cells and 30-35 percent ground parenchyma;
- fibre caps of equal size, equal fibre length, walls with polylamellate structure;
- ground parenchyma of small cells with thick polylamellate walls.

Thus, anatomical features can help to predict physical and mechanical properties. Especially, vascular bundle distribution and fibre dimensions, such as wall thickness, correlate with shear, compression parallel to grain and static bending.

This characterization can serve to analyse also hitherto unused species regarding their processing potential. Studies on anatomical properties of Papua New Guinea rattans have shown their market potential (Chung and Chen, 1994). Sample investigations of the rattans of West Africa have shown that the three endemic genera (*Eremospatha, Laccosperma, Oncocalamus*), with about 25 species, have the same basic structures as the well-used Asian rattans. This indicates a possible usage of these rattan genera, which have so far not been adequately processed for furniture (Weiner *et al.*, 1994). For Ghana, the five economic rattan species have recently been differentiated on characteristic anatomical features (Oteng-Amoako and Ebanyenle, 2000).

8.2 Further constraints

Major problems in rattan utilization relate to the availability of raw material, production technology, financing and marketing. There are a large number of small rattan processors at the village level, who work with very simple tools, old-fashioned designs and limited skills. Their market access is restricted by the poor quality of their products, so that the value-added by the processing industry is rather low (Belcher, 1999).

In many reports and discussions, the availability of raw material, especially of the valued *Calamus manan,* is defined as the most pressing problem for the furniture industry in the countries of origin, as well as in the main European furniture producers, such as France, Germany, Italy and Great Britain. As the rattan ban has affected the development of the furniture industry within non rattan-producing countries, other countries such as Myanmar, Viet Nam, Lao DPR and Papua New Guinea have increased harvesting of canes of varied quality. Furniture is now designed to require only smaller diameter poles. Factories cope by using low-grade poles, often stained and to be coloured. A major rattan company in Germany with large furniture factories in Java describes the shortfall, combined with the market price, as a decisive factor for further production so that, for example, early grading is neglected to obtain all possible canes at a reasonable price. For primary processing in the field, improved technologies for preservation and seasoning would reduce losses and improve the quality of the canes. Often labour-intensive methods and simple low-cost procedures are applied for peeling, splitting and bending. Technical improvements in the processing industry could increase the value of the products and thus also raw material prices. In reports on the rattan industry in Peninsular Malaysia with nearly 500 mills, the constraints of the industry and needs for further research and development have been discussed (Abd. Latif and Salleh, 1994; Abd. Latif and Shukri, 1989; Abd. Latif, 2000).

The shortage of rattan has also led to partial or even total replacement of rattan components for furniture by other materials such as plastic. "Original imitation" rattan is on the market, which is colourful, more economic in price and use with an attractive design.

9. Further needs and research expansion

In 1991, a study team headed by J.T. Williams on behalf of the International Fund for Agricultural Research, submitted to the IDRC their report on *Research needs for bamboo and rattan to the year 2000.* The general message in that report has hardly changed over the last decade, but certain developments have become obvious. It is therefore of interest to cite the relevant points on rattan in this presentation in order to realize some progress and to consider the areas where further activities are still needed. While a more detailed exploration on the gaps in utilization research for rattan is given in the above-mentioned report, the following priority needs were listed:

- Investigations on the properties of commercial and some neglected species are urgently required to assess more readily the utilization potential of the current "non-commercial" species.
- Since rattan products are susceptible to biological deterioration, protection with environmentally acceptable preservatives will widen the market.
- Improved processing technologies would lead to a greater diversity of products of better quality.

Of special importance is the development of better surface finishes for a pleasing visual appearance and greater wear resistance.
- Diversification of products according to species properties.
- Methods of colouring rattan for furniture.
- Development of panel and walling products.
- Waste utilization studies.
- Development of cost-effective designs appropriate to contemporary style.
- Development of hand tools and hardware.

REFERENCES

Abd. Latif Mohmod, 2000. Production and utilization of bamboo, rattan and related species: Management and research considerations. *XXI IUFRO World Congress, Kuala Lumpur, 2000, Proc. Subplenary Papers and Abstracts.* pp. 393–406.

Abd. Latif Mohmod & Salleh Mohmod Nor, 1994. *Priority areas of research on rattan and bamboo in Peninsular Malaysia.* FRIM Rep. No. 62. pp. 9–12.

Abd. Latif Mohmod & Shukri Mohamad, 1998, Distribution and current status of rattan manufactures in Peninsular Malaysia. In *The rattan industries in Peninsular Malaysia.* Rattan Inf. Centre, Occ. Paper No. 6. Kepong, FRIM. pp. 1–14.

Abd. Latif Mohmod & Siti Norralakmam Yahaya, 1993. Anatomical characteristics of five Malaysian canes and their relationships with physical and mechanical properties. In *Rattan management and utilization.* Proc. Rattan Sem. 1992, Kerala. pp. 207–213

Belcher, B., 1999. *The bamboo and rattan sectors in Asia: An analysis of production-to-consumption systems.* INBAR Working Paper No. 22, Beijing.

Bhat, K.M., 1992a. *Structure and properties of South Indian rattans.* Peechi, Kerala Forest Research Institute. 33 pp.

Bhat, K.M., 1992b Classification of canes (rattans) according to properties and potential end-uses. *J. Timb. Dev. Ass. (India)* 38: 23–32.

Bhat, K.M., 1996. *Grading rules for rattan – A survey of existing rules and proposals for standardization.* INBAR Working Paper No. 6, New Delhi. 44 pp.

Bhat, K.M. & Dhamodran, T.K., 1993. Rattan harvesting and processing technology in India: Present and future. In *Rattan management and utilization.* Proc. Rattan Sem. 1992, Kerala. pp. 233–243.

Bhat, K.M., Liese, W. & Schmitt, U., 1990. Structural variability of vascular bundles and cell wall in rattan stem. *Wood Sci. Technol.* 24: 211–224.

Bhat, K.M., Nasser, Mohamed & Thulasidas, P.K., 1993. Anatomy and identification of South Indian rattans (*Calamus* species). *IAWA Journ.* (14): 63–76.

Bhat, K.M. & Verghese, M., 1991. Anatomical basis for density and shrinkage behaviour of rattan stems. *J. Inst. Wood Sci.* 12(3): 123–130.

Chung, H.-H. & Chen, Y.-S., 1994. *Anatomical properties of Papua New Guinea rattans.* Taiwan Forest Res. Inst. 83 pp.

Dransfield, J. & Manokaran. N., eds., 1993. *Rattans – Plant resources of South-East Asia.* No. 6 Pudoc Sc. Publ. Wageningen. 137 pp.

Gnanaharan, R. & Mosteiro, A.P., 1997. *Local tools, equipment and technologies for processing bamboo and rattan.* INBAR Techn. Rep. No. 9. Beijing. 83 pp.

IPGRI & INBAR, 1998. *Priority species of bamboo and rattan*, by A.N. Rao, Ramanatha Rao & J.T. Williams, eds. Serdang, Malaysia. 95 pp.

Kumar, S., 1993. Protection of canes. *Proc. Rattan Management and Utilization, Trichur, India, 1992.* Kerala For. Res. Inst. pp. 304–308.

Liese, W., 1994. Biological aspects of bamboo and rattan for quality improvement by polymer impregnation. *Folia For. Pol. Ser. B.* 25: 43–56.

Mohanan, C., 1993. Biodeterioration of post-harvest rattans. *Proc. Rattan Management and Utilization, Trichur India, 1992.* Kerala For. Res. Inst. pp. 266–280.

Oteng-Amoako, A.A. & Ebanyenle, E., In press. The anatomy of five economic rattan species from Ghana. *Proc. Intern. Rattan Workshop held in Limbe, Cameroon Bot. Garden in 2000.*

Renuka, C., Bhat, K.M. & Nambiar, V.P.K., 1987. *Morphological, anatomical and physical properties of Calamus species of Kerala Forests.* KFRI Res. Rep. No. 46. Peechi, India. 58 pp.

Silitonga, T., 1989. The effect of several cooking oil compositions on manau (*Calamus manan* Miq.) canes. In *Recent Research on Rattans. Proc. Intern. Rattan Seminar 1987, Chiangmai.* Bangkok, Kasetsart Univ. pp. 187–181.

Tesoro, F., 1988. Rattan processing and utilization research in the Philippines. In *Proc. National Symposium on Rattan, Cebu, Philippines.* Laguna. pp. 41–54.

Weiner, G. & Liese, W., 1988. Anatomical structures and differences of rattan genera from Southeast Asia. *J. Trop. For. Sci.*: 122–132.

Weiner, G., 1992. *Zur Stammanatomie der Rattanpalmen.* Diss. Univ. Hamburg, 131 pp. + app.

Weiner, G. & Liese, W., 1990. *Rattan stem anatomy and taxonomic implications.* IAWA Bull. No. 11, pp. 61–70.

Weiner, G. & Liese, W., 1991. Anatomical comparison of commercial and non-commercial rattans. In *Proc. Oil Palm Trunk and Other Palmwood Utilization, Kuala Lumpur.* pp. 360–367.

Weiner, G. & Liese, W., 1992. Zellarten und Faserlängen innerhalb des Stammes verschiedener Rattangattungen. Holz Roh- Werkstoff 50. pp. 457–464.

Weiner, G. & Liese, W., 1993. Generic identification key to rattan palms based on stem anatomical characters. *IAWA J.* 14: 55–61.

Weiner, G. & Liese, W., 1994. Anatomische Untersuchungen an westafrikanischen Rattanpalmen (*Calamoidae*). *FLORA* 198: 51–61.

Weiner, G., Liese, W. & Schmitt, U., 1996. Cell wall thickening in fibres of the palms *Rhaphis excelsa* (Thunb.) Henry and *Calamus axillaris* Becc. In *Recent Advances in Wood Anatomy.* Rotorua, New Zealand, FRI. pp. 191-197.

Williams, J.T., Dransfield, J., Ganapathy, P.M., Liese, W., Salleh M. Nor & Sastry, C.B., 1991 *Research needs for bamboo and rattan to the year 2000.* Washington, DC, Intern. Fund Agric. Res.. 81 pp.

PRESENTATION OF INBAR'S PROGRAMME
ON RATTAN

I.R. Hunter and Y. Lou

1. Introduction

INBAR is a networking organisation. It has a very small staff in its Beijing Headquarters. It works mainly by providing seed money, partial finance or matching finance to enable activities in national programmes that meet its strategic vision, to come to fruition. Thus it is very difficult to draw a boundary around what is and what is not INBAR's programme. The difficulty is compounded by the fact that INBAR has assisted or contributed to several of the activities to be described in these sessions and repetition is best avoided. What we have chosen to do in this paper was therefore to extract several key aspects of the work we have facilitated in order to inform later discussions. We have ordered the material along the lines of INBAR's programme activities, considering ecological, developmental and industrial implications in turn.

2. Background

After timber, rattan is the second most important source of export earnings from tropical forests accounting for somewhere in the region of US$7,000 million annually. However, the way rattan moves in international trade means that such an estimate is only approximate and needs to be under-pinned by better and more thoroughly researched trade data. One of INBAR's ongoing research projects addresses just this.

Figure 1: Value of world exports of wood products

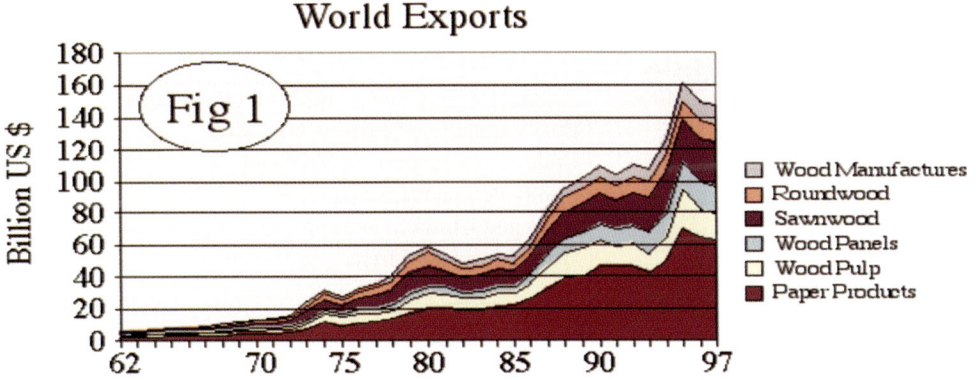

Moreover, we need to keep this figure in perspective. As Figure 1 shows (drawn from European Forest Institute Trade databases and published on the World Forests and Society website [http://www.metla.fi/wfse/]), trade in rattan is a rather trivial portion of world trade in forest products and only reaches 30% of wood exports from the Asia-Pacific region alone (Figure 2).

Figure 2: Wood exports from the Asia-Pacific Region

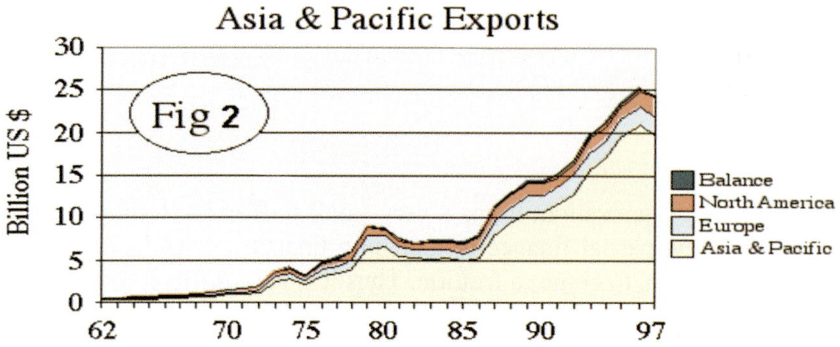

3. **Rattan as a Plant**

The distribution of rattan species around the world is relatively well described (See Table 1 taken from an Internet paper by T.C.H. Sunderland [http://www.africanrattanresearch.fsnet.co.uk]). INBAR assisted the work of the Africa Rattan Research Programme and so can perhaps, lay claim to part of this output. Flora exists for much of the distribution. However Sunderland is of the opinion that "despite the commercial importance of rattan, basic knowledge of the resource is somewhat limited and the flora of Africa and much of Southeast Asia and Malaysia remains poorly known". Sunderland is correct to observe, "This knowledge is essential in order to undertake meaningful inventories of commercially important taxa and to be able to assess the silvicultural potential of each species, based on sound ecological knowledge".

Table 1. The rattan genera, number of species and their distribution

Genus	Number of species	Distribution
Calamus L.	ca. 370-400	Tropical Africa, India and Sri Lanka, China, south and east to Fiji, Vanuatu and eastern Australia
Calospatha Becc.	1	Endemic to Peninsular Malaysia
Ceratolobus Bl.	6	Malay Peninsula, Sumatra, Borneo, Java
Daemonorops Bl.	ca. 115	India and China to Western-most New Guinea
Eremospatha (Mann & Wendl.) Wcndl.	ca. 13	Humid tropical Africa
Korthalsia Bl.	ca. 26	Indo-China and Burma to New Guinea
Laccosperma (Mann & Wendl.) Drude	ca. 6	Humid tropical Africa
Myrialepis Becc.	1	Indo-China, Thailand, Burma, Peninsular Malaysia and Sumatra
Oncocalamus (Mann & Wendl.) Mann & Wendl	ca. 4	Humid tropical Africa
Plectocomia Mart.	ca. 16	Himalayas and south China to western Malaysia
Plectocomiopsis Becc.	ca. 5	Thailand, Peninsular Malaysia, Borneo, Sumatra
Pogonotium J. Dransf.	3	Two species endemic to Borneo, one species in both Peninsular Malaysia and Borneo
Retispatha J. Dransf.	1	Endemic to Borneo

Rattan does not occur in the new world, yet there are many potential sites for it there. In a recently concluded experiment, INBAR introduced two rattan species to Cuba (*Calamus tetradactylus* and *Daemonorops*), where they established and grew satisfactorily (if not at a particularly exciting rate – the maximum height after four years being just over 2m).

Approximately 90% of rattan used commercially comes from the wild. Rattan, in the wild, occupies a position rather similar to ebony, greenheart or rose-wood. It is a relatively rare plant, preferentially sought and extracted. As with other similar plants, the degree of extraction exceeds the local regenerative capacity and resources as well as biodiversity are lost. Table 2 (drawn from T.C.H. Sunderland's Internet paper) summarises the situation for 20 commercial species and concludes that while three are threatened, for only five can it be clearly said that they are not threatened. In another estimate (Claire Coote – NRI, *pers. comm.*) it was thought that of the 200 plus species found in Malaysia over 98 are threatened or endangered.

Table 2. The major commercial species of rattan

Species	Distribution	Conservation status
Calamus caesius Bl.	Peninsular Malaysia, Sumatra, Borneo, Philippines and Thailand. Also introduced to China and south Pacific for planting	Unknown
Calamus egregius Burr.	Endemic to Hainan Island, China, but introduced to southern China for cultivation	Unknown
Calamus exilis Griffith	Peninsular Malaysia and Sumatra	Not threatened
Calamus javensis Bl.	Widespread in Southeast Asia	Not threatened
Calamus manan Miq.	Peninsula Malaysia and Sumatra	Threatened
Calamus merrillii Becc.	Philippines	Threatened
Calamus mindorensis Becc.	Philippines	Unknown
Calamus optimus Becc.	Borneo. Cultivated in Kalimantan	Unknown
Calamus ornatus Bl.	Thailand, Sumatra, Java, Borneo, Sulawesi, to the Philippines	Unknown
Calamus ovoideus Thwaites ex Trimen	Western Sri Lanka	Threatened
Calamus palustris Griffith	Burma, southern China, to Malaysia and the Andaman Islands	Unknown
Calamus pogonocanthus Becc. ex Winkler	Borneo	Unknown
Calamus scipionum Loureiro	Burma, Thailand, Peninsular Malaysia, Sumatra, Borneo to Palawan	Unknown
Calamus simplicifolius Wei	Endemic to Hainan Island, China, but introduced to southern China for cultivation	Unknown
Calamus subinermis (eddl. ex Becc.	Sabah, Sarawak, East Kalimamtan and Palawan	Unknown
Calamus tetradactylus Hance	Southern China. Introduced to Malaysia	Unknown
Calamus trachycoleus Becc.	South and Central Kalimantan. Introduced into Malaysia for cultivation	Not threatened
Calamus tumidus Furtado	Peninsular Malaysia and Sumatra	Unknown
Calamus wailong Pei & Chen	Southern China	Unknown
Calamus zollingeri Becc.	Sulawesi and the Moluccas	Unknown
Daemonorops jenkinsiana (Griff.) Mart.	Southern China	Unknown
Daemonorops robusta Warb.	Indonesia, Sulawesi and the Moluccas	Unknown
Daemonorops sabut Becc.	Peninsular Malaysia and Borneo	Unknown
Eremospatha macrocarpa (Mann & Wendl.) Mann & Wendl.	Tropical Africa from Sierra Leone to Angola	Not threatened
Laccosperma secundiflorum (P. Beauv.) Mann & Wendl.	Tropical Africa from Sierra Leone to Angola	Not threatened

Some of the main issues concerning rattan are accessibility of supplies, declining availability of secondary quality large diameter species and the different ways in which countries have reacted to this. Most producing countries have banned the export of poles and semi-processed rattan. This has led to the development of local processing capacity and also to smuggling of rattan poles.

Inventory of rattan, as with any rare plant, is a very difficult matter. Not only is it difficult to adapt normal forestry inventory methods to the estimation of infrequent plants, but the natural frequency is affected by man's activities, disturbing the estimate and requiring a further layer of stratification in the estimate. Resource quantification needs to be improved. Nevertheless in recent years INBAR has published a report that addresses these difficulties (Nur Supardi Md. Noor, Khali Aziz Hamzah, Wan Razali Mohd, 1998).

INBAR's programme in this resource area has therefore concentrated on ways to improve the size of the resource, in partnership with local peoples. For example, one of INBAR's Transfer of Technology Models (TOTEMS) describes the methods that are used by local communities in the Philippines to raise rattan seedlings. Local people either collect fresh seed from the canes or they collect naturally fallen seed that has germinated. Local knowledge of the phenology of the species is important to make this exercise efficient. They prefer not to collect ungerminated fallen seed because experience has taught them that subsequent germination is unpredictable and often low. Pre-germination treatment varies. Some have learnt that heating the seeds in the seedbed by burning weeds on the surface speeds germination. Others utilise orthodox seed scarification techniques. There is a market in recently germinated seedlings and also in more mature seedlings. The work of raising the seedlings is simple, provided adequate knowledge is available, profitable and can be fitted into other work. This TOTEM shows that, if a seed source is available, raising seedlings need not be a serious constraint.

Since the natural resource is under considerable pressure, amplification of resources, by planting, is essential. Another of INBAR's TOTEMS relates Malaysian experience in rattan planting. The TOTEM report stresses the importance of site selection. The mean annual temperature for rattan should be between 25 and 27 degrees centigrade. There should be over 2,000 mm of well-distributed rainfall and a high relative humidity. Deep and/or alluvial soils are preferred but waterlogged or shallow soils should be avoided. The report lists the 25 species that are considered potentially suitable in Malaysia.

Rattan seedlings of many of the commercial species do best in semi-shade. Thus logged-over forest and old plantations (rubber, oil palm or failed tree plantations) are ideal locations. Most tree species provide good support provided their lowest branches are not too high and provided they are strong enough. Inter-cropping with rubber can be an attractive option for small-holders. However, there is interruption to rubber tapping and sometimes a reduced rubber yield.

4. Rattan in Development

INBAR has analysed the market in two reports of production to consumption. In Java, which has a strong manufacturing base, INBAR found that the country has benefited from the policy of the Indonesian government which imposed prohibitive export taxes on raw and semi-finished rattan (Hariyatno Dwiprabowo, Setiasih Irawanti, Rahayu Supriadi, B.D. Nasendi, 1998). It has emerged as the main exporting region for finished rattan products. However, as Javanese forests are poor in rattan resources, this has created the island to be dependent on other islands for raw material. The Javanese rattan sector is important in terms of employment and cash income generation, and has made significant contributions to the national economy. A study shows that Indonesian rattan sector (Boen M. Purnama, Hendro Prahasto, B.D. Nasendi, 1998) is important because of two aspects: first, rattan plays a substantial role in the socio-economics of Indonesia; second, as the world's topmost rattan producer, Indonesia's policies on the sector have consequences that extend much beyond the country's border. Rattan is abundant in Indonesia, and about 90% of the rattan utilized comes from Indonesian forests. Rattan is also planted in Kalimantan, which is the first region in the world to establish commercial rattan plantations. It may, hence, seem paradoxical that the plant's utilization is yet to be optimized. To achieve optimum and sustainable utilization of rattan, the involvement of government and other related institutions is essential, points out this case study on the rattan sector of East and South Kalimantan. Efforts aimed at the development and improvement of rattan production-to-consumption system must involve financial and other types of incentives to the system participants, technical assistance and technology transfer, and improvement of trading and marketing practices.

INBAR has been funding a rattan project in Nepal since 1998 aiming at species identification and development of conservation and management methods. According to a research report and evaluation, after a new harvesting system – extraction of rattans at a five-year rotation basis were introduced and demonstrated - it shows that one block at a time is a good management practice for rattan brakes. In terms of the data from local communities, the yearly yield of harvested rattan has been doubled and the quality of rattan canes was improved. As a result, yearly incomes for the local people from the same area have been much raised.

INBAR has been active in a GTZ-financed development project in Hainan Island in China. The project has the overall aim of protecting the remaining forest on the island and seeks to achieve this by, *inter alia*, fostering alternative livelihoods. Hainan is one of the few parts of China that is warm enough for rattan. China is also a major importer of rattan that is processed into furniture for further export. Thus INBAR is assisting with the transfer of technology and with training to enable a rattan (and larger bamboo) industry to develop in the forest margins.

5. Rattan as a Product

Rattan fabrication typically occurs in two stages – cane extraction and then final product fabrication. The two stages are often conducted in different countries and the scale of activity differs. Rattan fabrication can be undertaken by small craft industries and can significantly supplement local incomes. Finished products, however, include composite furniture products, which tend to be combined in generalised categories in world trade data and bedevil efforts to get a clean and clear picture of world trade in rattan-based products.

An Internet report of "Japanese Market News" [http://www.wtco.osakawtc.or.jp/market/item/rattan.html] gives an interesting perspective on changes in and problems with the trade. *"In Japan, the traditional distribution route going from importer to wholesaler, retailer then end user has been thrown into disarray by the practice of buying directly by mail order or from supermarkets. Retail prices have also fallen, and the scope for wholesalers, retailers and importers to step in is shrinking. Department stores have ceased stocking rattan products. In addition, there are frequently complaints that rattan products is mouldy or contains bugs, and as it is expensive to wipe mould off and storing due to its bulk, low unit prices and the high cost of storage, trading in the products is now no longer a very attractive proposition. Following a peak in the latter half of 1980, trade in rattan products has fallen off and looks most unlikely to improve in the future. The number of companies that have dropped out of its trade and changed businesses or started importing other products is now increasing, and the fact that the product itself is a seasonal one is encouraging this trend."*

Difficulties with the supply chain; value/storage ratio and quality issues are all highlighted.

Many small-scale industries are involved in production of rattan products and they clearly need the best information to be able to make high quality reliable products. For example the bending of rattan poles for furniture in craft industry is often done by heating the side designated to be concave with a blowtorch. This technique puts scorch or burn marks on the cane and reduces quality, strength and price. One of INBAR's TOTEMS describes a method developed in the Philippines for steam bending rattan whereby canes can be bent to a minimum radius of three to four times their diameter. Thus design of the final item (e.g. furniture) is important because the shaping cannot be forced beyond a certain point. To achieve high quality results, the canes must themselves be of good quality and the moisture content around 18-20%. If drier the canes need longer steaming; if wetter they need longer to set in the new shape. Steaming should be at 100°C for just under 1 minute per millimeter cane diameter. The steamed pole is then clamped for 24 hours. The steaming equipment may cost in the region of US$10,000 but the quality improvement allows a higher proportion of export material.

The market report from Japan referred to mould and bugs as problems. Another of INBAR's TOTEMS therefore describes the Malaysian practice of curing and preserving rattan in boiling diesel oil. The practice is to immerse canes in diesel at between 60-150°C for 10 to 30 minutes. Excess diesel is scrubbed off and the canes dried. This process does to some extent prevent mould and insect attacks. The boiling tanks to carry out this operation are relatively cheap (<US$ 2,000). As the Japanese Market Report also highlighted, it matters crucially what is made from rattan. The report referred to bulky, low value products, which today, have difficulty finding shelf-space in supermarkets. Another of INBAR's TOTEMS describes the layout and organisation of a modern rattan furniture factory. The report gives informative photographs of each stage of the process. While it is true, as the report contends, that rattan furniture should have no problem when it comes to marketing as there are always markets for quality rattan furniture, locally or abroad", the Japanese Market Report reveals some worrying trends. The bulkiness of pre-fabricated rattan furniture is a problem. The tendency for rattan furniture to be thought of as an outside or conservatory item could be a problem in the longer term, if life-styles change. It is not yet part of INBAR's programme to work on design, but it should be.

REFERENCES

Hariyatno Dwiprabowo, Setiasih Irawanti, Rahayu Supriadi, B.D. Nasendi, 1998: Rattan in Java: A Case Study of the Production-to-Consumption Systems. INBAR Working Paper 13. 24 pages ISBN 81-86247-31-9

Boen M. Purnama, Hendro Prahasto, B.D. Nasendi, 1998: Rattan in East and South Kalimantan, Indonesia: A Case Study of the Production-to-Consumption Systems. INBAR Working Paper 20. ISBN 81-86247-42-4

Nur Supardi Md. Noor, Khali Aziz Hamzah, Wan Razali Mohd; 1998: Considerations in Rattan Inventory Practices in the Tropics. INBAR Technical Report Number 14. 50 pages ISBN 81-86247-24-6

Informal references

INBAR TOTEMS are currently in preparation and in press and will be released at the beginning of 2001.

The Japanese Market Report can be found at http://www.wtco.osakawtc.or.jp/market/item/ rattan.html

The Articles by T.C. H. Sunderland can be found at http://www.africanrattanresearch. fsnet.co.uk

The graphs of world trade in forest products can be found at the World Forests and Society web-site at http://www.metla.fi/wfse/

CIFOR RESEARCH : FOREST PRODUCTS AND PEOPLE
RATTAN ISSUES

Brian Belcher

Abstract

The Forest Products and People programme (FPP) of the Centre for International Forestry Research (CIFOR) undertakes research to better understand the true role and potential of non-timber forest products as tools to achieve development and conservation goals. Two main research thrusts are followed. First, the programme is undertaking an international comparative analysis of cases of forest product development. The project uses a systems focus, documenting the key elements of the product and the context under which it is produced, processed, and marketed. The second major thrust is a series of thematic case-based research projects designed to answer specific questions. An example is provided from a case study in East Kalimantan, Indonesia, where a traditional rattan cultivation system has been severely stressed by a combination of policy and economic factors. Government policies designed to encourage the domestic processing industry and monopsonistic manufacturing association have sharply depressed demand and prices. New developments in the region, in the form of roads, industrial plantations, mining, and other new economic activities, have both actively displaced existing rattan gardens (« push » factors) and offered attractive alternatives that have led some rattan farmers to shift to new activities (« pull » factors). And, recent widespread forest fires have destroyed large areas of rattan gardens, effectively forcing some rattan farmers out of business.

This set of conditions offers a good opportunity to study peoples' responses and to analyse whether and under what circumstances this particular intermediate forest product management system is a viable economic option now and in the future. Under current conditions, with low prevailing demand and prices, rattan gardens are a marginal activity in purely financial terms. However, rattan gardens remain important where competition for land is low because they fit well with the swidden cultivation system that is the economic mainstay in the region, because they have low establishment and maintenance costs, because they provide a mark of land « ownership », and because they still serve an important purpose in economic risk management, as a source of « savings ». As rattan remains an important commodity in Indonesia and internationally, and as the current farm-gate price for rattan appears to be artificially low, due in large part to the prevailing policy environment, the rattan garden system may remain viable, at least in the medium term. The paper concludes by briefly summarizing key points for consideration by the Expert Consultation.

1. Introduction

Over the past two decades there has been increased recognition of the many values of forests. This has led to new interest and effort to develop forest products (especially so-called "non-timber forest products or "NTFPs") as a means to achieve both development and conservation objectives. Governments, NGOs, community groups, and development agencies are actively seeking ways – through policy, investment, green marketing, and other interventions - to "develop" forest products. A new literature has emerged and significant investments have been made in numerous projects (see reviews such as Neuman and Hirsch, 2000; Townson, 1994, Ruiz-Perez and Arnold, 1996). Much of this interest is based on the premise that improving prices to producers, adding value locally through increasing post-harvest processing, and improving local organization, can lead to long-term economic and political gains for these groups. Some also argue that these kinds of interventions can lead to forest conservation. And yet, understanding of the true role and potential of forest-products development to contribute to human development or conservation, based as it is on untested theory and scattered and inconsistent case-based research, remains limited.

The Forest Products and People programme of CIFOR takes forest product management and use as its main point of departure in researching poverty alleviation, food security improvement, and environmental protection, especially in tropical forest areas. This programme has developed some questions and approaches that are relevant for this meeting; many issues and questions can be applied directly to the rattan sector. We also have some case-based research that involves an interesting system of rattan cultivation that has highlighted important policy lessons.

This paper will briefly summarize the key issues identified by the FPP programme. It will then present two research projects, highlighting issues that are considered relevant to this meeting. The first is an international comparison of cases of NTFP development. The second involves research on a rattan management system in Indonesia. A summary of the evolution of that system and its current state of development is provided, highlighting the policy lessons.

2. The Hope for NTFP-based Development

Over the past two decades, a growing number of people in development and conservation circles have been promoting NTFP-based development from different perspectives, based on three central propositions.

1. NTFPs, much more than timber, contribute to the livelihoods and welfare of people living in or near forests, offering them a variety of resources, employment and income, particularly in hard times.
2. Exploitation of NTFPs is ecologically more benign than timber harvesting and therefore provides a sounder basis for sustainable forest management.
3. Increased commercial harvest of NTFPs, and associated enterprise development, should add to the perceived value of tropical forests at both local and national levels, thereby increasing the incentive to retain forests rather than converting them to other land uses like agriculture or livestock.

Although these propositions remain largely untested, and have been strongly contested by some, they have attracted the interest of governments and other agents in tropical countries, and from a wide range of donor agencies, with associated increases in investment in forest product development. The experience to date has been mixed - various attempts at forest product development have produced conflicting results, and have triggered a debate over the viability of forest- (and especially NTFP-) based development and conservation.

We feel that investment of this kind needs a stronger rationale regarding both the development implications and the conservation implications of forest product development. There has been a tendency to overlook the fact that some of these products are important especially in situations where there are few or no alternative means of generating a minimum level of income. Indeed, some of the investments in developing and commercializing forest products may result from an exaggerated expectation of what they can contribute to growth in incomes and livelihood enhancement. Moreover, many NTFPs are produced in secondary forest, fallow land, or are cultivated in agricultural systems. Thus, the potential to create incentives for forest conservation may be over-estimated.

Another important weakness in the current debate is in the lack of recognition of dynamic aspects of forest product development. Interventions are intended to improve employment and income generating opportunities through enhancing the value and the capture of value at the community level without anticipating the possible social and economic impacts. But the process of increasing commercial links will bring many and often profound changes to communities, especially those that have been relatively isolated; increased incomes, improved market access, changing relative values of land and labour, and of the various products from the area. While it is common to ask questions about the potential for ecological sustainability, it is equally or more important to ask about economic sustainability of proposed interventions, and of the impacts they generate. Changing opportunities (through increased capital availability, access to markets, education, etc.) and changing tastes may mean that people quickly abandon low-yielding forest product production systems in favour of higher yielding alternatives. This may be highly desirable from the perspective of welfare improvement but it calls into question the conservation element. Indeed, the appropriate role for (some) forest products may be as stepping-stones to development rather than as the long-term integrated conservation and development options.

A well-focused research effort is required to improve understanding of the role and potential of forest products, and to articulate more clearly the kinds of interventions that are most likely to produce the desired outcomes in development and in conservation terms.

3. Research Needs Related to Problems and Trends

It is apparent that there is scope for the development of a wider range of products from forests, and also for new models of forestry management alongside industrial timber production. However, current understanding of the role, potential, and dynamics of forest products in development is insufficient. In order to realize the potential of forest products, work is needed to help target investments. Research is needed to advance understanding of:

- What kinds of forest products (characteristics) are suited to meeting development and conservation objectives?
- What conditions are needed to facilitate this development and, conversely, what conditions are likely to lead to failure?
- How does the role and potential of forest products change with economic development (dynamics)?
- What is the role of policy in determining sustainable and equitable use of forest products?

Such analyses would inform the formulation of policy options and guidelines for governments and development agencies that would in turn facilitate more effective investments in forest product conservation and development.

One of the key challenges facing us will be to develop a general framework within which to structure a comparison of numerous and diverse cases so as to use them effectively as an empirical basis for the advancement of forest-product-based development theory.

The FPP programme is organized into two main, inter-related, thrusts: (1) an international comparative analysis of a large number of cases of forest product development, and: (2) case-based research for detailed analysis of particular situations, in collaboration with national and international partners.

4. An International Comparative Analysis of Cases of Forest Product Development

CIFOR is working, with the support of the Department for International Development (DFID) of the United Kingdom, to improve understanding of the role and potential of NTFPs through a comparative analysis of a wide range of cases of forest products development. We aim to collate information from many cases that have already been studied, to document and describe the cases using a standardized set of descriptors, and to conduct a series of exploratory analyses.

The goal is to:

- create typologies of cases,
- identify conditions associated with particular kinds of development and conservation outcomes,
- develop and test hypotheses about forest product development.

This will provide guidance for action-oriented interventions based on forest products; that is, to identify conditions and "types' of cases that are amenable to development interventions, as well as to flag "types" of cases that may not be good investments.

There is a rich body of information on many aspects of commercial forest product development. This information includes numerous case-based studies of different elements of forest product systems, and results from development projects that have invested in forest product development. Many interventions have been tried at the project level, including various combinations of technical, institutional and financial support for forest product production, processing and marketing, with mixed success. As well, larger, crosscutting interventions have been attempted, including green markets, "fair trade" initiatives, and efforts to promote NTFP certification (Shanley *et al*, forthcoming).

However, it is difficult to build a theoretical framework from this basis. The information has been gathered using a range of methods, at different scales, and focusing on different elements of the forest product production, processing and marketing systems. Work is needed to document and compare cases using consistent terms and definitions for an appropriate range of variables.

This research will synthesize lessons from approximately 60 cases (20 each in Latin America, Africa and Asia) that have already been researched and analysed, applying a standard comparative analytical approach. A range of descriptors (variables) has been identified based on a review of the literature and the authors' experience. These variables have been recognized in the literature as being important in some aspect of forest product development. The following categories of information are addressed:

- Geographic setting
- Biological and physical characteristics of the product
- Characteristics of the raw material production system
- Ecological implications of production
- Socio-economic characteristics of the raw material production system
- Institutional characteristics of raw material producers
- Policies affecting raw material production
- Characteristics of the processing industry
- Characteristics of the trade and marketing system
- Outside interventions
- Outcomes of forest product commercialization

The objective is to find out which characteristics tend to be associated with which other characteristics, and so create a typology of cases. We also want to discover which sets of characteristics, or "types" of cases, tend to be associated with what kinds of human development and conservation outcomes. We are hopeful that this information will be a valuable addition to the management and policy debate.

The comparative methodology is based on that developed by Ruiz-Perez and Byron (1999). It uses exploratory statistical analyses (including multivariate analysis) to find patterns, to develop typologies, to identify key context variables and to analyze their relationship with observed outcomes. The goal is to develop a useful typology of cases, identify key variables (those with maximum explanatory power), and investigate relationships between particular classes of forest products production-to-consumption systems and their development and conservation outcomes.

In addition to the statistical analysis, the relatively large database of cases, characterized using a standard format, will offer a unique opportunity for qualitative analyses. It is intended to publish summaries of the case descriptions in three volumes (by region) for ease of use by other researchers.

In order to capture all of the relevant variability, the analysis will be based on a production-to-consumption systems approach. That is, the case descriptions and the comparative analysis will consider the whole system, from production of raw material through to final market, including social, economic, technological and ecological aspects of the production systems, of the products, and of the market (Belcher, 1997, 1998).

5. Thematic Case-based Research – An Example Focusing on Rattan

The FPP programme is also undertaking a series of case-based research projects. The approach is illustrated with an example from Kalimantan, Indonesia, where rattan is the key commercial NTFP.

As with most CIFOR research, the project involves collaboration with local partners (in this case, the Centre for Social Forestry at the University of Mulawarman, Samarinda) and two international partners

(The Centre for Earth Observation Science, University of Manitoba, Canada[2], and the EU-supported FORRESASIA project). The study focuses on the Pasir and Kutai Kabupaten (Districts) of the Indonesian Province of East Kalimantan. The people in the area are mainly indigenous people (Dayak tribes) who live in scattered villages accessible by river and, increasingly with road access, who practise swidden agriculture. Rice is the mainstay, with several other field crops grown, supplemented by hunting, fishing and collecting from the forest, and increasing integration in the cash economy. The area was selected because:

1. there is a high level of traditional forest use by people living in the area;
2. the traditional rattan gardens of the area represent an interesting and important intermediate-intensity forest product production system;
3. the area is currently undergoing rapid externally-generated changes (new roads, large-scale establishment of oil palm and pulpwood plantations), leading to new pressures and opportunities for people living in the area.

This combination of factors makes the area very interesting for a study of the changing role and importance of forest products.

The ongoing work includes village and household level socio-economic surveys, ecological surveys, economic research, and spatial and land cover analyses based on community mapping and time-series remote sensing (optical and radar satellite) imagery[3].

The rattan cultivation system in Kalimantan has been described frequently in the literature (Weinstock, 1983; Mayer, 1989; Godoy, 1990; Peluso, 1992; Fried and Mustafa, 1992; Boen et al., 1996; Belcher, 1997; Eghenter and Sellato, 1999). Farmers start the swidden cycle in May by slashing undergrowth vegetation, followed by felling the trees in a selected area of primary or secondary forest. In August, after a drying period of a month or so, the field is burned and by September farmers start planting the hill rice that will be harvested in February. Rattan seeds or seedlings are planted either after slashing or at the same time as rice. After two years of rice production most of the agricultural activities cease (except for harvesting of longer-maturing root-crops and fruits) and the rattan grows up with the regenerating forest. The rattan can be harvested seven to ten years after planting, and regularly after that.

The origins of the rattan cultivation system in Kalimantan are not well documented. It probably dates back to the mid-19th century (Van Tuil 1929; Tan, C. F. 1992). Rattan was originally used mainly for subsistence purposes, but over time gained commercial importance. We can only speculate about the domestication process, but it is a relatively small step from wild gathering to planting within a 'ladang' (rice-swidden system). Rattan seeds or seedlings can be established simultaneously with the rice crop at very low extra cost. Our studies show that it requires and extra 7 or 8 man-days in the first year, and small inputs for weeding and protecting the young rattan plants afterwards. Once they are established, the rattan plants can be harvested periodically, using simple technology, over a long period of time, for just the cost of the harvesting labour (cutting and carrying). Most likely an intensification of the system to the current situation occurred with the entrance of rattan in the international trade in mid-19th century.

The details of the current approach vary from farmer to farmer and place to place, but the basic elements are consistent. Farmers plant rattan seeds, wildings or seedlings, in a newly created agricultural field (or "ladang") as part of a shifting cultivation system. The main agricultural crop is upland rice, along with maize, cassava and banana among other food crops. The main rattan species used is *Calamus caesius* (known locally as "rotan sega"), with several other species also grown. The young rattan plants are protected in the ladang and, when the farmer shifts to a new swidden plot one to two years later, the rattan left to grow up with the secondary forest vegetation, to create a "kebun rotan" or rattan garden. The average size of a rattan garden is 1.4 ha. Density of rattan clumps ranges from about 50/ha up to 350/ha, with a mean of around 170/ha (García-Fernández, forthcoming).

---

[2] With support from the Canadian International Development Agency
[3] Detailed accounts of the research are available in Belcher et al, forthcoming.

Harvesting of *C. caesius* typically commences eight to ten years after planting. Some of the others mature more quickly. *C. caesius*, and most of the other cultivated species, has multiple stems and can sustain repeated harvests. Thus, the rattan gardens can be harvested periodically over time. Farmers report that production peaks between 24 to 30 years and begins to decline between 37 and 43 years after planting (García-Fernández, forthcoming).

The rattan stems are cut, and cleaned and dried for sale through a network of traders. The main market for the primary cultivated species has been the 'lampit' (rattan mat) industry in South Kalimantan (though this industry has largely collapsed - discussed below) and the furniture and handicrafts industry, primarily located in Java. A substantial portion has also been smuggled to Malaysia (Haury and Seragih, 1996, 1997) and on to other countries with large rattan furniture manufacturing industries (especially the Philippines and China).

The village elders report that rattan cultivation gained importance after independence, when rattan prices reached high levels. Rattan became a main economic crop at the end of the 1960's with the increased motorization of river transportation and an increasing number of traders and exporters. The main driving force was the regular increase in prices of rattan. At the same time, alternative sources of income were lost as forest products that had been important, such as resins and gums, became less valuable. The rapid development in Malaysia and Indonesia of hevea rubber plantations in the 1920's and 1930's meant reduced importance for the gums. Resins followed the same path with the development of synthetic substitutes around the time of the Second World War. Locally, village elders lay the blame on logging companies, who removed the big resin producing dipterocarps. By the end of the 1970's, rattan became the main source of income of most villages of the study area (the exceptions being a few villages in the very south of Pasir), with many farmers concentrating on rattan cultivation and purchasing rice to meet their requirements.

The economic role of rattan was exaggerated in the 1980s with the rapid development of the 'lampit' (rattan mat) industry in South Kalimantan. In 1984 there were just 21 'lampit' manufacturing enterprises in Amuntai, the centre of the industry, making 64,000 m² of lampit. By 1987 the industry was at its peak, having swollen to 435 units producing over 1 million m² (Figure 1). The industry used cultivated *C. ceasius,* and demand and prices reached unprecedented highs (Figure 2). Farmers report that competition from buyers was very high. Traders would come to the villages, offering advances of cash and consumer goods to secure rattan supplies. But, good things do not last, and this boom was short-lived.

There has been a tradition in Indonesia of heavy government intervention in resource industries, often in collusion with powerful private interests (de Jong *et al*, forthcoming). The boom in the rattan sector in the 1980s attracted the attention of some of these people, and a series of regulations (Box 1) was swiftly put in place to try to capture some of the profits being generated.

Box 1: Policy instruments affecting rattan in Indonesia

- a ban on the export of unprocessed (raw) rattan in October 1986
- a ban on the export of semi-finished rattan in January 1989 (replaced in 1992 with a prohibitive export tax)
- the reclassification of rattan webbing as a semi-finished product (from finished product) in 1992, further reducing demand for cultivated rattan species used for this product
- regulation of the rattan processing industry, with restrictions on the investment in the area. For example, in 1989 all foreign and domestic investment in raw rattan processing and semi-finished rattan production was closed, and foreign investment in finished products manufacturing was also closed. Later this restriction was relaxed to allow investment in rattan processing outside of Java. This policy was finally fully relaxed in 1995, but in the meantime it has probably kept rattan-processing capacity below what it would otherwise have been.
- establishment of a Joint Marketing Board (ASMINDO), an approved exporters system, and an export quota system for *lampit*, by a Ministry of Trade Decree.

These measures were ostensibly aimed at protecting the resource and encouraging the domestic processing industry. The ban on the export of unprocessed rattan and on semi-processed rattan acted as a subsidy for domestic processors by artificially reducing demand for raw material (NRMP, 1996). In this respect the policy was successful; the rattan processing industry in Indonesia has grown substantially. However, the bans had a strong depressing effect on raw material prices, at great cost to the people involved in raw material extraction and cultivation.

One of the most important changes for the rattan growers of Kalimantan was the move to establish the Association of Furniture and Handicraft Industry (Indonesia) (ASMINDO). This was done to "prevent unhealthy competition" among lampit exporters, following the same approach used by the *Asosiasi Panel Kayu Indonesia* (Indonesian Wood Panel Association) (APKINDO) to control the plywood industry. Indeed, ASMINDO was effectively controlled by the same man who controlled APKINDO. ASMINDO imposed export restrictions on its membership in order to manage supply, in an effort to control quality and to increase unit prices. This strategy was based on the reasoning that, as the main supplier of lampit, Indonesia could control the market. Individual manufacturers reported that the quota was assigned based on political connections and payments.

These measures led to severe reductions in manufacturing and export of 'lampit' (Fig. 2). There were also big fluctuations in value-added, as the unit price changed (in nominal terms) from US$6.38 down to as low as US$1.22 and back up to US$8.39 in 1987, 1990, and 1995 respectively. The total number of enterprises had dropped to 20, and now, according to anecdotal evidence, the industry is almost completely destroyed, with only one 'lampit' factory and a number of home-based manufacturers producing for the domestic market. ASMINDO officials lay the blame for this situation on changing tastes and decreased demand in the main importing country, Japan. In fact, Chinese manufacturers developed a substitute for rattan 'lampit', made from bamboo. This product was exported to Japan, beginning in the early 1980s, but exports expanded dramatically in 1995 to fill the gap created when the Indonesian prices increased and quantities decreased.

The drastic reduction in output has likewise reduced demand, and prices, for raw material. Raw material prices have changed little in nominal terms since 1987, and have decreased in real terms. Researchers in other rattan farming areas in Kalimantan report similar though more pronounced trends. In more remote areas, with higher transport and other transactions costs, there have been no buyers for several years.

The price slump following the introduction of restrictions on exports was a hard blow to all rattan farmers. Most farmers were not aware of the reasons for the price slump. They had already experienced ups and downs in prices of rattan, so they were waiting for the good times to come back. As the situation did not improve over time, more and more farmers have begun to seek alternative sources of cash income. Villages with better access to alternative opportunities started to set themselves apart from the dominant rattan based model. These villages were mainly located in the eastern part of our survey area in Kutai and in Pasir as a whole.

A series of events have brought major changes to the region, with tremendous impact on the rattan farmers in the area.

6. Oil Palm

First and perhaps most important has been the rapid expansion of oil palm plantations in the province. Industrial oil palm plantations typically cover several thousand hectares, often in rattan growing areas. By 1998, an estimated 70,000 ha were planted to oil palm in East Kalimantan (with substantially larger areas in neighboring provinces). Nearly 4 million ha have been designated for conversion in East Kalimantan and applications had been approved for more than 450,000 ha to be released by 1999 (Casson, 2000). In many cases there is direct competition for land, with oil palm concessions given on land that has been used and

Fig. 1 Rattan Lampit Industry in Amuntai, South Kalimantan, 1984-1993

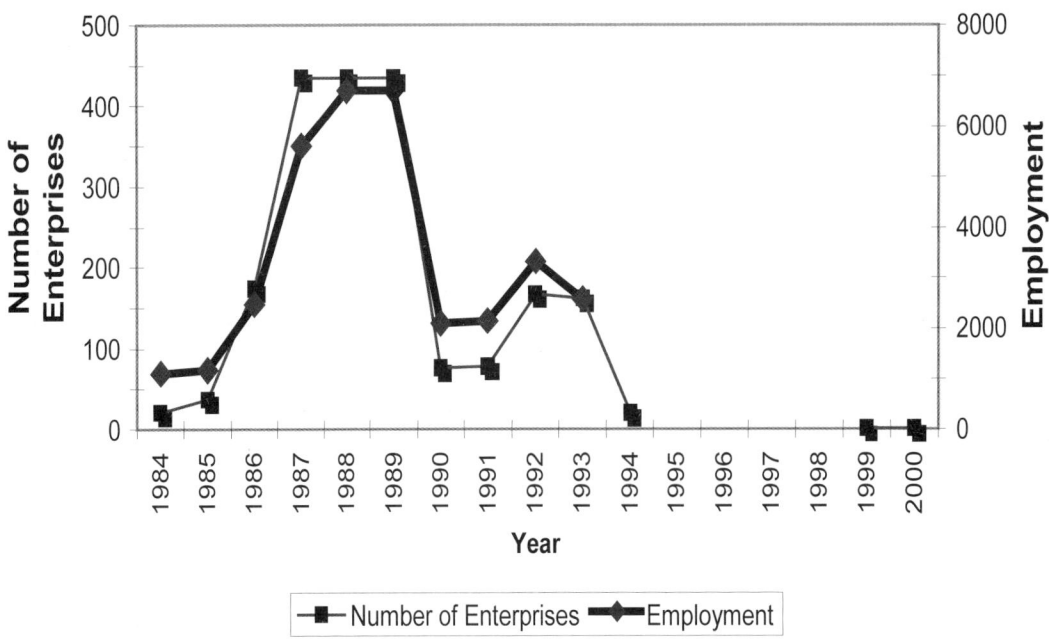

Fig. 2. Exports to Japan (net weight in Kg) of Lampit from Indonesia & Bamboo Mat from China 1984-1999

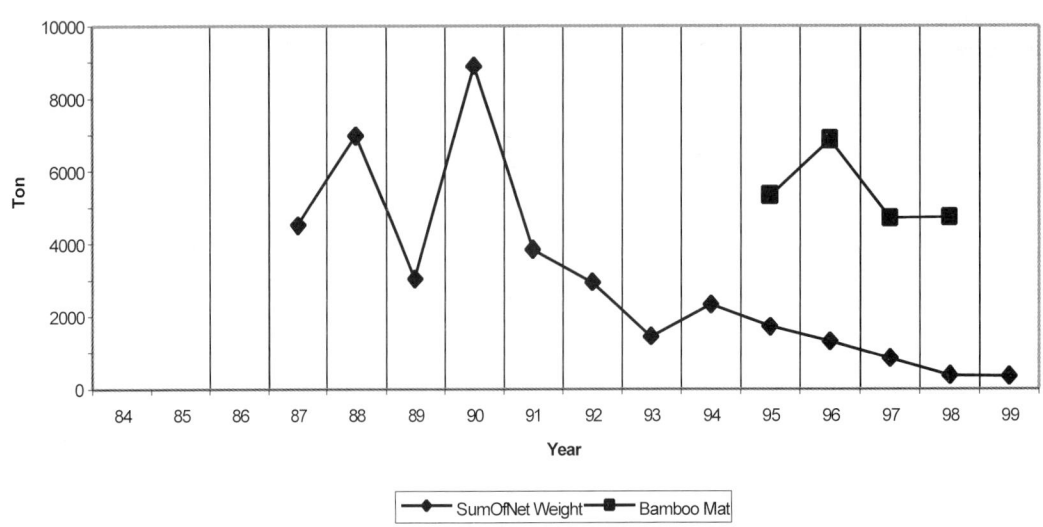

managed by indigenous people for swidden agriculture, including rattan gardens. In one of the study villages, Modang, the establishment of a large oil palm plantation in the early 1980s resulted in many people being displaced, and large areas of productive rattan gardens being destroyed. More recent attempts to establish oil palm plantations have led to severe, sometimes armed, conflict between villagers and company employees. For example, in Lempunah village there is a severe conflict currently underway between the company, P.T. London Sumatra, and villagers, that has involved malicious destruction of rattan gardens and forest on the one side, and burning of vehicles and buildings, and up-rooting newly planted oil palm plants on the other (C. Gonner, *pers. com.*).

But oil palm also has a "pull' effect. Oil palm growing is seen as an interesting new opportunity by local people who appreciate benefits such as regular cash income (palm kernels can be harvested every week), guaranteed market, and a more "modern" lifestyle. Indeed, the main reasons for people's resistance seems to be the lack of adequate compensation for land that they consider to belong to them, and the wish to maintain a broad portfolio of economic activities. People do not want to limit their options. The oil palm companies, in contrast, want to encourage (or force) people to concentrate their efforts on oil palm growing, partly to ensure more efficient production and to ensure sufficient raw material to run their processing factories at capacity and partly, no doubt, to foster a dependence among growers. These issues notwithstanding, there is a strong desire among people in the area to get involved in oil palm growing for several reasons.

7. Pulpwood Plantations (HTI)

The other big land-use change has been large-scale planting of pulpwood plantations (known by the Indonesian acronym HTI). Many of these have been situated on "degraded lands". Under the definition of degraded lands that has been used by the government of Indonesia, this applies to rattan gardens, which are seen as degraded forests. Indeed, our spatial analysis showed a very high correlation of rattan growing areas with HTI.

8. Forest Fires

The other major impact on the rattan gardens was the fires of 1997. During a period of prolonged drought associated with an *El niño* event, several million hectares of Kalimantan were burned by wildfires. The hardest hit areas were logged over forests and areas of new plantation (oil palm, HTI) establishment; these often coincide. In many places, fire was used as a weapon in land conflicts. For example, in Lempunah, where P.T. London Sumatra was trying to establish an oil palm plantation and where local people have been resisting having their land taken from them, large areas of rattan gardens were burned (C. Gonner, *pers. com.*).

The fires did not affect all the villages of the area with the same intensity. The Eastern-most villages of Kutai and Pasir as a whole were the hardest hit. As these villages were also the ones with the best access to other opportunities, the trend towards change was reinforced.

In some villages, fires destroyed up to 90% of the rattan gardens. Beyond the physical damage, this event had a very traumatic effect on local people. Rattan gardens had been seen as a source of security. While prices might fluctuate, the rattan could always be sold for cash if and when it was needed. The rattan kept growing, and in many ways people used their rattan gardens like a savings account. Many respondents use the analogy themselves, saying that a rattan garden is like having money in the bank. All of a sudden, with the widespread burning of rattan gardens, the sense of security was replaced by the recognition that rattan gardens too are vulnerable. This new reality, combined with the low prevailing prices, had a determining effect in many villages to abandon rattan cultivation.

In other areas the response was different. In the west part of Kutai, some villages were spared the fires, while others were as severely hit as Pasir villages. People from villages in both categories seem to retain a high interest in rattan growing. Some have decided to convert from "sega" cultivation to "pulut merah" cultivation. This small-diameter species is relatively fast growing (compared to "sega") and current prices are high. Farmers are able to harvest quicker and to reduce the risks of total loss by fire. Furthermore "pulut merah" thrives in wetter areas along rivers, which are less prone to fires. The shift to this new species is so popular that "pulut merah" seeds are in high demand all over the area.

Other villages, especially those dominated by Benuaq and Bentian ethnic groups, still maintain their interest in rattan gardens, even after the price slump and the destructive fires. They still hope that the prices will soar again. But this may be due to their limited choice. In these remote villages the only source of cash is rattan. No other commodity is traded in the area. Even with very low prices they need to sell rattan, if they are lucky enough to have a buyer. But they no longer invest in establishing large rattan gardens. They only cut small amounts on a regular basis, in order to meet their basic subsistence needs. In villages closer to the primary forest, farmers look for wild rattans still in high demand by traders for the furniture industry. Provided that there are traders willing to buy timber, illegal logging is a favourite occupation for local people in need of cash all over the area.

9. Krismon

One other important factor came into play with the Krismon (from Krisis moniter) or monetary crisis associated with the Asian financial collapse. With a massive devaluation in the local currency, the relative value of export commodities soared. In Indonesia, agricultural commodities such as coffee, cocoa, pepper, rubber and palm oil, and mineral resources (oil, coal, gold) appreciated in value, as did any labour-intensive industry. In our study area the impact was seen in Pasir along the trans-Kalimantan road, where numerous immigrants from South-Kalimantan started gold panning on a large scale with motorized equipment, and in a trend toward increased coffee growing. Though not directly linked to the slump in rattan prices, the development of gold panning (with very high returns to labour) had a large impact by increasing the opportunity cost of labour. There was also a short-lived boom in the rattan furniture industry, but the raw material demands did not result in much price increase for the small diameter canes grown in the study area.

Figure 3: Stem of *Deamonorops cristata*, East Kalimantan (van Valkenburg)

10. The Future for Rattan Cultivation

This is a very interesting case to consider when evaluating ideas to develop intensified rattan management systems, or other NTFPs. This system was developed to fit with the traditional « ladang » (swidden) system. It offers the advantage of low cost establishment and maintenance with relatively high yields. The traditional system is highly diversified, and the rattan element fits well. Harvesting is very flexible – the rattan continues to grow for years, so there is no penalty for delaying harvesting to coincide with labour availability or higher prices. Many people note that it functions like a bank account – rattan can be harvested to respond to urgent needs for cash, to respond to medical emergencies, for example, or for ceremonial requirements. However, it has been seriously stressed by several factors, including:

- Low prices, in this case driven by the policy environment
- Fires and competing land uses leading to reduced rattan garden area
- New, financially superior alternative opportunities for land use (oil palm) and labour (wage jobs, gold panning).

In fact, the rattan gardens in East Kalimantan tend to be resilient, especially in areas where there are limited other opportunities. While this may seem obvious, there are some important lessons in the reasons for their resilience. These systems:

- Offer a valuable risk management tool in which the rattan is available as long-lived, low-maintenance source of savings/income. This is especially important in systems without other, well-developed risk management institutions (people do not have bank accounts, let alone insurance policies).
- Play an important "marker" function for property "ownership". Within the traditional system, rattan gardens are respected as a sign of occupation. Under the present circumstances, with large-scale state-sanctioned land appropriation by oil palm, HTI and mining companies, rattan gardens have been used successfully to demonstrate ownership and claim financial compensation from the company (however meagre).
- Provide a source of cash income in areas where there are few other opportunities to earn cash.
- Provide other valuable forest products and services as the rattan gardens function as secondary forests, giving habitat for medicinal plants, ritual plants, and plants and animals valued for food.
- Retain important cultural values; rattan gardens represent important traditions and provide links to ancestors (many rattan gardens have been inherited from fathers and grandfathers).
- Live long, with little input required. Thus they have a high degree of inertia.

Many of these benefits will be true of other intermediate systems, in other places.

The question arises as to whether this system should be subsidized or otherwise supported, and if so, how? Clearly, as discussed above, the rattan gardens are very important to significant number of, as an integral part of their livelihood systems. The stresses placed on the system have been, for the most part, generated from outside. Rattan trade policies have been designed to keep raw material prices low. Large-scale plantation agriculture has been pursued at the expense of people already living in the area. And the fires were largely human induced, many deliberately targeted to rattan gardens, even if they were facilitated by a natural period of drought. On this count, it seems that the system could be economically competitive if provided with a level playing field.

There are also other benefits to be considered. The rattan garden system offers important ecological benefits, in terms of biodiversity, forest cover, carbon sink, climate. Essentially, the financial value of the rattan makes a long fallow period feasible. During the long fallow, the forest can regenerate and increasingly provide these ecological services.

From a national perspective, the strongest argument for removing barriers and even for actively supporting the rattan cultivation system is that it supplies a valuable export industry.

There are several policy options that could be pursued simultaneously. Simple measures include reducing trade barriers that depress domestic raw material prices (including internal barriers, such as the ubiquitous illegal fees charged to traders, and official export taxes). Industry has resisted this, fearing that higher raw material prices would threaten their competitiveness. Additional measures then would be needed to assist industry to become more competitive. This could be achieved through more efficient raw material production (through research and extension to improve the cultivation system) and trade (especially through improved market information) and through improved design, quality, efficiency and marketing of manufactured products. Combined with these measures, there is a strong case in favour of more careful land-use planning to ensure that important rattan growing areas are not planted to industrial estate crops.

Under the current circumstances, the young people interviewed in our surveys place their hopes on plantation crops. They acknowledge that their low level of education and know-how prevents them from being hired as salaried workers by large companies and even from migrating. Condemned to stay in the village, they long for the regular incomes from plantation crops: oil palm or rubber. Rattan is seen as a thing from the past, something rather backwards, inherited from their forefathers. But such negative perception may easily be overridden if prices go up and if returns to labour become favourable again.

11. Conclusion

In many ways, rattan is a quintessential NTFP. It is typically produced under extensive conditions – the bulk of the world's supply still comes from wild resources – by people with relatively low economic and political power (though in the case described here it is cultivated under relatively extensive conditions). Individual producers tend to harvest small quantities in remote and highly dispersed production areas, often under open access property regimes. Product quality varies enormously. All of these things lead to high transactions costs for trade and relatively low bargaining power for producers, low commodity prices, and low incentives for sustainable management (Belcher, 1997). As a result, resource depletion is widespread. And yet, rattan (and many other NTFPs) offers great potential as a point of entry to improve opportunities for poor people to improve their livelihoods and at the same time to support an industry that creates many jobs and earns valuable foreign exchange for the producing countries.

This paper has raised a number of questions/issues that, it is hoped, will help focus efforts to achieve this potential. One of the main messages is that it is important to consider whole systems in order to identify the real problems. While rattan resources (both wild and managed) may be depleting, the underlying cause are social, economic and political, not technical. The case in East Kalimantan is of course unique in that the rattan is cultivated. But the same forces that have led people to abandon their rattan gardens in this case have also led to unsustainable management (insecure property rights; low raw material prices) and resource depletion (deforestation; wide-spread land use change) elsewhere.

Research and development for rattan must consider these issues. Sustainable management of NTFPs generally, and rattan specifically, will require improved institutional mechanisms – secure property rights for rattan managers/producers, transparent markets and reduced transaction costs to reduce the risk and increase the efficiency of the trade. People can only be expected to manage a resource sustainably if they are able to capture sufficient benefits now and have a reasonable expectation that they will continue to do so in the future. The case in Kalimantan demonstrates that, even with a system that can (technically) be managed sustainably can be seriously undermined by outside forces (and, not least, misguided policies). While there will be a need for some technical research (improved treatment to prevent post-harvest losses, for example), such research should take place in the context of an understanding of the system and the real constraints and opportunities.

REFERENCES

Belcher, B.M., and M. Ruiz-Perez, forthcoming. An International Comparison of Cases of Forest Product Development: Overview, Description and Data Requirements. CIFOR Working Paper.

Belcher, B.M., P. Levang; C. García-Fernández, S. Dewi, R. Achdiawan, J. Tarigan, W. F. Riva, I. Kurniawan, S. Sitorus, and R.Mustikasari, forthcoming. Resilience and Evolution in a Managed NTFP System: Evidence from the Rattan Gardens of Kalimantan. Paper presented at the workshop "Cultivating Forests: Intermediate Systems for NTFP Production, Lofoten Islands, Norway, June 2000.

Belcher, B.M., 1998 A production-to-consumption systems approach: lessons from the bamboo and rattan sectors in Asia. *In:* Wollenberg, E. and Ingles, A. (eds.). Incomes from the forest: methods for the development and conservation of forest products for local communities, 57-84. Center for International Forestry Research, Bogor, Indonesia.

Belcher, B.M., 1997. Commercialization of Forest Products as a Tool for Sustainable Development: Lesson from the Asian Rattan Sector. Ph.D. Thesis, University of Minnesota.

Boen, P.M., P. Hendro, and A. Satria, 1996. Study on the Socio-Economic Aspects of the Rattan Production to Consumption System in Indonesia: A Case Study in Kalimantan. Draft Report.

Casson, A., 2000. The Hesitant Boom: Indonesia's oil palm sub-sector in an era of economic crisis and political change. CIFOR Occasional paper No. 29, Bogor, Indonesia.

Haury, D. and B. Saragih, 1996. Processing and Marketing Rattan. Ministry of Forestry in Cooperation with Deutsche Gesellschaft für Technische Zusammenarbeit (GTZ). SFMP Document No. 6a.

Eghenter, C. and B. Sellato, 1999. Kebudayaan dan Pelestarian Alam: Penelitian Interdisipliner di Pedalaman Kalimantan. WWF Indonesia, Jakarta.

Tan, C. F., 1992. The History of Rattan Cultivation. Malayan Forest Record no. 35, pp 51-55.

Fried, S.T. and Mustofa Agung Sardjono, 1992. Social and Economic Aspects of Rattan Production, Middle Mahakam Region: A Preliminary Survey. GFG Report No. 21, pp. 63-72.

de Jong, W., D. Rohadi, R. Mustikasari, B. Belcher and P. Levang, forthcoming. The Political Economy of Forest Products in Indonesia: A History of Changing Adversaries. Centre for International Forestry Research. Bogor, Indonesia.

García-Fernández, C., Title to be announced. Forthcoming Ph.D. thesis.

Godoy, R.A., 1990. The Economics of Traditional Rattan Cultivation. Agroforestry System, Vol. 12, pp. 163-172.

Haury, D. and B. Saragih, 1996. Processing and Marketing Rattan. Ministry of Forestry in Cooperation with Deutsche Gesellschaft für Technische Zusammenarbeit (GTZ). SFMP Document No. 6a.

Haury, D. and B. Saragih, 1997. Low Rattan Farmgate Prices in East Kalimantan. Causes and Implications. Ministry of Forestry in Cooperation with Deutsche Gesellschaft für Technische Zusammenarbeit (GTZ). SFMP Document No. 12.

Mayer, J., 1989. Rattan Cultivation, Family Economy and Land Use: A Case from Pasir, East Kalimantan. German Forestry Group (GFG) Report no. 13, pp 39-53.

McNeely, J.A., 1988. Economics and Biological Diversity. International Union for Conservation of Nature and Natural Resources. Gland, Switzerland.

Neumann, R.P. and Hirsch, E., 2000. Commercialization of non-timber forest products: review and analysis of research. Center for International Forestry Research, Bogor, Indonesia.

Peluso, N.L., 1992. The Rattan Trade in East Kalimantan, Indonesia. In Nepstad, D.C., and S. Schartzman (eds.), 1992. Non-Timber Products from Tropical Forest: Evaluation of a Conservation and Development Strategy. Volume 9, Advances in Economic Botany, The New York Botanical Garden, Bronx, New York, USA, pp. 115-127

Pompa, A.G. and A. Kaus, 1999. From Pre-Hispanic to Future Conservation Alternatives: Lessons from Mexico. Proc. Natl. Acad. Sci. USA Vol 96, pp. 5982-5986.

Ruiz-Perez, M. and N. Byron, 1999. A methodology to Analyze Divergent Case Studies of Non-Timber Forest Products and Their Development Potential. Forest Science 45(1) pp. 1-14.

Ruiz-Perez, M. and Arnold, M., (eds.). 1997 Current issues in non-timber forest products. Center for International Forestry Research, Bogor, Indonesia.

Shanley, P., Pierce, A.R., Laird, S.A. and Guillen, A., forthcoming. Tapping the green market. Earthscan Press.

Townson, I.M., 1994. Forest products and household incomes: a review and annotated bibliography. Oxford Forestry Institute.

Van Tuil, J.H., 1929. Handel en Cultuur van Rotan in de Zuideren Oosterafdeeling van Borneo (Trade and Cultivation of Rattan in the Southern and Eastern Divisions of Borneo). Tectona 22, pp 695-717. In Dutch.

Weinstock, J. A., 1983. Rattan: Ecological Balance in a Borneo Rainforest Swidden. Economic Botany, Vol. 37(1) pp 58-68.

Wiersum, K.F., 1997. Indigenous Exploitation and Management of Tropical Forest Resources: An Evolutionary Continuum in Forest-People Interactions. Agriculture, Ecosystems and Environment 63, pp 1-16.

RATTAN GENETIC RESOURCES CONSERVATION AND USE
IPGRI'S PERSPECTIVE AND STRATEGY

L.T. Hong, V. Ramanatha Rao and W. Amaral

1. Introduction

The International Plant Genetic Resources Institute (IPGRI) is one of 16 centres of the Consultative Group on International Agricultural Research (CGIAR). IPGRI's mission is to encourage, support and undertake activities to improve the management of genetic resources worldwide so as to help eradicate poverty, increase food security and protect the environment. IPGRI focuses on the conservation and use of plant genetic resources important to developing countries and has an explicit commitment to specific crops. IPGRI works in partnership with other organizations, undertakes research and training, and provides scientific and technical advice and information. IPGRI operates in five geographical areas: Sub-Saharan Africa, Europe, Central and West Asia and North Africa, and Asia, the Pacific and Oceania.

Prior to 1993 there was little concerted effort by organizations in focusing on biodiversity or genetic resources conservation and genetic improvement of bamboo or rattan even though the International Network on Bamboo and Rattan (INBAR) was working tirelessly on various aspects of bamboo and rattan research. The urgent need to generate information for the effective conservation and sustainable use of rattan (and bamboo) was then recognised. At that time, INBAR had requested IPGRI to take the lead in activities related to research on the genetic resources conservation of rattan (and bamboo). Thus the INBAR-IPGRI Biodiversity and Conservation Working Group was constituted in 1993 to address these issues. It was opportune that Japan then provided the financial assistance to IPGRI to initiate the programme (Ramanatha Rao *et al*). The planning and implementation of the research activities on rattan (and bamboo) are conducted from the Asia-Pacific Regional Office of IPGRI, which is located in the vicinity of the campus of Universiti Putra Malaysia, Serdang, Malaysia.

The objective of this information paper is to provide a synopsis on IPGRI's role and a sample of its activities in rattan genetic resources conservation for use.

2. Rattan

Rattan palms are found only in the old world distributed in equatorial Africa, south Asia, southern China, the Malay Archipelago, Australia, the western Pacific as far as Fiji. The greatest diversity of rattan genera and species is found in the Southeast Asian region. In Africa three of the four genera recorded are endemic. *Calamus*, with 370-400 species is the largest genus and is distributed throughout the geographical range of rattans (Dransfield & Manokaran, 1993).

A unique feature of rattans is the abundance and diversity of species, sometimes as many as 30 which occur in one locality in what is apparently rather uniform vegetation. However there could be habitat differences and subtle breeding barriers between species that are not yet understood. In addition, knowledge in the genetic diversity within and between species is still scarce for rattan. With the fast depletion of the tropical forests it is imperative to obtain this knowledge for the sustainable management of the remaining rattan resources. For the commercially more popular species like *Calamus manan*, the problem is more acute as the rate of "regeneration" is dependent on seedling survival as compared to other canes like *C. caesius*.

3. The Objectives

IPGRI does not carry out research on genetic resources conservation for use by itself. It works in partnership with national research institutes and universities to execute the research. The genetic variability and diversity of rattans are two areas of research with little inputs. However, there is a need to have a focus for research in the conservation of rattan genetic resources and this calls for a strategy to maximise financial and manpower resources. The objectives formulated by IPGRI for rattan (and bamboo) activities include: to identify priority bamboo and rattan species for conservation and use; to assess diversity of selected bamboo and rattan genetic resources; to develop complementary conservation and sustainable use strategies for these resources; and to establish information base in the region and to strengthen capacity of national programmes through research and training.

In the last few years significant amount of information have been generated, compiled and distributed through IPGRI activities on rattan. Nevertheless there is still a substantial gap in information needed for effective conservation of the resource in many countries. Results from a number of rattan activities in the region that IPGRI has supported have benefited the countries concerned and have also improved national capacity to address the conservation of genetic resources of rattan.

4. The Strategy

The work accomplished so far on identification and diversity of the genetic resources of rattan in various countries is expected to assist in maximising the utilization of the species and thus enhancing conservation for sustainable management. The results obtained are already creating awareness among the national research institutes not only in Asia, but also in Africa and Central and South America.

The need to focus on specific priority areas necessitated IPGRI to formulate a strategy to ensure effective conservation of rattan genetic resources for use (Ramanatha Rao *et al*, undated). Activities identified and executed under this strategy would direct efforts of national research organisations to achieve the set objectives. However, it is noted here that not all of the areas identified are of equal importance for countries in the region. Activities are being developed according to the priority and needs of each country. The areas of focus have been grouped under four headings and these are:

1. *Assessment and inventory* – *In situ* conservation actions are required and will take priority while complementary conservation strategies are being developed for the rattan resources. Therefore the assessment on the current status of rattan resources is a vital activity for successful *in-situ* (and *ex-situ*) conservation efforts. Activities to be undertaken under this area will include assessment and inventories, distribution patterns and size of populations, rates of extraction etc.;

2. *Development and implementation of conservation procedures* - There is a need to implement different procedures for conservation to ensure sustainable management of the rattan genetic resources for use. This will include development of *in situ* and *ex situ* conservation plans, assessment of seed viability and seed storage and *in vitro* conservation protocols, establishment and management of field genebanks and guidelines for safe movement of germplasm;

3. *Rates of extraction and human impact* – The long-term (detrimental) impact of over exploitation on natural rattan resources in some countries has been felt. There is still a lack of information on the natural regeneration and socioeconomic impact of exploitation and conservation. This information is needed to establish proper *in situ* (and *ex situ*) measures, for sustainable utilization of the resource to ensure the socio-economic benefits are maintained; and

4. *Development of methods for sustainable conservation and use* – It must be recognized that many rural poor are dependent on non-timber forest products such as rattan. Efforts to conserve should not interfere with extraction and use of this resource for their daily needs as well as the income generation activities of these people and other forest dwellers. Understanding the preference for extraction by the forest and forest-fringe dwellers, especially when alternative means of livelihood become available, is of significance to sustainable conservation of the rattan in the natural habitats. Activities would include, assessment of economic gains through extraction of rattan, identification and selection of rattan material that performs well under different environment and ecosystems, and identification and selection of species suitable for cultivation to reduce pressure on naturally occurring stands.

5. Highlights of Some Achievements

IPGRI has undertaken rattan projects with partners in a number of countries in the region stretching from Nepal to China. Some achievements generated through IPGRI's activities on rattan are highlighted here for information (Appendix 1).

5.1 Prioritization of species for genetic resources conservation

The importance of correct identification of rattans when establishing priorities for conservation and use strategies cannot be over emphasized. A good taxonomy also provides the means for reliable transfer of information and for predicting the properties of rattan (Dransfield, 2000). The taxonomic identification of some of the commercial species is still uncertain. In view of the large number of species and their diverse geographical ranges and ecologies, focus has to be directed to the conservation of genepools of more useful species. This is also relevant in that only a small number of the total species is used or have a commercial value. IPGRI together with INBAR published a list of priority species based on a broad criteria (Williams & Ramanatha Rao, 1994). Criteria used for selection included information on: utilization, cultivation, products and processing; germplasm and genetic resources and agro-ecology. Nine species were identified as priority. Acceding to the needs and feedback of the countries in the region this priority list was later expanded to include 21 species (Rao *et al.*, 1998). The priority list is a useful guide for countries in focusing their research on rattan (Appendix 2).

5.2 Assessment and inventory

Studies have been carried out on rattan genetic resources and identification of commercially important species in countries such as Bangladesh, China, the Western Ghats in India, Indonesia, Laos, Malaysia, Myanmar, Nepal, the Philippines, Sri Lanka, Thailand and Vietnam (Rao & Ramanatha Rao, 1999; Vivekanandan *et al.*, 1998; Xu *et al.*, 2000). The surveys carried out in the collection of data have also helped countries to quantify the extent of depletion of rattan resources and have assisted in identifying most suitable areas for conservation. For example the studies in Vietnam have shown that taxonomic descriptions and species identification are incomplete for most rattans, especially those in the Central and Southern regions. *Calamus platyacanthus,* a big sized cane similar to the *C. manan* of Southeast Asia, was found to extend from Yunnan province in China to various provinces in Vietnam.

5.3 Patterns of genetic variation

This area of rattan research has just begun to generate interest among researchers. Therefore only a few studies are available to come to an understanding of genetic diversity within and between populations.

The population and genetic diversity of three *Calamus spp.* each in the Andaman and Nicobar Islands, and Malaysia that were studied showed significant phenotypic variation. The study of 13 populations from seven provinces of *Calamus palustris* in Thailand showed that approximately 18% of the total diversity was due to differences among populations. Work in comparing random amplified polymorphic DNA (RAPD) and isozyme methods for identifying diversity in *C. palustris* in Thailand revealed highly polymorphic isozyme gene loci in this species. This meant that isozyme analysis alone was sufficient to assess genetic diversity of *C. palustris*. A study to evaluate the status of genetic diversity of rattan to construct spatial and temporal patterns of loss of their populations in the Western Ghats in India has identified the presence of 27 species of rattan. Population genetic variability assessed using *C. thwaitesii* has shown a lack of population differentiation. Another related research on the identification of genetic markers for gender-determination in two dioecious species of *Calamus* has just been initiated by the National University of Singapore.

5.4 Processes regulating genetic diversity

A study on socioeconomic aspects of loss of rattan resources in Karnataka, India, was carried out. The objectives were to determine the degree of extraction of and economic reliance on the resource at the local and state level, to identify the social and economic factors responsible for the decline in the resource and to examine the social and economic consequences of the decline of the resource. This project has just been completed and results are being analysed and suggestions on how to mitigate the impacts of extraction and land use changes are being proposed.

5.5 Human resource development

One of the constraints in executing research is availability of skilled manpower. This has been a concern of IPGRI since the rattan research project was started. Over the past few years, efforts were made by IPGRI to promote and assist in the training for conservation and use of rattan resources on a sustainable basis. Collaboration with INBAR and other organisations has increased the skills of partners to carry out work in this area. Various courses on taxonomy, conservation, ecology, silviculture, molecular approaches in plant population genetics have been held, and relevant workshops have been conducted for this purpose.

6. The Future Ahead

For sustainable conservation and use of rattan resources, it is imperative that the practice of uncontrolled exploitation should be abandoned and replaced by effective measures of conservation, cultivation and sustainable management that would also help the rural poor in the long term. The work done so far on the identification of available genetic resources in various countries would assist in the optimum utilization of rattan, including the expansion of the number of species to be brought under cultivation. IPGRI would continue to support studies on the conservation of rattan genetic resources for use under the four strategic areas identified. Its efforts will be focussed on specific areas where information is still lacking to ensure effective conservation of the species for sustainable management. It has been and will always be the policy of IPGRI to collaborate with organisations (at national and international levels) concerned with the conservation and use of rattan to bring about more effective sustainable management of these resources for improving the economic status (especially) of the rural population.

Figure 4: A rattan harvester in East Kalimantan (van Valkenburg)

REFERENCES

Dransfield, J., 2000. General introduction to rattan - The biological background to exploitation and the history of rattan research. FAO Expert Consultation on Rattan Development. 5-7December 2000, Rome, Italy. 12pp.

Dransfield, J. and Manokaran, N. eds., 1993. Plant Resources of South-East Asia, No. 6. Rattans. Pudoc Scientific Publ., Wageningen, the Netherlands. 137 pp.

Ramanatha Rao,V., Rao, A.N. and Ouedrago, A.S. (undated). Sustainable conservation and use of bamboo and rattan resources: Elements of a strategy.

Rao, A.N., Ramanatha Rao, V. and Williams, J.T. eds., 1998. Priority species of bamboo and rattan. IPGRI-APO, Serdang, Malaysia. 95 pp.

Rao, A.N. and Ramanatha Rao, V., eds. 1999. Bamboo and rattan genetic resources and use. Proceedings of the 3rd INBAR-IPGRI biodiversity, genetic resources and conservation working group meeting. 24-27 August 1997. IPGRI-APO, Serdang., Malaysia. 203 pp.

Sastry, C.B., 2000. Rattan in the twenty-first century – An outlook. FAO Expert Consultation on Rattan Development. 5-7 December 2000, Rome, Italy.

Vivekanandan, K., Rao, A.N. and Ramanatha Rao, V., eds., 1998. Bamboo and rattan genetic resources in certain Asian countries. IPGRI-APO, Serdang, Malaysia. 229 pp.

Williams, J.T. and Ramanatha Rao, V., eds., 1994. Priority sspecies of bamboo and rattan. INBAR Technical Report No. 1. INBAR and IBPGR, New Delhi.

Xu, H.C., Rao, A.N., Zeng, B.S. and Yin, G.T., eds. 2000. Research on rattans in China. Conservation, cultivation, distribution, ecology, growth, phenology, silviculture, systematic anatomy and tissue culture. IPGRI-APO, Serdang, Malaysia. 148 pp.

Annex 1

Current rattan projects supported by IPGRI

	ON-GOING STUDIES	Organization/Country
1	Distribution and status of rattan in Bardiya district of Nepal	Institute of Foresters, Nepal
2	Distribution, population status and genetic diversity of *Calamus manan* in Sumatra	Indonesian Institute of Sciences, R&D Centre for Biotechnology, Indonesia
3	Studies on rattans of Dakshina, Kannada and Kodagn districts of Karnataka with particular reference to species diversity, density of population, seed viability and germination	Mangalore University, India
	COMPLETED STUDIES	
1	Herbarium survey to determine the distribution of certain rattan species in China	Research Institute of Tropical Forestry, Guangzhou, China
2	Evaluation of *ex situ* and *in situ* conservation of rattan germplasm in China	Research Institute of Tropical Forestry, Guangzhou, China
3	Genetic assessment of three rattan species	FRIM, Malaysia
4	Identification of patterns of genetic variation among three selected rattans	Universiti Malaya, Malaysia
5	The distribution and conservation of bamboo and rattan species in Northern Thailand	Chiang Mai University, Thailand
6	Genetic diversity of *Calamus* species	Royal Forest Department, Thailand
7	Mapping genetic diversity of rattan in Western Ghats of India	ATREE, India
8	Genetic diversity and conservation of certain rattan species in Andaman Nicobar Islands and Western Ghats, South India	Kerala Forest Research Institute, Peechi, India
9	Estimation of nuclear DNA content of various rattan species	National University of Singapore
10	Distribution, phenology and conditions suitable for seed germination of certain rattans in Vietnam	Forest Science Institute of Vietnam
11	Ecogeographic survey and phenology of rattan in Nepal	Forest Research Centre, Nepal

Annex 2

List of priority rattan species for R & D
(Rao *et al.*, 1998)

Calamus manan
Calamus caesius
Calamus trachycoleus
Calamus sect. Podocephalus
Calamus andamanicus
Calamus burckianus
Calanus erinaceus
Calamus foxworthyi
Calamus merrillii
Calamus nagbettai
Calamus ovoideus
Calamus polystachys

Calamus warburghii
Calamus zeylanicus
Calamus zollingeri
Calamus palustris and relatives
Calamus inermis
Calamus nambariensis
Calamuc deeratus
Calamus tetradactylus
Calamus hollrungii and relatives

ITTO's EXPERIENCES IN PROMOTING TROPICAL NON-TIMBER FOREST PRODUCTS

Hwan Ok Ma

1. Introduction

The International Tropical Timber Organization (ITTO) was established by the International Tropical Timber Agreement (ITTA), 1983 and operates under the successor agreement to the ITTA, 1983, which was negotiated in 1994 and came into force on 1 January 1997. The ITTO headquarters are in Yokohama, Japan and the organization had a staff of 32 people as of November 2000.

The mission of ITTO is "to facilitate discussion, consultation and international cooperation on issues relating to the international trade and utilization of tropical timber and the sustainable management of its resource base".

More information on the organization can be found on the ITTO web site at www.itto.or.jp.

1.1 ITTO Structure and Functions

The governing body of ITTO is the International Tropical Timber Council (ITTC), which includes all members and meets twice a year. ITTO has 55 members, representing over 75% of the world's tropical forests and almost 90% of world trade in tropical timber products.

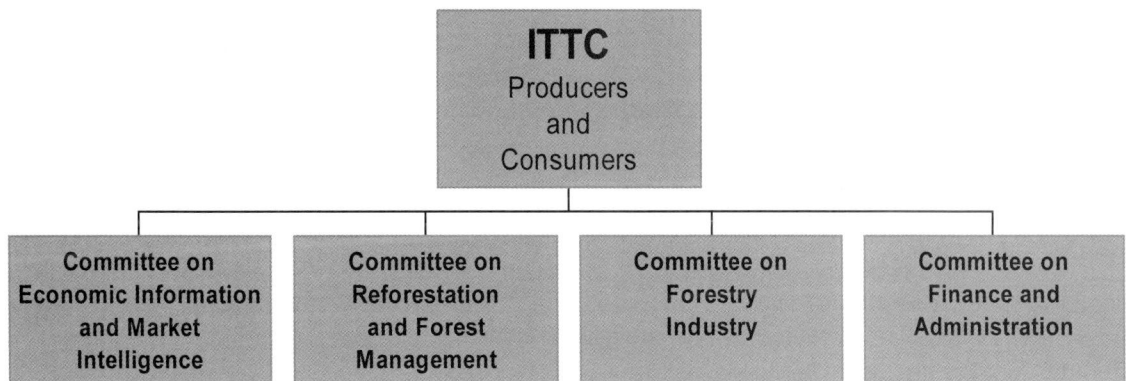

1.2 The "ITTO Objective 2000"

This is a strategy adopted in 1991 by which its members would progress towards achieving trade in tropical timber from sustainably managed forests by the year 2000. In November 2000, ITTO reaffirmed its full commitment to "moving as rapidly as possible towards achieving exports of tropical timber and timber products from sustainably managed sources".

1.3 Policy Work

Development of a series of ITTO guidelines covering:
- Guidelines for the sustainable management of natural tropical forests (1990);
- Criteria for the Measurement of Sustainable Tropical Forest Management (1992);
- Guidelines for the Establishment and Sustainable Management of Planted Tropical Forests (1993);
- Guidelines for the Conservation of Biological Diversity in Tropical Production Forests (1993);
- Guidelines on Fire Management of Tropical Forests (1997);
- Criteria and Indicators for the Sustainable Management of Natural Tropical Forests (1998);
- Manual for the Application of Criteria and Indicators for Sustainable Management of Natural Tropical Forest – national level and forest management unit level (1999).

1.4 Project Funding in ITTO: (in million US$)

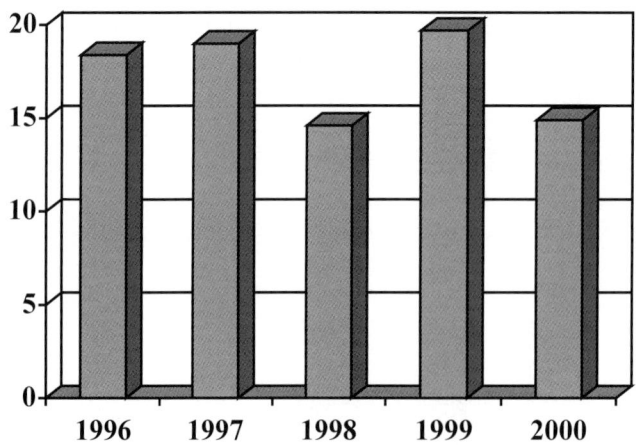

2. ITTO's Involvement in NTFPs

Since ITTO started operations in 1987, the promotion of Sustainable Forest Management (SFM) has been at the core of its activities. Achievement of SFM implies that all forest resources, including NTFPs, are taken into consideration. ITTO believes that NTFP development is an important tool for local community development and resource conservation, and as a strategy for rural poverty alleviation.

2.1 Main Lessons Learned from ITTO Projects

All forest communities use NTFPs intensively. However, not much emphasis has been put on NTFPs in the past, and NTFP-related activities are poorly documented. Guidelines for sustainable use of NTFPs are non-existent and even difficult to establish. Further, technical and financial resources and market information are not easily available to rural communities. However, NTFPs have a great potential to raise income levels of rural communities and to contribute to SFM.

On-going ITTO Projects in NTFPs

Title	Implementing Agency	ITTO Budget
Sustainable Management and Utilization of Sympodial Bamboos in South China	Research Institute of Subtropical Forestry, CAF, China	US$482,452
Non-timber Production and Sustainable Development in the Amazon	University of Brasilia Brazil	US$387,185
Processing and Utilization of Almaciga Resin as Source of Industrial Chemicals	Forest Products Research and Development Institute (FPRDI), Philippines	US$342,743
Management and Utilization of Paca (*Guadua Sacocapa*)	National Institute for Natural Resources, Peru	US$601,900
Promotion of the Utilization of Bamboo from Sustainable Sources in Thailand	Forest Research Office, RFD Thailand	US$452,996
Improvement of Sustainable Management and Utilization of Tropical NTFPs in Cambodia	Department of Forestry and Wildlife, Cambodia	US$77,648
Promotion of Sustainable Utilization of Rattan from Plantation in Thailand	Forest Research Office, RFD Thailand	US$292,457
Utilization, Collection, and Trade of Tropical NWFPs in the Philippines	FPRDI Philippines	US$345,196

3. ITTO project on Rattan

<u>ITTO PD 24/00, "Promotion of sustainable utilization of rattan from plantation in Thailand"</u>

This is a 3-year project (from 2001 to 2003) to be implemented by the Forest Products Research and Silvicultural Research Divisions of the Forest Research Office, Royal Forest Department, Thailand.

The objectives of the project are:
- to study and develop techniques for managing rattan plantations for sustainable production with a view to developing and disseminating guidelines and technologies on plantation management and harvesting of rattan. .
- to promote the efficient utilization of rattan shoots and canes for value-added products by developing guidelines for utilization of rattan and transferring technologies to support socio-economic development.
- to establish community-owned enterprises.

Problems to be addressed by the project are:

for Rattan canes:
- Shortage of rattan canes
- Lack of knowledge on managing rattan plantations for sustainable cane production
- Lack of knowledge on production of value-added rattan products
- Need for research on basic properties of important rattan species

for Rattan shoots:
- Demand for rattan shoots
- Lack of knowledge on cultivating and managing rattan plantations for sustainable shoot production
- Lack of knowledge on processing of rattan shoots

Programmes and Operational Activities

Location	Sakon Nakhon (North)	Krabi (South)
Area of plantation	2 ha	2.2 ha (in existing 5 and 10-year old plantations)
Species	for shoot production *Calamus siamensis* (bitter taste): 1ha *Calamus sp.* (sweet taste): 1 ha	for cane production *Calamus longisetus* (Kumpuan) *Calamus latifolius* (Pong) *Calamus caecius* (Tha kha thong)
Treatments	Spacing/Fertilizer applications Harvesting methods	Fertilizer applications Number of canes being harvested

Expected Outputs:
- Two demonstration plots on management of rattan plantations
- Guidelines for sustainable management of rattan
- Study on physical and working properties
- Techniques for preservation, processing (bending and bleaching of rattan canes), harvesting and edible shoot processing
- A cottage industry through small cooperatives for the production of rattan shoots and for the production of rattan furniture parts to develop value-added rattan products.

4. Concluding Remarks

Since ITTO began operations in 1987, it has funded more than 500 projects, pre-projects and activities valued at some US$200 million and the promotion of Sustainable Forest Management (SFM) has been at the core of its activities. Achievement of SFM implies that all forest resources, including NTFPs, are taken into consideration. Promotion of NTFPs is identified as one of the priority actions of ITTO to improve the tropical timber resource base in the ITTO Libreville Action Plan currently implemented by the Organization.

Figure 1: Map of Thailand with situation of field sites of rattan project

One of the lessons learnt from ITTO projects is that, in most cases, the immediate and most pressing problem faced by rural communities is low-income levels and inadequate social facilities. Noting that poverty alleviation is one of the most urgent tasks in developing countries, promotion of the sustainable use of NTFPs, based on the needs of the community, can provide income-generation activities in rural communities. This will enhance the implementation of economically and ecologically sound SFM practices.

Recognizing that NTFPs have a great potential to alleviate rural poverty and to contribute to SFM, ITTO looks forward to further collaboration and joint activities with FAO, INBAR, CIFOR and other organizations involved in the promotion of NTFPs.

STATUS OF RATTAN RESOURCES AND USES IN AFRICA AND ASIA

STATUS OF RATTAN RESOURCES AND USE IN WEST AND CENTRAL AFRICA

Terry C.H. Sunderland

Abstract

Four genera of rattan palms, represented by 20 species, occur in West and Central Africa. In common with their Asian relatives, the rattans of Africa form an integral part of subsistence strategies for many rural populations as well as providing the basis of a thriving cottage industry. Although many of the African rattan species are used locally for a multiplicity of purposes, the commercial trade concentrates on the bulk harvest of only a few widespread and relatively common species.

African rattans have long been recognised by donor agencies and national governments as having a potential role to play on the world market as well as a great role within the regional Non-Timber Forest Products (NTFPs) sector of Africa. However, the development of the rattan resource in Africa has until recently, been hindered by a lack of basic knowledge about the exact species used, their ecological requirements and the social context of their utilisation. Hence it has not been possible to design appropriate management strategies that might be implemented to ensure their sustainable, and equitable, exploitation. As increased interest is being shown in the potential role of high value NTFPs to contribute to the conservation and development paradigm, rattan has been one of the frequently mentioned products that could be developed and promoted in a meaningful way. Recent research has concentrated on the provision of information on the taxonomy, ecology and utilisation of these taxa. Now that this baseline information is available, rattan research in Africa is now concentrating of the development and promotion of the rattan resource from both ecological and socio-economic perspectives.

1. A brief introduction to the biology of African rattans

1.1 Morphological distinctness from the Asian rattans

In Africa, there are 20 species of rattan, representing four genera that are relatively easy to differentiate, particularly through the morphology of their climbing organs (Sunderland, 2000). Calamoid palms climb with the aid of two main organs; they may either have a flagellum or possess a cirrus. Flagella only occur in certain species of *Calamus*, including *C. deërratus*, the sole representative of *Calamus* in Africa. The flagellum arises directly from the sheath and is regarded as a modified inflorescence (Fisher and Dransfield, 1977; Dransfield, 1978; Baker *et al.*, 1999). Indeed, inflorescences of *C. deërratus* are flagellate.

The remaining taxa within the Calamoideae, particularly those of Asian origin, climb with the aid of a cirrus, a whip-like extension at the distal end of the leaf rachis armed with short, recurved thorns that often resemble a cat's claw (Tomlinson, 1990). However, the three rattan genera endemic to Africa, *Laccosperma, Eremospatha* and *Oncocalamus*, possess a vegetative morphology unique within the Calamoideae in that the cirrus is actually a marked extension between the distal leaflets rather than beyond them. The leaflets are present as reduced, reflexed thorn-like organs termed acanthophylls. This structure is also present in some members of the unrelated genera present only in the new world: *Chamaedorea* (sub-family Ceroxyloxideae; tribe Hyophorbeae) and *Desmoncus* (sub-family Arecoideae; tribe Cocoeae) (Uhl and Dransfield, 1987).

In common with other members of the genus, *Calamus deërratus* possesses female inflorescences where the flowers are arranged in pairs comprising a fertile female flower and a sterile male flower, while the male infloresecnce has rows of solitary flowers. The inflorescence units of the endemic rattan genera of Africa are quite distinct. For example the genera *Eremospatha* and *Laccosperma* have pairs of hermaphroditic flowers[4]. Although pairs of unisexual flowers are a common feature within the Calamoideae, the dyad composed of hermaphroditic flowers is unique to *Eremospatha* and *Laccosperma* within the Palmae and is considered to be an unspecialised form of flower arrangement (Uhl and Dransfield, 1987; Baker *et al.*, 1999).

In addition, the flower cluster of *Oncocalamus* is distinctive and complex, not only within the Calamoideae, but also within the Palmae as a whole. *Oncocalamus* is monoecious, and the flower cluster consists of a central 1-3 female flowers with two lateral groups subtended by a single bract, with each group bearing basal 1-3 female flowers and 3-5 distal male flowers. The unusual flower cluster of the African taxa, and *Oncocalamus*, in particular, suggests that a complex evolution of the Calamoideae has occurred on the African with much extinction, caused by dramatic climatic upheaval, leaving only isolated lineages. The speciation patterns exhibited today by African palms, in that they have a distinct Guineo-Congolian centre of diversity, probably due to the maintenance and later speciation of forest refugia during periods of these climatic changes, supports this assertion.

1.2 Cane anatomy

Anatomical studies of three of the four (*Laccosperma, Eremospatha* and *Calamus*) African genera have recently been undertaken in Ghana (Oteng-Amoako and Ebanyele, in press). The initial results of the study suggest that the thickness of the fibre walls, the proportion of fibre tissues and metaxylem vessel diameter differ significantly between the genera, hence influencing the relative utility of the members of each.

The relatively higher proportion of thick-walled fibres and narrower diameter of metaxylem vessels suggests that the genus Laccosperma has a greater density and hence strength properties than the canes of *Eremospatha* and *Calamus deërratus*. These latter taxa have a relatively higher proportion of thinner wall fibres and larger metaxylem vessels, which contribute to greater void volume of the stems resulting in lower density and strength (*ibid.*). The study also revealed that fibre wall thickness, fibre proportion and metaxylem diameter, which are the likely determinants of rattan quality, do not differ significantly between *Calamus* and *Eremospatha* and therefore these genera are included within the same density and strength groupings. These findings concur with those of Wiener and Liese (1994) who additionally examined material of *Oncocalamus*. This latter genus was found to have very thin fibre walls and very large metaxylem vessels and possessed the least desirable properties of density and strength of any of the African rattans.

These anatomical conclusions generally correspond with those of researchers of rattan utilisation in Africa. It is generally accepted that the large-diameter species of *Laccosperma* are particularly durable, whilst *Oncocalamus* is particularly weak and brittle and, as such, is not commonly valued as a source of cane (Profizi, 1986; Defo, 1997; Defo, 1999; Sunderland, 1999a; 1999b). However, it is surprising that *Eremospatha* and *Calamus* are found to be anatomically similar, and hence share similar cane properties, as most workers note that *Calamus deërratus* is considered of inferior quality to that of the desired species of *Eremospatha* and is only utilised in the absence of other species. Further anatomical studies, which are currently under way, might shed more light on this anomaly.

---

[4] Less commonly, *Laccosperma* may also possess triads of flowers.

2. Ecology and distribution

Rattans in Africa are widespread throughout West and Central Africa and are a common component of the forest flora. Some species, such as *Laccosperma secundiflorum* and *Eremospatha macrocarpa*, have large ranges and occur from Liberia to Angola, whilst *Calamus deërratus* is particularly widely distributed and occurs from the Gambia, across to Kenya and southwards to Zambia. In terms of diversity, the greatest concentration of rattan species, along with the highest levels of endemism, are found in the Guineo-Congolian forests of Central Africa. Eighteen of the 20 known African rattan species occur in Cameroon; the "hinge" of Africa. The diversity of rattans in the Upper Guinea forests, by comparison, is somewhat poor with only seven species, none of which are endemic to that region.

Within this forest zone, rattans occur in a wide range of ecological conditions. The majority of the species occur naturally in closed tropical forests and are early gap colonisers. Because of this, many of the taxa are extremely light demanding and respond well to a limited reduction in the forest canopy. Increases in forest disturbance, such as through selective logging activity encourages the regeneration of rattans and these palms are often a common feature along logging roads and skid trails. For some taxa such as some species of *Oncocalamus*, their light-demanding nature is such that they are often the earliest colonisers of heavily disturbed areas. Other species of rattan, notably *Calamus deërratus*, grow in permanently and seasonally inundated forest or swamps, whilst other species, such as *Laccosperma opacum* and *L. laeve*, are highly shade-tolerant and prefer to grow under the forest canopy.

The seed of most rattans in Africa are dispersed predominantly by hornbills (Whitney *et al.*, 1998). However, primates, predominantly the drill (*Mandrillus leucophaeus*) and mandrill (*Mandrillus sphinx*), chimpanzees (*Pan troglodytes*) and gorillas (*Gorilla gorilla*) along with elephants (Gartlan, *pers. comm.*; White and Abernethy, 1997; Sunderland, 2000) are also key dispersal agents. The seeds are often scattered far from the mother plant. Limited predation, and sometimes catching by rodents accounts for some additional, although limited, dispersal. Interestingly, significant germination also occurs near to the parent plants through natural fruit fall, particularly in areas where over-hunting has led to a significant decline in faunal dispersal agents. After germination, rattan seedlings can remain on the forest floor for some time waiting for the optimum light conditions needed to begin the long journey to the canopy. Interestingly, despite intensive field work and herbarium collection in the past three years, there appears to be no obvious phenological pattern to flower development and seed production for the majority of the species.

3. The resource base

Although numerous studies have concentrated on evaluating the local importance of rattan in Africa, very few have attempted to adequately define the resource base. However, it is now known that the utilisation of rattan to supply the thriving cottage industry is limited to a few species (Sunderland, in press). Table 1 (below) presents the major commercial species of rattan utilised in each region.

4. Conservation Status of African Rattans

It is reported that the demand for rattan is increasing and much greater amount of cane is being processed in many areas of Africa today than was being worked five or ten years ago (Morakinyo, 1995; Ndoye, 1994; Falconer, 1994; Townson, 1995; Trefon and Defo, 1998; Defo, 1997; Sunderland, 1998; Defo, 1999; Sunderland 1999a; 1999b; Kialo, 1999; Minga, in press; Holbech, 2000; Sunderland *et al.*, in press; Oteng-Amoako and Obiri-Darko, in press). This has led to a

Table 1. Commercially important rattan species by region

Region	Commercially utilised species
West Africa (Senegal, Côte d'Ivoire, Ghana, Benin, W. Nigeria)	*Laccosperma secundiflorum* (P. Beauv.) Kuntze *Eremospatha macrocarpa* (G. Mann & H. Wendl.) H. Wendl. *Eremospatha hookeri* (G. Mann & H. Wendl.) H. Wendl. *Calamus deërratus* (G. Mann & H. Wendl.)
West/Central Africa (E. Nigeria, Cameroon, Congo, Gabon, E. Guinea)	*Laccosperma secundiflorum* (P. Beauv.) Kuntze *Laccosperma robustum* (Burr.) J. Dransf. *Eremospatha macrocarpa* (G. Mann & H. Wendl.) H. Wendl.
Central Africa (DR Congo, Central African Republic)	*Laccosperma robustum* (Burr.) J. Dransf. *Eremospatha haullevilleana* de Wild. *Eremospatha macrocarpa* (G. Mann & H. Wendl.) H. Wendl.
Southern/Eastern Africa (Zambia, Uganda, Kenya, Tanzania)	*Calamus deërratus* G. Mann & H. Wendl. *Eremospatha haullevilleana* de. Wild.

* Indicates primary commercial species

Table 2. The conservation status of African rattan species

Species	Geographical range (km²)	IUCN Category
Calamus deërratus G. Mann & H. Wendl.	8,049,170	Not threatened
Eremospatha barendii sp. nov.	One collection only	Endangered
E. cabrae de Wild.	1,918,050	Not threatened
E. cuspidata (G. Mann & H. Wendl.) H. Wendl.	1,891,190	Not threatened
E. haullevilleana de Wild.	2,703,930	Not threatened
E. hookeri (G. Mann & H. Wendl.) H. Wendl.	1,102,420	Not threatened
E. laurentii de Wild.	2,731,880	Not threatened
E. macrocarpa (G. Mann & H. Wendl.) H. Wendl.	4,259,660	Not threatened
E. quinquecostulata Becc.	9,276	Vulnerable
E. tessmanniana Becc.	5,899	Vulnerable
E. wendlandiana Dammer ex Becc.	604,086	Not threatened
Laccosperma acutiflorum (Becc.) J. Dransf.	1,485,230	Not threatened
L. laeve (G. Mann & H. Wendl.) H. Wendl.	1,226,210	Not threatened
L. opacum (G. Mann & H. Wendl.) Drude	1,807,940	Not threatened
L. robustum (Burr.) J. Dransf.	1,537,390	Not threatened
L. secundiflorum (P. Beauv.) Kuntze	3,195390	Not threatened
Oncocalamus macrospathus Burr.	701,830	Not threatened
O. mannii (H. Wendl.) H. Wendl.	129,432	Not threatened
O. sp. nov. ('Vuley')	18,423	Vulnerable
O. wrightianus Hutch.	2,872	Endangered

significant decline in wild stocks and has resulted in considerable local scarcity. This scarcity and the associated irregular supply of unprocessed rattan have been identified as one of the major constraints to the continued development of the industry.

In this regard, now that the taxonomic base of the African rattan sector has been established (Sunderland, 2000) it is now possible to determine the global conservation status of each species.

This has been calculated using the standard IUCN conservation categories where among other criteria, geographical limits in species distribution are used to determine the conservation status.

5. The African Rattan Trade

During the colonial period, there existed a significant trade in cane and cane products in Africa. In particular, Cameroon and Gabon supplied France and its colonies (Hédin, 1929), and Ghana (formerly the Gold Coast) supplied a significant proportion of the large UK market during the inter-war period (Anon, 1934). The export industry was not restricted to raw cane and in 1928 alone over 250,000 FF worth of finished cane furniture was exported from Cameroon to Senegal for the expatriate community there (Hédin, 1929). More recently, an initiative promoted by UNIDO in Senegal was exploiting wild cane for a large-scale production and export (Douglas, 1974), although this enterprise folded not long after its establishment due to problems securing a regular supply of raw material.

Table 3. Raw rattan cane exports from Douala and Kribi to France 1926 to 1928
(modified from Hédin, 1929)

Year	Tonnes Exported	Value (FF)
1926	100	250,000
1927	58	137,000
1928 (Douala)	32	80,000
1928 (Kribi)	34	85,000

6. The nature of the trade

The conditions and circumstances under which rattan is harvested and transported in Africa are remarkably consistent throughout its range. The majority of the harvesting for commercial trading is undertaken by individuals usually farmers or hunters, or other rural people primarily involved in other occupations. Rattan harvesting provides many of these individuals with extra revenue, particularly in times of need such as for medical expenses or the payment of annual school fees (Trefon and Defo, 1998; Sunderland, 1998). Many cash-crop farmers also harvest rattan to obtain extra capital to purchase chemicals, planting stock and other necessary items for their primary occupation (*ibid.*). Despite the recognised capital returns of rattan harvest and sale, the unpleasant and difficult nature of rattan harvesting means that most harvesters state they would prefer to concentrate on their primary occupations given the opportunity.

In general, rattan harvesters tend to work in the same forest area, and return each time they need to cut cane. If the harvester is not an indigene of the area, the chief of the local village is paid a small retainer for providing access to the forest. The harvesters usually prefer to collect as close to a motorable road as possible to avoid head-portering the bundled canes too far. However, local scarcity near many urban centres now forces many harvesters further into the forest (Sunderland, 1998; Defo, 1999; Profizi, 1999). The added porterage resulting from this increased range is slowly generating an increase in the raw cane prices, which is being felt at the market level.

Village-based harvesters transport the harvested rattan to the urban markets themselves, or they may sell at the village to a local trader who then transport the cane for sale to urban artisans. Some urban-based artisans harvest rattan themselves, although this is often only the case where there is close proximity to the wild resource. Falconer (1994), and Oteng-Amoako and Obiri-Darko (in press) provide a good overview of the production to consumption system of rattan in Ghana, as do Defo (1999) and Sunderland *et al.* (in press) for Cameroon.

Although many of the commercial species of rattan respond well to selective logging activities, logging has also resulted in increased rattan exploitation. The development of a wide network of logging roads throughout many forest areas in West and Central Africa has enabled greater access to otherwise inaccessible areas of forest. Indeed, the logging trucks themselves are often known to be responsible for the transport of harvested rattan (Defo, 1997; Sunderland, 1998).

Indigenous management systems for the rattan resource in Africa are unknown, and, throughout its range, rattan is considered an "open-access" resource; there are very few, if any customary laws regulating the harvest of rattan from the wild. This is also mirrored in the national legislation for most countries. Those States that require the exploitation of forest products to be governed by the issue of licenses and permits, often do not adequately monitor the exploitation of these resources, nor receive the full forestry taxes related to that exploitation. In general though, many national forestry codes still do not include the exploitation of non-timber forest products in their regulations and the over-harvesting of many commercially important products, including rattan, continues unabated and uncontrolled. However, as will be discussed, these legislative and institutional constraints to sustainability are currently being addressed.

Figure 1. Generalized "production-to-consumption" system for rattan in Africa

Production → →   →   →   →   →   →   → Consumption

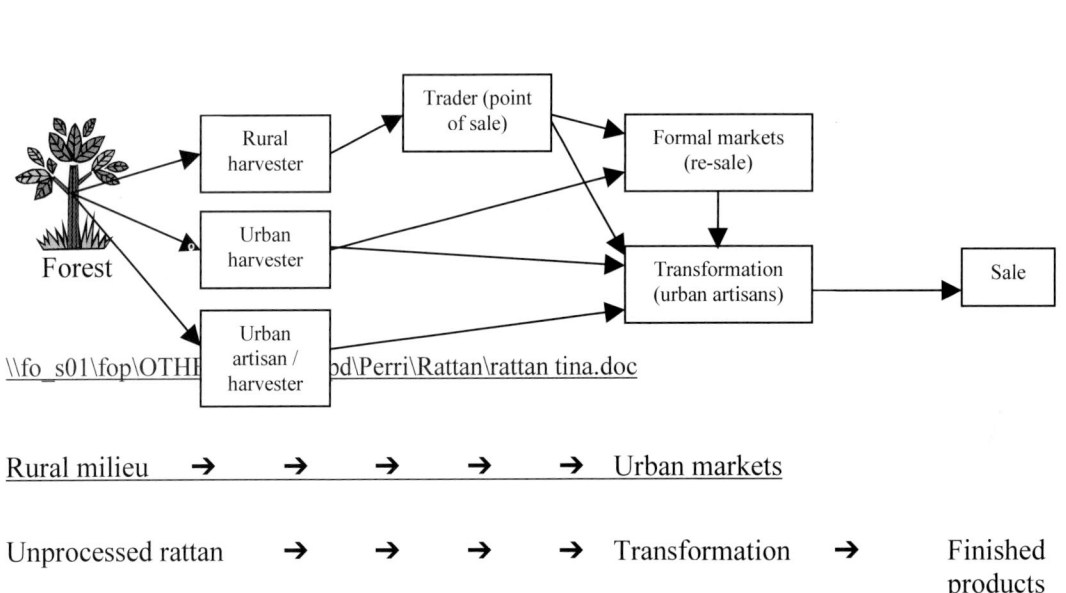

\\fo_s01\fop\OTHE   d\Perri\Rattan\rattan tina.doc

Rural milieu   →   →   →   →   → Urban markets

Unprocessed rattan   →   →   →   → Transformation   →   Finished products

7. Amount and Value of the Trade

Large quantities of raw cane enter the urban centres of West and Central Africa each day (Morakinyo, 1995; Ndoye, 1994; Falconer, 1994; Townson, 1995; Trefon and Defo, 1998; Defo, 1997; Sunderland, 1998; Defo, 1999; Sunderland, 1999a; 1999b; Kialo, 1999; Minga, in press; Holbech, 2000; Sunderland *et al.*, in press; Oteng-Amoako and Obiri-Darko, in press). Table 4 (below) summarises the findings of some of these studies where quantification of field data has been possible.

Table 4. The scale and value of the African rattan trade in selected urban markets

City (country)	Population (sample size)	Estimated amount of cane used / month (m)	Estimated mean annual value (US$)	Reference
Lagos (Nigeria)	10,712,800 (not known)	180,000 m	1,141,180	Morakinyo (1994)
Accra (Ghana)	1,512,800 (27 enterprises)	not known	64,080	Falconer (1994)
Kumasi (Ghana)	602,000 (11 enterprises)	not known	95,475	Falconer (1994)
Ankasa (Ghana)	Not known (12 markets)	4,300m (all species)	62,000	Holbech (2000)
Bata (Equatorial Guinea)	80,000 (15 enterprises)	20,550m (all species)	27,400	Sunderland (1998)
Douala (Cameroon)	1,262,000 (25 enterprises)	26,955m (large dia.) 28,875m (small dia.)	127,405	Sunderland *et al.*, (in press)
Yaounde (Cameroon)	1,157,400 (31 enterprises)	23,165m (large dia.) 29,765m (small dia.)	103,500	Sunderland *et al.*, (in press)
Kinshasa (DR Congo)	2,456,000 (114 enterprises)	13,760m (large dia.) 14,448m (small dia.)	56,600	Minga (in press)

8. Processing and Transformation

Processing of raw cane essentially entails the removal of the epidermis (skin) from the stem and the drying of the raw cane prior to its use. Immature stems, or the very apex of mature stems, where the leaf sheath is also present are not used, and are often left or discarded at the time of harvest. The processing of raw cane throughout much of Africa is undertaken manually, with the stems being scraped with kitchen knives to remove the skin followed by drying, usually undertaken in the open air. This rudimentary means of processing is not only labour intensive, but also results in inferior quality cane being available for artisan use and hence limits the value of the finished products. This inferior quality has also led to speculation that the quality of cane in Africa inherently poor (Dransfield, *pers.comm.*). However, this speculation has not been supported by thorough anatomical studies and it is possible that if processed and transformed more efficiently cane from Africa could, in terms of quality, rival that of Asia.

There are also long-term conservation benefits from improved methods of processing and transformation. Most notable of these is the fact that a more durable and longer lasting product will ensure that less cane needs to be continually harvested from the wild. In addition, from the social perspective the advantages of urban artisans producing better quality products are clear and directly relate to the current DFID initiative of ensuring the valorisation of forest products contribute to overall poverty alleviation.

In this respect, there are currently initiatives to introduce appropriate processing and transformation technology from Asia that are suitable for the African milieu (Sunderland and Nkefor, 1999). A model processing unit has recently been constructed in Limbe, Cameroon and will be used primarily as a training and will function as a demonstration unit. Similar units will be established in Ghana and Nigeria over the course of the next two years.

9. Discussion: Manifold Routes to Sustainability

The sustainable harvesting and management of the African rattan resource is primarily hindered by a paucity of a sound information on stocking, growth, yield and harvest intensity. In addition, the lack of adequate land and resource tenure precludes many attempts at long-term and sustainable harvesting and the fact that rattan is considered an "open-access" resource throughout much of its range mitigates the prospects for long-term sustainable management. However, a number of research strategies are currently being developed to address these shortfalls in baseline information and institutional and social constraints.

The harvesting techniques employed in the extraction of rattan in Africa, and which are generally the same despite the considerable geographic variation on the continent, have an impact on potential sustainability. Particularly for clustering species, the mature stems selected for harvest are those without lower leaves (i.e. where the leaf sheaths have sloughed off) and usually only the basal 10-20m is harvested; the upper "green" part of the cane is too soft and inflexible for transformation and is often left in the canopy. In many instances, all the stems in a cluster may be cut in order to obtain access to the mature stems; even those that are not yet mature enough for exploitation and sale. This is particularly an issue where resource tenure is uncertain or weak.

However, where resource tenure is somewhat more clearly defined, younger stems, are not removed and are left to regenerate and provide future sources of cane, usually on a 2-3 year rotation. Despite the fact that this example of better "stool management" relies on adequate land and resource tenure, there is considerable reason for optimism in the African context. Currently there is a significant paradigm shift from the management of forest resources being controlled by the State to those of community-based management regimes. Formal legislation in this regard is now being developed and implemented in many rattan-producing countries. Through the empowerment of forest communities in this way there is significant potential to ensure the long-term sustainable, and equitable, exploitation of a wide range of forest resources, not only rattan.

Rattans are currently harvested exclusively from wild populations in Africa. Unlike some areas of Southeast Asia where rattan is traditionally cultivated as part of mixed gardens by sedentary cultivators, or is planted in recently-burned forest by shifting cultivators (Godoy, 1992), no known similar cultivation practices exist in West and Central Africa. However, the ecological and social factors prevalent here are favourable to the development of a cultivated and managed rattan resource.

In this regard, recent research by the African Rattan Research Programme has concentrated on aspects of seed storage and pre-treatments. The material made available by these trials has led to the recent establishment of an experimental silvicultural trial. The trial consists of a 1-hectare plot of *Laccosperma secundiflorum*, planted beneath obsolete rubber and has been undertaken in collaboration with the Cameroon Development Corporation (CDC). Further community-based trials, within the legislative context of community forest management, have also recently been established in Cameroon, soon to be followed by similar initiatives in Ghana and Nigeria. These latter trials are concentrating on the introduction of rattans into agroforestry systems and enrichment planting of farmbush and secondary forest. Annual growth rates as well as the economic viability of these cultivation systems are currently being monitored and assessed.

Recent initiatives aimed at introducing certification schemes for NTFP resources as well as for timber, also have potential for the sustainable management of the rattan resource in both Africa and Asia. Recent guidelines for the criteria for the certification of rattans have recently been developed by Sunderland and Dransfield (in press).

10. Conclusion

As essential biological, ecological and socio-economic information on the African rattan resource becomes available and suitable strategies to ensure sustainability are implemented, there is significant potential for the rattans of Africa to contribute significantly not only to the regional development of the resource, but also to the thriving global market. Through applied forest management regimes, and through the development of community-based management supported by appropriate legislation, African rattans could provide a real opportunity for the meaningful, and sustainable, development of rural areas and, potentially, for forest conservation through extractive management.

Figure 5: Rattan stems for river transport by rafts (Siebert)

Acknowledgements

I would like to thank the Central African Regional Programme for the Environment (CARPE), the United States Forest Service and the International Network for Bamboo and Rattan for funding the first phase (1996-2000) of the African Rattan Research Programme. The programme is now currently funded by the Forestry Research Programme (FRP) of the UK's Department for International Development (DFID) under grant number R7636 ZF0139.

REFERENCES

Anon, 1934. Shipments of rattans to the UK and USA from the Gold Coast. Gold Coast Forestry Department Annual Report.

Baker, W.J., J. Dransfield, M.M. Harley & A. Bruneau, 1999. Morphology and cladistic analysis of sub-family Calamoideae (Palmae). In: A. Henderson & F. Borchsenius (eds.) Evolution, variation and classification of palms. *Memoirs of the New York Botanical Garden.* 83: 307-324

Defo, L., 1997. *La filière des produits forestiers non ligneux: l'exemple du rotin au Sud Cameroun.* Dept. of Geography, University of Yaounde.

Defo, L., 1999. Rattan or porcupine? Benefits and limitations of a high value non-wood forest product for conservation in the Yaounde region of Cameroon. (In): T.C.H. Sunderland & L.E. Clark (eds.). *The non-wood forest products of Central Africa: current research issues and prospects for conservation and development.* Food and Agriculture Organization. [in press].

Douglas, J.S., 1974. *Utilisation and Industrial Treatment of Rattan Cane in Casamance, Senegal (Return Mission).* United Nations Industrial Development Organization. New York.

Dransfield, J., 1978. The growth forms of rainforest palms. In: P.B. Tomlinson & M. Zimmerman (eds). *Tropical trees as living systems.* Cambridge University Press. pp 247-268

Falconer, J., 1994. *Non-timber forest products in southern Ghana. Main report.* Natural Resources Institute. UK.

Fisher, J.B. & J. Dransfield, 1977. Comparative morphology and development of inflorescence adnation in rattan palms. *Bot. J. Linn. Soc.* 75: 119-140

Godoy, R., 1992. Raw Materials for Craft Industries: The Case of Rattan. In: T. Panyatou & P. Ashton. *Not by Timber alone: Economics and Ecology for Sustaining Tropical Forests.* Island Press. UK.

Hedin, L., 1929. Les rotins au Cameroun. *Rev. Bot. Appl.* Vol.9: 502-507.

Holbech, L.H., 2000. *Non-timber forest products survey: market survey and trade route assessment around the Ankasa Protected area.* (Unpubl.) Report for the Protected Area Development Programme, Western Region, Ghana.

Kialo, P., 1999. *Le marché du rotin à Libreville: stratégies pour sa formalisation.* Paft-Gabon Information. No. 6

Minga, M.D. [in press]. L'impact de l'exploitation du rotin sur la préservation de la forêt à Kinshasa, République Démocratique du Congo. In: T.C.H. Sunderland & J.P. Profizi (eds). *New research on African rattans.* INBAR. Beijing.

Morakinyo, A.B., 1994. *The ecology and silviculture of rattans in Africa: a management strategy for Cross River State and Edo State, Nigeria.* MSc Dissertation. University College of North Wales, Bangor.

Morakinyo, A.B., 1995. Profiles and pan-African distributions of the rattan species (Calmoideae) recorded in Nigeria. *Principes.* 39(4), pp 127-209.

Ndoye, O., 1994. *New employment opportunities for farmers in the humid forest zone of Cameroon: The case of palm wine and rattan.* Paper presented to the Rockefeller Fellow Meeting, Addis Ababa, Ethiopia. November 14-18.

Profizi, J.-P., 1999. The management of forest resources by local people and the state in Gabon. (In): T.C.H. Sunderland & L.E. Clark (eds.). *The non-wood forest products of Central Africa: current research issues and prospects for conservation and development.* Food and Agriculture Organization. [in press].

Oteng-Amoako, A.A. & E. Ebanyele. [in press]. The anatomy of five economic rattan species from Ghana. In: T.C.H. Sunderland & J.P. Profizi (eds). *New research on African rattans.* INBAR. Beijing.

Oteng-Amoako, A.A. & B. Obiri-Darko. [in press]. Rattan as a sustainable cottage industry in Ghana: the need for development interventions. In: T.C.H. Sunderland & J.P. Profizi (eds). *New research on African rattans.* INBAR. Beijing.

Profizi, J.P., 1986. Notes on West African rattans. *RIC Bulletin.* 5(1): 1-3

Sunderland, T.C.H., 1998. *The rattans of Rio Muni, Equatorial Guinea: utilisation, biology and distribution.* A report for the European Union Project No.6 ACP-EG-020: Proyecto Conservación y Utilización Racional de los Ecosistemas Forestales de Guinea Ecuatorial (CUREF).

Sunderland, T.C.H., 1999a. The rattans of Africa. In: R. Bacilieri & S. Appanah (eds.) 1999. *Rattan cultivation: Achievements, Problems and Prospects.* CIRAD-Forêt & FRIM, Malaysia. pp 237-236

Sunderland, T.C.H., 1999b. New research on African rattans: an important non-wood forest product from the forests of Central Africa. In: T.C.H. Sunderland, L.E. Clark & P. Vantomme (eds). *The non-wood forest products of Central Africa: current research issues and prospects for conservation and development.* Food and Agriculture Organization. Rome. pp 87-98

Sunderland, T.C.H., 2000. *The taxonomy, ecology and utilisation of African rattans (Palmae: Calamoideae).* PhD Thesis. University College, London and the Royal Botanic Gardens, Kew.

Sunderland, T.C.H. & J.P. Nkefor, 1999. *Technology transfer between Asia and Africa: rattan cultivation and processing.* African Rattan Research Programme. Technical Note No. 5.

Sunderland, T.C.H. [in press]. Indigenous nomenclature, classification and utilisation of African rattans. In: L. Maffi & T. Carlson (eds.). *Ethnobotany, and Conservation of Biocultural Diversity.* Advances in Economic Botany. New York Botanical Garden

Sunderland, T.C.H. & J. Dransfield. [in press]. Certification guidelines for rattans. In: P. Shanley, S. Laird, A. Pierce & A. Guillen (eds). *The Management and Marketing of Non-Timber Forest Products: Certification as a Tool to Promote Sustainability.* RBG Kew/WWF/UNESCO People and Plants Series no. 5.

Sunderland, T.C.H., L. Defo, N. Ndam, M. Abwe & I. Tamnjong. [in press]. A socio-economic profile of the rattan trade in Cameroon. In: T.C.H. Sunderland & J.P. Profizi (eds). *New research on African rattans.* INBAR. Beijing.

Tomlinson, P.B., 1990. *The structural biology of palms.* Oxford University Press.

Townson, I., 1995. *Incomes from non-timber forest products: patterns of enterprise activity in the forest zone of southern Ghana.* ODA Forestry Research Programme.

Trefon, T. & L. Defo, 1998. *Can rattan help save wildlife?* APFT Briefing Note No. 10.

Uhl, N.W. & Dransfield, J., 1987. *Genera palmarum: a classification of palms based on the work of H.E.Moore Jr.* pp 610. The International Palm Society & the Bailey Hortorium, Kansas.

Weiner, G. and Liese, W., 1994. Anatomische untersuchungen an westafrikanischen rattanpalmen (Calamoideae). *Flora.* 189: 51-61

White, L. & K. Abernethy, 1997. *A guide to the vegetation of the Lopé Reserve, Gabon.* Wildlife Conservation Society. New York. USA.

Whitney, K.D., M.K. Fogiel, A.M. Lamperti, K.M. Holbrook, D.M. Stauffer, B.D. Hardesty, V.T. Parker, and T.B. Smith, 1998. Seed dispersal by *Ceratogymna* hornbills in the Dja Reserve, Cameroon. *J. Trop. Ecol.* 14: 351-371

RATTAN AS SUSTAINABLE INDUSTRY IN ARICA: THE NEED FOR TECHNOLOGICAL INTERVENTIONS

A. Oteng-Amoako and B. Obiri-Darko

Abstract

Rattans have enormous potential to rural economies in Africa, however, this sector has long been neglected. Consequently, African's share of the world's US$6.5 billion annual trade in rattan does not exist. The industry is threatened by over-harvesting, poor quality raw rattan stems, inconsistent quality of rattan products and national policies which appear to stifle the industry and dampen the aspirations of collectors, weavers and traders alike. A call is made for effective and adaptable technologies in the areas of rattan management, plantation development, primary processing, product development and competitive marketing. Primary stakeholders should be empowered to take up the challenge and advocate for positive changes to outmoded policies in the sector.

1. Introduction

The consequences of rapid depletion of tropical forest resources are quite enormous: global warming, soil erosion, endangered biodiversity and shortage of timber trees are but some of the many problems. Non-timber forest products such as rattan and bamboo have a potential role in solving some of these environmental and developmental concerns. Rattan, sustainably developed, can be a supplement for timber revenues and have potential to alleviate poverty among marginalised groups and serve as a source of income for rural people.

In spite of its enormous potential, the development of rattan sector in tropical Africa seems to have been neglected, a situation in contrast to many countries in Southeast Asia. There is therefore an urgent need for appropriate technologies to develop the sector and fulfil its economic potential.

This paper gives a brief account of the current rattan production to consumption system in Africa, examplified by a baseline study in Ghana by Oteng-Amoako and Obiri-Darko (2000a) and a review of similar studies in other countries by Sunderland in Cameroon and Equatorial Guinea (1998a, 1998b), Morakinyo (1995a, 1995b) in Nigeria, and Defo (1997) in Cameroon, Equatorial Guinea and Congo. The problems and constraints are highlighted and recommendation is made on development interventions to sustain the sector and bring economic prosperity to collectors, processors, weavers and traders of raw rattan and rattan products.

2. Overview of Rattan Production to Consumption System in Africa

The rattan production to consumption system in most producing countries of Africa is a low input and labour intensive system, characterised by extraction of rattan from the natural forest, processing them into different products at village and urban centres using simple tools. The products are then sold in urban and rural domestic markets with very few quantities being exported to neighbouring and European countries (Fig. 1).

Rattans are extracted with simple tools from forests and cleared of spines and leaves and then bundled in groups of 10 to 100 sticks of 10 metres long for sale at rural and urban markets. They are processed manually by rural and urban weavers using simple tools and technology. Furniture, shopping and laundry baskets and serving trays are major products from urban areas in Ghana and Nigeria while carrier and storage baskets are the main products at the rural level in many countries. The quality of finished products varies widely. In general, products from Ghana, Nigeria and Cameroon are of better quality than other producing countries like Equatorial Guinea where the industry is very much rudimentary (Sunderland, 1988b).

The urban products are sold in the open by roadsides and very rarely from display centres. Consumers are both the poor, middle class locals and expatriate tourists. However, rattan products are very much a poor man's furniture in many countries and a considerable part of the urban market is geared towards local, low level consumption. Only a small quantity is exported to Europe and other countries. At the rural level, carrier and storage baskets are sold to farmers and market traders.

A review of various studies on rattan sector in Africa by Oteng-Amoako and Obiri-Darko (2000a; 2000b), Morakinyo (1995a; 1995b), Sunderland and Nkefor (1999), Sunderland (1999), Defo (1998) and Falconer (1994) identifies problems in the sector which may be summarised into four areas as follows:

- Lack of inventory data on rattan resource;
- Inadequate supply of quality rattan cane;
- Inefficient processing and poor quality of rattan raw materials and finished products;
- Lack of adequate overseas markets for rattan cane and finished products especially in Ghana.

To address these constrains in Ghana, the following development interventions which are also applicable to other countries in Africa have been suggested by Oteng-Amoako and Obiri-Darko (2000b):

- Increasing the quantity and quality of rattan resource base;
- Improved processing efficiency for quality raw rattan and rattan products;
- Enhanced marketing of raw rattan and rattan products in an enhanced local and international competitive market;
- Empowerment of primary rattan stakeholders for financial, social and political enhancement (to reflect changes in forestry legislation to include community forest management);
- Effective policy to promote sustainable industry.

3. A Sustainable Resource Base – How?

The transfer of technology needed to sustain the rattan resource should be undertaken with the involvement of primary stakeholders in the planning and execution of the interventions. This approach is likely to increase security of introduced technology and demonstrate its effectiveness to the stakeholders who will ultimately have to determine which interventions to be implemented. For example, Stockdale (1994) has reviewed research methodologies for the sustainable management of natural rattan stands which include the need for adequate inventory, the determination of sustainable harvesting levels and appropriate management practices.

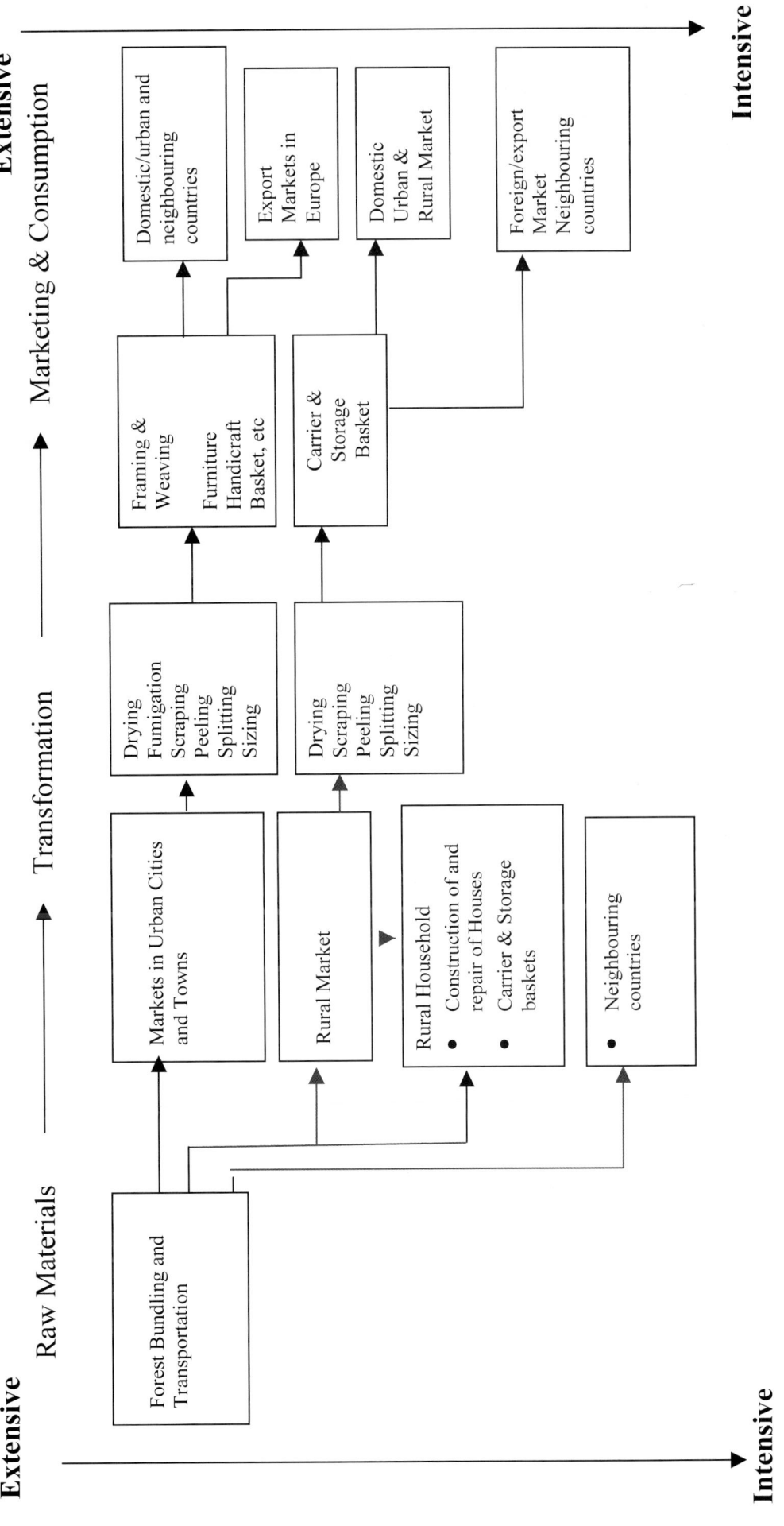

Fig. 1: Overview of Rattan. Production to consumption system in Africa

3.1 Distribution

Rattan occurs in the tropical forest of Africa and Asia (Fig.2). In Africa, rattan occurs predominantly in lowland equatorial rainforests and are distributed from Senegambia in the Northwest to Angola and Zambia in South-Central Africa Tanzania in the East. *Calamus deerratus* is the most widespread rattan species extending from Senegal to Tanzania but the most commercially utilised species are reported to be *Laccosperma secundiflorum, L.robustum and Eremospatha macrocarpa*. However, thorough assessment of the rattan resource is needed in the rattan-producing countries of Africa.

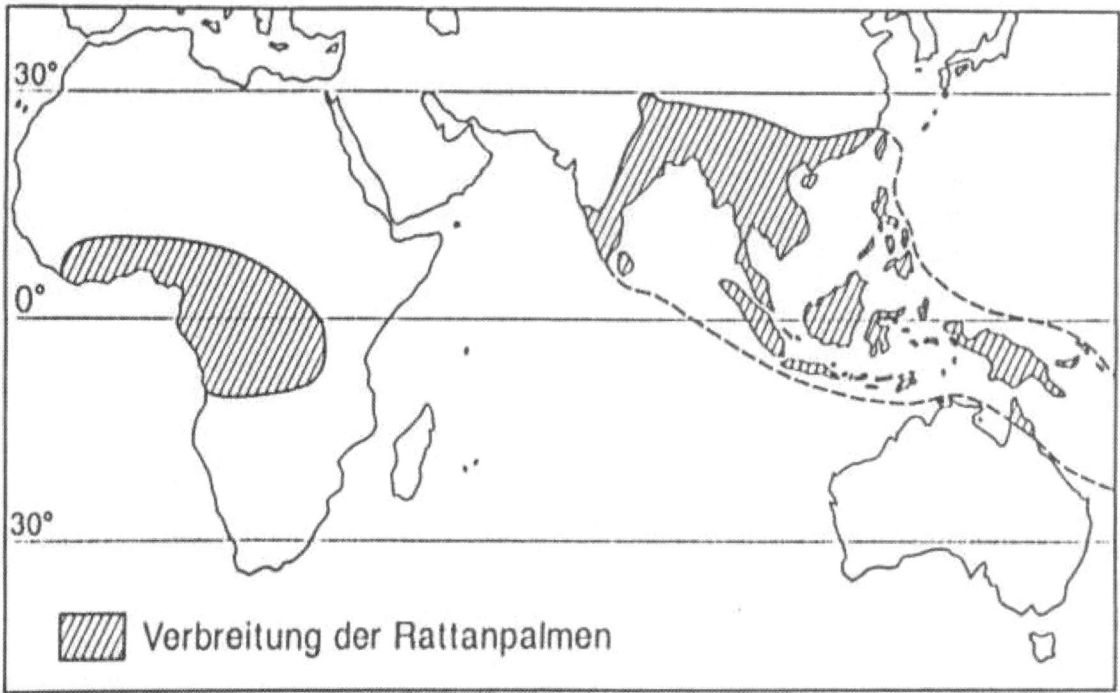

Fig. 2: Distribution of rattan in the world
(from Wiener and Liese, 1993)

3.2 Inventory

Little is known about the stocking level and the quality of rattans in the forests of Africa. Recent work by the Cross River Forestry Project in Nigeria (CRFP, 1994), the Tropenbos programme in Cameroon (Van-Dijk, 1995) and in Ghana (Wong, 1998) have attempted to address the inventory problem with little success. Therefore, the first means to attain a sustainable resource base is the development of a reliable cost effective inventory method to assess the stocking of the commercial species of African rattans. Stockdale (1994) suggests random stratified method using rectangular 0,1 ha plots perpendicular to the slope as the most cost effective means of gathering quantitative information on wild rattan resource. The measurement of stem length is a crucial parameter in undertaking rattan inventories. Therefore, common inventory systems of measuring stem length should be used to allow for comparison of results between countries. Inventory of rattans should be accompanied by a thorough understanding of rattan taxonomy which is essential in determining the resource base of an area. Without reference to an adequate taxonomy and defining the resource base, inventories are of little or no use in providing information for management decisions.

3.3 Taxonomy and Anatomical Identification

There are conflicting reports as to the number of rattan species found in Africa. Morakinyo (1995b) recognizes 13 "good" species, 17 by Sunderland (1997a) while Wiener and Liese (1993) report of 25 species. However, a new taxonomy on African rattans was presented at the "International Expert Workshop on African Rattans", held in Limbe in February 2000, in which 20 species from Africa, including two new taxa, are described, listed and illustrated by Sunderland (2000).

In spite of the above differences, all experts agree to four genera including three endemic genera of *Laccosperma, Eremospatha* and *Oncocalamus,* each consisting of at least two species. *Calamus* which is represented by only one variable species in Africa, has many species in Southeast Asia. Sunderland (1997a) has recently provided a field guide for taxonomic identification of mature and immature rattans at species level. Likewise, identification of Africa rattan stems to species level using anatomical features initiated by Wiener and Liese (1994) and now being pursued by Oteng-Amoako and Ebanyenle (2000) should be supported. Correct identification of rattans will enhance their effective processing and utilisation. Knowledge on anatomical features could assist to predict physical properties of the species, including density, strength and other working properties.

3.4 Sustainable Harvesting

Harvesting of rattan in Africa is undertaken manually and it can be a very difficult activity particularly when rattans become entangled with each other and in the canopies of adjacent trees. It often results in various forms of injury to collectors and calls for a better and efficient method of harvesting such as the use of simple but effective tools (manual harvesting is still practised throughout Southeast Asia as well) and wearing of adequate protective clothing.

In spite of the above problems, over-harvesting and poor management of rattans in their natural habitat have contributed to acute shortage of raw rattan species (especially *Eremospatha)* in some producing countries of Africa, including Ghana (Oteng-Amoako and Obiri-Darko 2000a, 2000b; Falconer, 1994) and Nigeria (Morakinyo 1995a, 1995b). As a consequence, village collectors in Ghana for example, have to walk a distance of about ten kilometres to and from their village to collect rattans (Oteng-Amoako and Obiri-Darko, 2000). This shortage of raw material should be addressed expeditiously through implementation of sustainable management practices with the involvement of rattan collectors and weavers to avoid risks of over-harvesting, aggravating even more the local scarcity of the species.

Inefficient harvesting methods may influence the survival, growth and vegetative or sexual reproduction of other stems in a cluster. Rattans may regenerate naturally if judicious harvesting methods are implemented. Many variables which have to be investigated for suitable harvesting methods include the number of mature stems removed from a cluster (i.e. harvesting intensity), the height at which a rattan cane is cut at the stump, the harvesting cycle, the maturity of the stem, the removal of entangled stems to create gaps and the length of harvested stem. Indicators for stem maturity are: the peeling of the dry leaf sheaths, the yellowish or green colour of the basal portion and stems free of thorns (Sunderland and Nkefor, 1999). Intervention to these research needs could assist in making judicious decisions on the required management regimes and silvicultural intervention for both cultivated rattan plantation and for rattan management in natural forests, and on planting distance, harvest period and intensity and rotation cycle, for formulation of a sustainable yield model.

3.5 Seed Technology and Nursery Practices

In Africa, raw material for artificial or natural regeneration of rattan will have to be met from seed source in the foreseeable future until the science of biotechnology including tissue culture for production of rattan seedlings is considerably developed. There is therefore a need to improve upon stakeholders' knowledge of seed collection, storage and raising of seedlings through nursery establishment. This should be followed by technology for transplanting seedlings in the field. Research should aim at seedling establishment and at reducing dormancy period and increasing the percentage of seed germination (this work is currently being undertaken in Cameroon). Nursery and planting methods of rattans in gardens and plantations are summarised by K. Awang (1994), and K. Awang, J. Dransfield and N. Manokaran (eds) (1992).

3.6 Enrichment Planting

Enrichment planting, which involves planting and cultivation of rattans in gaps in natural forest stands, should be encouraged because the technology is relatively simple with limited capital input. Enrichment planting using nursery seedlings or wildings should be undertaken in gaps using line planting techniques in logged-over forests, fallow or degraded lands by bushfire, shifting cultivation or mining. A success story of enrichment planting in Malaysia where more than 17,000 hectares of logged over forest has been successfully planted with *Calamus species* can serve as a guide (Manokaran, 1987). The shortfalls of rattan cultivation have recently been reviewed by Bacilieri & Appanah (eds.), 1999. Research needs should include studies on microhabitat preferences as advanced by Dransfield (1979), Aminuddin (1990) and Siebert (1993).

3.7 Silviculture

There is no record on silvicultural practices of natural rattan stands in Africa. Consequently, silvicultural practices of weeding, fertilising, peeling of dried sheaths off the stem to prevent attack by insects which also opens up the forest canopy for increased light can be considered (although it will entail high labour costs) to access their effect on yield and stem quality. Previous studies in Southeast Asia by Tan (1992), Aminuddin and Nur Supardi (1992), Manokaran (1985, 1987) and Bacilieri and Appanah (1999) could provide useful background for the study.

3.8 Plantation Establishment

Establishment of rattan plantation is another option for which technology is urgently needed in Africa. Rattan plantation has been successfully developed in association with rubber trees *(Hevea brasilensis)* in Malaysia since the 1980's for *Calamus manan*, a solitary cane. Sunderland *et al* (1999) have recently established a one-hectare trial plantation of *Laccosperma secundiflorum* under an obsolete rubber plantation in Cameroon. Caution should however be exercised in selection of planting distance for our clustering rattan species to avoid possible hindrance of access to established trees (in case they might be tapped for rubber). The establishment of community rattan plantations will also serve as a source for collection of (superior) seeds material and as a demonstration farm for prospective growers.

Provenance trials of different rattan species on degraded lands should follow once the basic technology for plantation establishment has been established and adopted by small-holders. An estimated harvest yield of 4,000 and 7,000 sticks per hectare for large and small diameter cane respectively, has been recorded in Southeast Asia which makes rattan plantation a possible profitable venture (Sunderland and Nkefor, 1999). A long-term possibility of introducing exotic rattan species into plantations of Africa should not be overlooked. (This is not possible for all species, as for example there is currently a ban on the export of *Calamus manan* seeds from all Asian producer countries as a mean to protect their domestic industry.)

3.9 Rattans in Agroforestry

Agroforestry can offer a rattan farmer the benefit to inter-crop trees with seasonal crops so that the farmer can reap the crops while he waits for trees to mature. Inter-cropping rattans with seasonal crops will enable the rattan farmer to secure additional income from the same land until the rattans are harvested after 7 to 10 years. Alternatively, rattans planted with timber trees will give income for the farmer after 7 to 10 years when species are matured and before the trees mature in 30 plus years. Although the concept of agroforestry has long been introduced in Africa, there is a need to know what crop types should be planted with what rattan species or which tree and rattan species should be inter planted. For example, can the practice of inter-cropping rattans with rice practised in some part of Indonesia (Weinstock, 1983) be used in Africa?

4. Efficient Processing for Quality Raw Materials and Finished Products

The quality of rattan products depend mostly on the quality of raw materials used, the ingenuity of the weaver, the efficiency of tools and equipment used and other inputs such as finishes and varnishes.

4.1 Quality of Raw Material

Transformation of primary processed rattan stems into quality finished rattan products depends mostly on the quality of rattan stems. For quality finished products, rattan weavers prefer matured stem with long internodes, devoid of discolouration, scorch marks, splits fungi and insect and damage and seasoning defects such as abnormal shrinkages, cracks, splits (Oteng-Amoako and Obiri-Darko, 2000b). Other quality indicators identified by Sunderland and Nkefor (1999) include non-tapered or uniform stem diameter and stems with glossy or bright colour.

4.2 Primary Processing

Processing of raw rattan in African is manually done using simple domestic knives to remove the skin (epidermis), followed by drying in the open air with little or no preservative treatment (Oteng-Amoako and Obiri-Darko, 2000a; Sunderland and Nkefor, 1999). This labour intensive activity results in inferior quality of raw materials often infested with fungi and borer insects. Therefore, the technology needed in processing is the adoption of simple but efficient scrapping tools and effective preservative methods to prevent raw rattan from attack by biological hazards before they are used. The Southeast Asian method of boiling green rattan in diesel or coconut oil which gives superior raw material (Latif, 1991) should be investigated alongside other innovative methods for possible adoption in Africa.

4.3 Transformation (Weaving) of Rattan

Research into desired physical properties such as ease of bending, sanding, glue bonding, drying and bleaching are essential for the production of quality finished products. Furthermore, the use of steam instead of blow gun (fire) to bend rattan which avoids scorch marks; the use of staples and dowels instead of nails and proper application of varnish on finished products need further evaluation on our indigenous species. Transfer of technology on product designs and use of modern processing machines from Southeast Asia may be appropriate. To facilitate technology transfer, there will be a need to improve upon the few existing training centres in African countries and establish new ones at selected processing sites. These training centres should be manned by master craftsmen who will be capable of organising periodic workshops to introduce innovative designs to weavers. Equally important is the establishment of rattan processing centres where weavers can share at affordable cost the use of modern and efficient machines to boost quality and quantity of rattan products.

5. Product Promotion and Marketing

As previously noted, the international trade in rattan is currently worth some US$ 6.5 billion a year (ITTO, 1997) while the annual domestic market in Southeast Asia is conservatively estimated by Manokaran (1984) to be worth US$ 2.5 billion. Furthermore, an estimated US$ 0.7 billion of the world's populations are reportedly involved in rattan processing and trade (Dransfield and Manokaran, 1993). International trade of rattan is dominated by countries of Southeast Asia and considerable efforts should be made to improve Africa's share of international market.

Very limited data are available on rattan exports from countries in Africa. Komolafe (1992) reports of a limited export of finished products from West Africa to Europe and raw rattans to Asia, including China and Korea. Recently, Sunderland (1999) has reported export of raw cane from Ghana and Nigeria to Southeast Asia, and Morakinyo, (1995) has reported of a flourishing export trade from Nigeria to Korea. According to Defo (1997), Trefon and Defo (1998) and Morakinyo (1995), demand for rattan products has increased significantly in West and Central Africa. Notwithstanding, many processors have complained of poor price for finished products. For example, 50% of weavers in Ghana complain about poor price and irregular demand for finished products as one of the most important constraints in marketing of rattan products (Oteng-Amoako and Obiri-Darko, 2000a; 2000b)

The need to increase Africa's share of international trade in raw and finished rattan products calls for effective promotion of raw rattan and rattan products through better processing and transformation. Appropriate grading and standardisation of raw rattans and rattan products should be introduced and private capability developed to improve the quality of rattan products. Better packaging of graded products is also essential, as is bulk freight of goods to reduce the cost of transportation. Display or promotion centres to sell raw rattan and exhibit finished products are equally important. Special promotional fairs and advertisements should be organised by national rattan associations in collaboration with export promotion councils and small-scale industries of rattan producing countries to increase public awareness. Francophone and anglophone countries of Africa should take advantage of their historical trade links with France and United Kingdom respectively to increase export to these countries. In addition, trade within and between countries of West and Central Africa reported by Falconer (1994), Morakinyo (1999) and Sunderland (1995) should be improved.

Trade restrictions and export levies should be relaxed to serve as incentives for primary stakeholders to boost their market supply. Rattan entrepreneurs should be encouraged by way of sponsorship by national governments to attend international trade fairs to exhibit their products. Market intelligence for rattan products should be conducted periodically to ensure fair and equitable prices for the commodity on the international market. It may be necessary to establish direct links between producers in Africa countries and foreign buyers to eliminate middlemen. The upsurge of hotels, restaurants and other service centres, the increasing number of expatriate community and foreign businesses in emerging economies in the region could increase possible demand provided quality finished products are produced and promoted.

6. Effective Policy and Legislation

In most countries of tropical Africa, past forest policies and legislations have emphasised the supply of timber for wood industry with little or no recognition of non-timber forest products including rattans and bamboo. Consequently, non-timber forest products were not considered in forest management plans. However, current forest policies in some countries of Africa encourage involvement of private particularly rural people in forestry decision making. They encourage rural people living on the fringes of the forest, organisations and communities, to grow, protect, manage and utilise their own forest resources. Notwithstanding these new initiatives, rattan collectors in some countries, like Ghana, still have to pay permit and other fees for collection (Oteng-Amoako and Obir-Darko, 2000a; 2000b). Local collectors should be allowed to collect rattan on a sustainable harvesting basis provided it can be demonstrated to decision-makers that appropriate management regimes are developed and implemented by communities which recognized access rights to the resource. Involvement of the local people in decision making and management of the reserves will motivate them to see themselves as part owners of the forest reserves. They will then harbour no apprehension to manage, protect and cultivate it through enrichment planting or plantation establishment. Therefore, all stakeholders should be allowed to participate in all policy formulations that affect them.

7. Empowerment of Primary Stakeholders

The production to consumption study of Ghana and Cameroon indicate that only few of rattan collectors, primary processors and weavers belong to some form of rattan associations. Although most of these associations were formed to address the financial, political and social needs of their members, these in most cases have not been achieved. Therefore, an urgent intervention is to empower the primary stakeholders by organising them into formidable and dynamic associations. The associations will promote sustainable harvesting, enrichment planting and plantation establishment among rattan workers in response to the much needed technological interventions where applicable, they will seek loans, grants and subsidies for their members and lobby for flexible permits, lower royalties and conveyance fees for harvesting of rattan. The associations will co-ordinate technical interventions in processing, marketing and establishment of more training and product display centres. A well-organised association could rally resources from members for acquisition of inputs such as machines and other equipment for proposed processing centres and financial institutions for use by its members. It will promote bulk haulage of rattan and rattan products to reduce high cost of transportation which accounts for about 50 percent of the selling price of rattan products in Ghana (Oteng-Amoako and Obiri-Darko, 2000a and 2000b). The task of empowerment could be taken up by a non-governmental organisation with a mandate to promote non-timber forest products to alleviate rural poverty (Fig. 3).

8. Conclusions and Recommendations

The rattan industry sector of Africa, in spite of its economic potential, has long been neglected. Consequently, Africa's share of the international trade in rattan, worth some US$6.5 billion a year, is negligible. The following recommendations are needed to address the situation and hopefully realise Africa's economic potential in the sector:

- Development interventions to sustain the resource base, improve upon the quality of rattan products and effectively market them in local and international markets;
- Transfer of adaptable and effective technologies to stakeholders of Africa in the areas of resource management, efficient and innovative processing and competitive marketing of products;
- Empowerment of primary stakeholders of collectors, weavers and traders for access to loans and grants and advocacy for positive policy changes to accelerate developmental needs of the sector.

These interventions need the cooperation of all stakeholders. While the task ahead may look daunting it is not insurmountable.

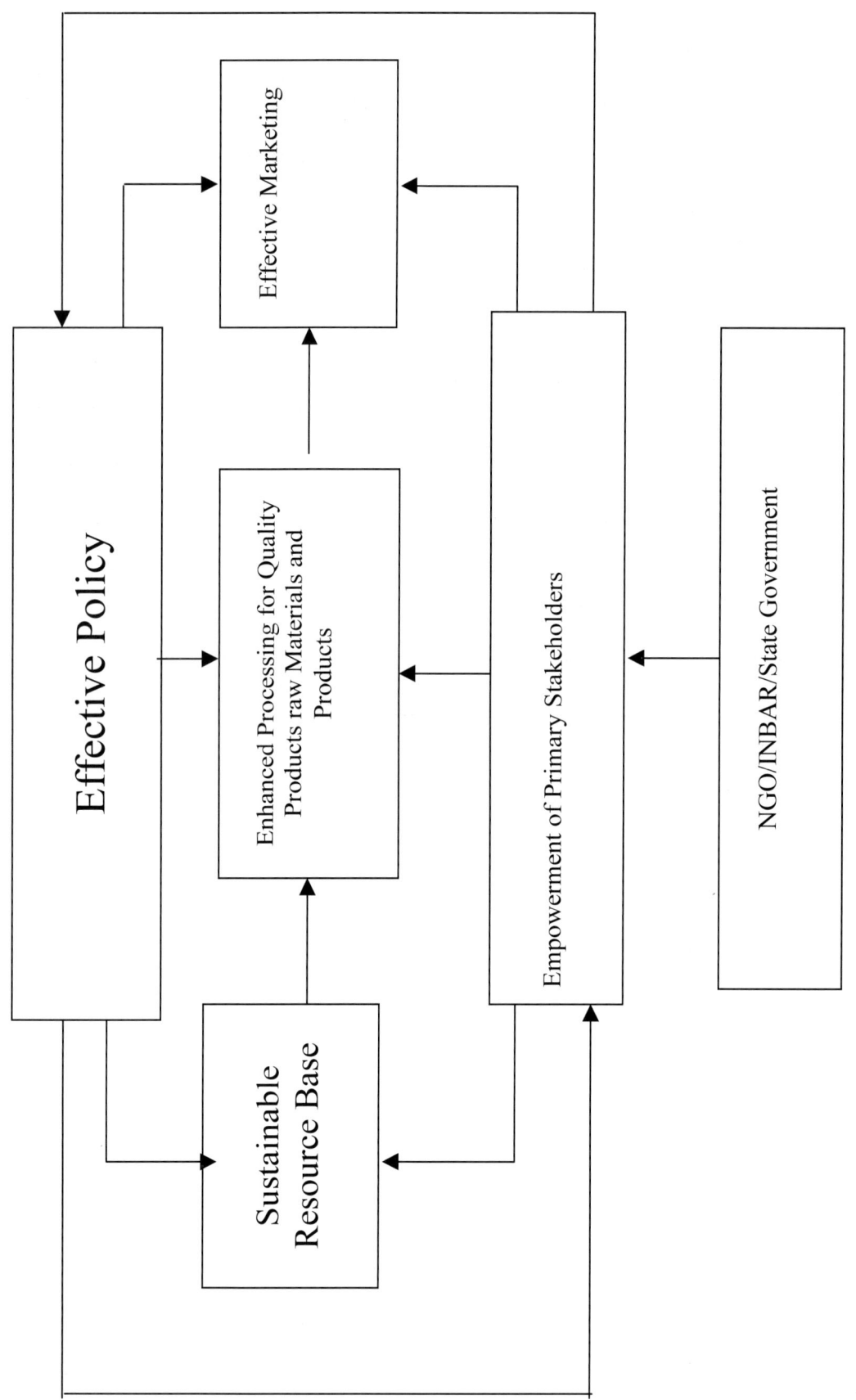

Figure 3: Technology Needs for Rattans: Which Way in Africa?

REFERENCES

Aminuddin M., 1995: Rattan in Malaysia: Conservation status: biodiversity base and its strategic programme. FRIM Report No 67.

Aminuddin, M. and M. N. Nur Supardi, 1992: Field preparation, planting and maintenance: large-diameter canes. In: A guide to the cultivation of rattan, Malaysian Forest Record No.35, Forest Research Institute, Malaysia, Kepong, pp. 205 – 238.

Aminuddin, M., 1990: Ecology and silviculture of *Calamus manan* in Peninsula Malaysia. Ph.D. thesis, University of Wales, Bangor, UK

Awang, K., 1994: A guide for standardised species and provenance trials of rattan: In: Methodologies for trials of bamboo and rattan. INBAR Technical Report No. 4: pp. 55-73

Awang, K., J. Dransfield and N. Monokaran (eds), 1994. A guide to the cultivation of rattan. Forest Research Institute, Forest Record No. 35, Kuala Lumpur, Malaysia.

Bacilieri, R and S. Appanah, 1999: Rattan Cultivation: Achievements, problems and prospects CIRAD–Forêt and FRIM, Malaysia.

Cross-River State Forestry Project (CRFP), 1994: Report of the recognisance inventory of high forest and swamp Forest Areas in Cross River State, Nigeria. Cross-River State, State Forestry Dept/ODA.

Defo, L., 1997: La filière des produits forestiers non ligneux: L'exemple du rotin au sud-Cameroun. Dept. of Geography, University of Yaounde.

Dijk, H. Van, 1995: Assessment of the abundance and distribution of non-timber forest product species. Intermediate report. Tropenbos – Cameroon Programme.

Dransfield, J. and Manokaran N. (eds)., 1993: Rattan: PROSEA Volume 6. Wageningen. pp 137

Wan R.W.M., Dransfield, J & Manokaran, N. (eds). 1992: A guide to the cultivation of rattan. Malayan Forest Record No. 35. Forestry Research Institute, Malaysia. pp 293.

Dransfield, J., 1979: A manual of the rattans of the Malay Peninsula. Malayan Forest Records No. 26, Kuala Lumpur.

Falconer, J.,1994: Non-timber forest products in Southern Ghana. Main Report. Natural Resources Institute, United Kingdom.

International Tropical Timber Organisation, 1997: Bamboo and rattan resource for the 21st Century. Tropical Forest Update, Vol. 7 No. 4.

Komolafe, D., 1992: Cane makers face raw material scarcity. Daily Times (Nigeria), August 5, Page 30.

Latif, M. A. 1991: Guidelines for selection and preparation of rattan for industrial use. RIC Handbook Nᵒ 2

Manokaran, N., 1990: The state of the rattan and bamboo trade. RIC occasional Paper No. 7. The Rattan Information Centre, Forest Research Institute Malaysia, Kepong.

Manokaran, N., 1987: Survival and growth of the economically important rattan species (*Calamus manan*) in Ulu Langart, Selangor. Malay. Forester, 40: 192 – 196.

Manokaran, N., 1985: Biological and ecological considerations pertinent to the silviculture of rattans. In: Proceedings of the Rattan Seminar, October 2-4, 1984; Kuala Lumpur, edited by K.M. Wong and Manokaran, N. The Rattan Information Centre, Forest Research Institute Malaysia, Kepong, pp. 95 – 105.

Morakinyo, A. B., 1995a: The commercial rattan trade in Nigeria forests. Trees and People Newsletter, No. 25

Morakinyo, A. B., 1995b: Profiles and Pan-African distributions of the rattan species (Calamoideae) recorded in Nigeria. Principles 39(4): 127-209

Oteng-Amoako, A. and B. Obiri-Darko, 2000a: Socio-economic survey of rattan industry of Ghana. Paper presented at the International Workshop of Bamboo and Rattan, 2-7 April 2000, Nairobi, Kenya.

Oteng-Amoako, A.A. and B. Obiri-Darko, 2000b: Rattan as a sustainable cottage industry in Ghana: A need for development interventions. Paper presented at international rattan seminar in Limbe, Cameroon, 1– 4 February 2000.

Sieberk, S., 1993: The abundance and site preferences of rattan (*Calamus exilis* and C. *zollingeri*) in two Indonesian national parks. Society and Natural Resources.

Stockdale, M.C.,1994: Appropriate methodologies in research for sustainable management of natural stands of rattan. In: Constraints to Production of Bamboo and Rattan. INBAR Technical Report No. 5, 209 – 245.

Sunderland, T.C.H. ,2000: A new taxonomy of African rattan. Paper presented at the International Expert Workship on African rattans, Limbe, Cameroon, February 2000.

Sunderland, T.C.H., 1999: New research on African rattans: an important non-wood forest product from the forests Central Africa. In: T.C.H. Sunderland, L. E. Clark & P. Vantomme (eds). The non-wood forest products of Central Africa: Current research issues and prospects for conservation and development. Food and Agriculture Organization. Rome. pp 87-98

Sunderland, T.C.H, Nkefor J. and P.C. Black, 1998a: The establishment and maintenance of a silvicultural trial of *Laccosperma secundiflorum* in Cameroon. Technical Note No. 3 African Research Programme.

Sunderland, T.C.H., 1998b: The Rattans of Rio Muni Equatorial Guinea: Utilisation, biology and distribution, African Rattan Research Programme.

Sunderland , T.C.H., 1997a: The abundance and distribution of rattan palms in the Campo Founal Reserve, Cameroon and an estimate of market value: Technical Note No. 2 African Rattan Research Programme

Sunderland, T.C.H., 1997: Guide to rattan and rattan collecting in Africa. African Rattan Research Programme. Technical Note No.1.

Sunderland, T.C.H. and J. P. Nkefor, 1999: Technology transfer between Asia and Africa: Rattan cultivation and processing. Afrirattan Technical Note No. 5.

Sunderland, T.C.H., J. P. Nkefor and P. Blackmore, 1999: The establishment of a silvicultural trial of *Laccosperma secundiflorum* in Cameroon. African Rattan Research Programme, Technical Note No. 3

Tan, C.F. 1992: Field preparation, planting and maintenance of small-diameter canes. In: A guide to the cultivation of rattan, Malaysian Forest Record No. 35, Forest Research Institute, Malaysia, Kepong, pp. 99 – 106

Trefon, T. and L. Defo, 1998: Can rattan help save wildlife? APFT Briefing No. 10

Weinstock, J. A. 1983: Rattan: Ecological balance in a Borneo rainforest Swidden. Econ. Botany, 37: 58 – 68.

Wiener G. and W. Liese, 1993: Anatomische Untersuchungen an Westafrikanischen Rattan Palmen (Calamoideae). Flora (1994) 189: 51 – 61

Wong. J., 1998: The state of Ghana's forests 1986-1997: non-timber forest products. Report to ODA/FRR.

STATUS OF RATTAN RESOURCES
AND USES IN SOUTH ASIA

C. Renuka

1. Introduction

Rattans, one of the important forest products after timber, form an integral part of rural and tribal populace of many of the tropical countries. They are not only the chief raw material for industries in various parts of the world, but they also hold great social significance as a source of livelihood for the people residing near the forest areas. Although economically important, rattans remained as a neglected natural resource till recent times. With the rampant destruction of forests and habitats, their stock, at present, is highly depleted. Over extraction has compounded this problem, so much so that various organizations and institutions round the world are now undertaking scientific studies to conserve and cultivate them. During the last two decades there was an upsurge of research activities that has led to an appreciation of the importance of rattan and the need for its conservation.

A comprehensive account of the mainland Asian rattans was first given by Beccari (1908, 1911, 1913, 1918). After him not much original work was done in the field till 1975. Then a period of active research started in Asian countries. In mainland Asia, India, China and Sri Lanka progressed in this field much more than the other countries because of the already available infrastructure facilities and scientific personnel. This paper is a status report of rattans in South Asia. India, China, Nepal, Bangladesh, Myanmar and Sri Lanka are included in this report.

2. INDIA

India, with an area of 3.287 million square kilometers is marked with remarkable ecological, biological and cultural diversity. As per the forest resource assessment 1990 (FAO, 1995) India has 51.73 million ha. of natural forests. Endowed with magnificent forests ranging from evergreen to moist deciduous, which cover 17.4% of the land area, India's economy is strongly linked with its forests.

In India, forests are Government properties and managed by the State Governments. In Northeast India Government owns only up to 10% of the forestland, the rest remaining as private forests.

Rattans are recognized as one of the most useful forest products in India and the resource plays an important role in the rural economy, employing many people in the remote areas who earn their living through extraction and cleaning of rattans. Urban people are employed in the small-scale industries. Apart from its importance as a commodity for the furniture and handicraft industries, there are great many other uses also known to the Indian people since ancient times. In the ancient books, the medicinal uses of rattans have been reported. Rattans are used for rafting, house construction, for making baskets, as poles for carrying goods etc. Rattan leaves are used extensively as a thatching material. In Nicobar the spiny sheath is used for scrapping coconut. The tribal people of northeastern India make extensive use of long canes for making cane bridges. Some species of rattans are used in tribal rituals and festivals.

2.1 Rattan resource

In India there are about 60 species of rattans under four genera, *Calamus, Daemonorops, Korthalsia* and *Plectocomia*. They are mainly distributed in three major geographic regions, the Western Ghats of Peninsular India, Sub-Himalayan hills and valleys of eastern and northeastern India and Andaman

& Nicobar Islands. The rattans comprise more than fifty percent of the total palm taxa found in India (Basu, 1992).

One genus and 21 species have been so far reported from Western Ghats; 3 genera and 18 species from Andaman & Nicobar Islands and 3 genera and 17 species and two varieties from North Eastern States (Lakshmana, 1993; Renuka, 1992, 1995, 1999; Sunny Thomas *et al.*, 1998). Each region has its own specific rattan flora and the species distribution does not overlap. Out of the reported species only 25 per cent are economically important.

– Peninsular India

Western Ghats of Peninsular India, with its tropical evergreen rain forests, form one of the ideal habitats of rattans. They are also seen in the Nilgiris and in the Ghat forests of Andhra Pradesh. Depending on the species they are distributed in the evergreen, semi-evergreen and moist deciduous forests.

Rattans occur from almost sea level to 2000m elevation, most showing altitudinal preferences. With the exception of *C. rotang*, a cane of the plains, all others are plants of the hills and mountains. Most of the species are distributed below 1000m and only four species are seen above this level. More species occur towards the southern part of Western Ghats.

– Northeast India

Rattans are found in the evergreen, sub montane or the sub Himalayan mixed forests of northeastern states. Rattans are also distributed in the moist deciduous forests in Orissa and Bihar and in the coastal swamp forests of W. Bengal and Orissa. They have their range of distribution from alluvial plains to the moist hill forests up to 2000m altitude. Most of the species are found below 500m while *C. acanthospathus* and *P. himalayana* are seen at much higher altitudes, around 2000m.

– Andaman & Nicobar Islands

While the uninhabited islands are much richer, the inhabited ones also harbour several taxa of rattans along the boundaries of farms, roadsides and in fallow lands. The species are not evenly distributed in the islands. While *C. andamanicus* and *Korthalsia laciniosa* occur both in the Andaman and Nicobar group of islands, most others have restricted distribution in certain parts only. Eleven species are confined to the Andaman group and five to the Nicobar group.

– Present status

An analysis of distribution of rattans in the three different major areas in India shows that much change has taken place in their distribution over the years because of the shrinkage of the natural forest cover. In the northeastern states shifting cultivation had been degrading and denuding the forests since long ago. Many of the species reported earlier from certain localities are absent now (Renuka, 1999). The growing popularity of rattan furniture resulted in overexploitation of this important forest resource. In many regions commercial species have been seriously depleted as the rapid exploitation continues unabated. This situation, if left unresolved, will bring about severe economic and social repercussions. *Calamus travancoricus, C. rotang, C. dransfieldii* and *C. nambariensis* have become extremely rare in their original localities.

2.2 Resource management

– *In situ* conservation

In India, there has been no serious effort so far to conserve rattans *in situ* . Even though National Parks and Bio-reserves are helpful in promoting *in situ* conservation, illicit harvesting cannot be controlled efficiently. For conserving the natural populations, some of the State Forest Departments have introduced extraction rules. Generally the extraction is carried out on a 4-year rotation. The Government has also banned the export of the raw material .

Rattans are planted and protected in sacred groves. There are about 80 rattan bearing sacred groves in Kerala alone (Mohanan & Muraleedharan, 1988).

– *Ex situ* conservation

State forest departments of Kerala, Karnataka, Tamil Nadu and Goa have started large-scale plantations of rattans. Certain species are cultivated in homesteads. But only three or four economically important species are protected like this.

A live collection consisting of about 30 species is maintained in the Kerala Forest Research Institute campus. Seed stands of 12 species have been raised in Thrissur Forest Division. The State Forest Research Institute in Arunachal Pradesh has also started germplasm conservation.

– Propagation techniques

Phenological details of almost all the species are available. Seed germination and nursery techniques have been standardized for India (Renuka, 1991). Suckers can also be used after treating them with growth regulating substances like NAA (1-naphthaleneacetic acid) and IBA (indole-3-butyric acid) for better rooting. Suckers extracted during July-September after a treatment with NAA at 2000 ppm have shown good rooting (Seethalakshmi, 1993).

– Tissue Culture

The earliest reports of tissue culture studies on Indian rattans were those by Padmanabhan and Krishnan (1989), Padmanabhan and Sudhersan (1989), Padmanabhan and Illangovan (1989, 1994). Research on *in vitro* culture of rattans is currently being carried out in India at the Kerala Forest Research Institute (KFRI), Peechi and Tropical Botanical Gardens and Research Institute, Palode. At KFRI, work was initiated in 1992 and micro-propagation protocols have been developed for seven species of *Calamus* of Western Ghats and the Andaman and Nicobar Islands, using immature or mature embryos or shoot tips of (1-2 year old) seedlings for initiating the culture (Muralidhran, 1994; Valsala & Muralidhran, 1998, 1999). Plantlets have been planted out in a forest plot to assess their performance. An interesting finding was that plantlets with multiple stems could also be induced in culture and established in soil (Muralidhran, *pers. comm.*). Since seedlings of rattan normally take 3-4 years to form suckers, the use of such micro-propagated plants in plantations has the potential of increasing the yield of stems in the normal rotation. It is however necessary to carry out field trials to demonstrate this potential.

Regeneration of plantlets in some of the species has been achieved through organogenesis in callus cultures as well as from *in vitro* leaf segments. Somatic embryogenesis and plantlet formation were also achieved from callus cultures (Valsala and Muralidharan, 1998).

– Identification keys

Correct identification of the species is essential for management practices. Since rattans are extracted even before flowering and fruiting, identification of rattans based on floral characters becomes difficult. Hence identification keys were prepared based on vegetative characters and easily distinguishable field characters (Renuka, 1992, 2000). Identification keys based on anatomical characters were also prepared (Bhat *et al.,* 1993).

2.3 Rattan processing and product development

– Harvesting

The State Forest departments lease out the right of collection to private parties. In Kerala the right of collection is vested with the Kerala State Federation of Scheduled Castes and Scheduled Tribes Development Co-operative Ltd. The major policy objectives are: (i) to eliminate intermediaries in collection, and (ii) to increase the income and employment opportunities of the tribal people. Even though the right of collection and marketing of cane was assigned to the co-operative societies, illegal harvesting continues. In other parts of the country rattans are harvested by contractors under permits issued by the Forest Department. They are to pay the royalty to the Government (Haridasan, 1997).

– Harvesting methods

The traditional method of harvesting is simple, by manual means. The mature rattans are pulled out from the support and cleaned. The soft uppermost part is discarded. The remaining portion is cut into suitable lengths for bundling and then transported to the depot.

At present no grading is done. Bhat (1996) has formulated certain grading rules. Furniture and handicraft items are the main products and private manufacturers or co-operative societies carry out production.

– Processing and preservation techniques

In India air-drying is the most common method used for reducing the moisture content of rattans. Recently oil curing methods have been introduced (Yekantappa *et al.,* 1990, Dhamodaran & Bhat, 1992). The standardized curing method uses diesel – coconut oil mixture (9:1 ratio) or kerosene.

Although rattan curing has a desired effect in controlling fungal infection and discoloration, it may not be economically feasible always to cure rattan at the felling site. One alternative is to give the canes a prophylactic treatment in extraction sites by dipping them in plastic lined pits filled with preservative chemicals such as sodium pentachloro phenoxide, copper sulphate and sulphur azide (Mohanan, 1993).

– Socio-economics

Socio-economic studies on rattans are mainly confined to Kerala. Harvesting of cane is under taken by the Kerala State Scheduled Caste and Scheduled Tribe Development Federation (Federation), aiming to provide more employment to tribes. But the absence of a permanent setup for collection in the Federation resulted in middlemen dominance in the harvesting scene. Consequently, a significant part of collection charge offered by the Federation is taken away by them, thereby reducing the share of actual collectors. Rattan based industry in the State depends heavily on import of cane from the Northeastern Region. This has caused an adverse income and cost relationship in the industry, which provides fewer incentives to carry out the production, resulting in decline of units in the State. Similarly, problems are prevalent in marketing section also. Most of the units in the industry have no marketing set up of their own. They unduly depend on private traders for marketing their products. Due to lack of proper marketing set up and inefficiency in marketing, rattan-processing units in the State receive only a very low marketing margin (Muraleedharan & Anjana, 1994; Muraleedharan *et al,* 1996 a, b; Muraleedharan & Anitha, 1999).

3. CHINA

China has an area of 9.561 million square kilometers, covered with mountains, hills and plateaus. With such a diversity of physical features, the climatic conditions also vary a great deal.

In China rattan research was initiated in 1963 by the Research Institute of Tropical Forestry (RITF) under the Chinese Academy of Forestry. In the early sixties, considerable attention was paid to the conservation of the diminishing rattan resources. To increase local supplies, rattan research was reactivated in 1985.

3.1 Rattan resource

About 40 species and 21 varieties under three genera have been reported so far (Pei & Chen, 1991). The genus *Daemonorops* has only one species, *Calamus* 35 species and 21 varieties and *Plectocomia,* four species (Xu *et al.*, 2000). It has been recorded that rattan is naturally distributed in more than eleven provinces of south China. Hainan Island and Xishuangbanna are two centres where the diversity and productivity are the highest. Significant differences in geographical and climatic conditions between southeast and southwest China influence rattan distribution in the two main centres. Rattan is usually found from low lands to the hills below 1800m altitude depending on the species. It occurs in tropical montane rain forests, tropical evergreen monsoon forests or secondary tropical forests, but a few species can be found growing in tropical semi-deciduous monsoon forests and sub-tropical evergreen broad-leaved forests.

Rattan resources in China have been well utilized except some shrubby species of *Plectocomia.* Hainan Island and Xishuangbana, Yunan are the two major centres for rattan production, with an estimated annual yield of 4000-6000 tons. This is about 90% of the country's total, which only meets 10-20% of the domestic market demand (Xu *et al,* 2000).

- **Present status**

The natural rattan resources have been heavily depleted because of unlimited exploitation of canes in the last few years and increasing destruction of forests. Extraction of commercial species before flowering also attributes to the low concentration of rattans in the forests.

3.2 Resource management

- *In situ* conservation

In Hainan Island, rattans are still frequent in natural forests, while in the continental main land, overexploitation has resulted in serious depletion of the resource (Zeng *et al.*, 1999).

- *Ex situ* conservation

The work on genetic conservation of rattan resources in China started in the mid 1970s at The RITF and the Yunnan Tropical Botanical Institute (YNTBI). A live collection of rattans was established in RITF in 1985. YNTBI also established *ex situ* conservation plots. Eight species of rattans have been successfully conserved *in vitro (Xu et al.*, 1999).

Xu *et al.* (1999) evaluated the conservation status of rattans in China. According to them the status of *ex situ* germplasm stocks is not satisfactory. Nearly 61.9% of them are in poor condition, 53.6%, with no more than five plants in one garden, 19% with only one plant each in one garden. Five species or varieties are represented by only one plant. Of all *ex situ* collections in China, only 42.9% are producing seeds. Inter-specific hybridization occurs naturally in gardens. Apomictic species also cannot be ruled out. Much work still needs to be done on the genetic conservation of rattans (Yin *et al.*, 2000).

– Growth and silviculture

Growth and phenological data are available for certain species (Xu *et al.*, 2000). Yin and Xu (2000) worked out the suitable storage condition for seeds of *Daemonorops jenkinsiana* (as *D. margaritae*). Studies to estimate the sex ratio in plantations are underway. Li *et al.* (2000) studied the relationship between rattan community and the environment. Much work has been carried out on the mineral nutrition requirement, cultivation methods, effectiveness of VA mycorrhizal fungi and biomass accumulation and nutrient recycle in rattan plantations (Xu *et al.*, 2000).

– Tissue culture

Rattan tissue culture work in China started in the late eighties in the Kunming Institute of Botany (Chengji, 1987); with time and experience, the techniques for *in vitro* culture work and mass propagation improved. Till 1998, tissue culture studies were conducted on 16 species (Zeng, 2000). Most of the explants used were embryos and collar regions from seedlings. For mass propagation of 5 species, this technique was used. Plantlets with fibrous roots survived better. But the key factor for best performance during out planting is the height of the plantlets. No indication could yet be found as whether these studies can be considered successful.

– Rattan processing and product development

Rattan industry in China is comparatively well developed and a variety of products are available and sold not only locally but exported also. The estimated value of rattan products in 1993 was more than US$100 million, of which about US$60 million from export per year. The annual demand for rattans in China may reach 30 thousand tons. Most of the raw material is imported from countries of Southeast Asia. Since Indonesia has banned the export or raw rattan, China is exploring the possibility of getting supplies from Myanmar, Vietnam and Laos (Zeng & Yin, 1997).

4. NEPAL

Nepal has a total area of 14.7 million ha. and has a rich floral and faunal biodiversity because of its diverse land configuration and altitudinal ranges, i.e. 75m to 8848m. Geographically, Nepal is divided into five regions, Terai, Siwalik (foothills), Mid hills, High hills and High Himalayas (Annex 1) .

Rattan is one of the economically important native species found in the deciduous forests of Nepal. This has been used locally by rural people for a long time. Besides local uses, rattans have great cultural value in Nepal. The Tharu (an ethnic group) people, for example, use rattan sticks in temples. They also believe that the rattan stick is holy and no evil spirit will come near it. They keep a rattan stick with them while attending religious functions. Rattans have been protected in the temple compounds where people cannot harvest them.

4.1 Rattan resource

Two genera and seven species have been reported from Nepal. Rattans are distributed from eastern to western region of Nepal at the altitudinal range of 100- 1000m, mostly in the terai and mid hill region. The Terai is the flat land following the foothills, and it has rich sal forests. This is the area where rattan used to be found in large quantities. Rattans grow profusely in the Terai regions. In the hilly regions, the distribution is scattered (Paudel, 1997). In the mid hills, rattan is found on the mild and well-drained soils. *C. tenuis* is present as cane brakes in some areas of mid and far-western regions (Paudel, 1997).

4.2 Resource management

Rattan brakes (patches of rattan stands in natural forests) are well protected in the community forests and are extracted on a five-year rotation. Extraction is done by clear felling the rattans. After removal of mature stems, the community set fire in the plot, which helps for the regeneration.

Most of the harvested rattan is smuggled out to India from the western Nepal where more than 800 ha of rattan brakes have been reported to exist (Dixit, 1998). As a result of this the smooth functioning of the rattan-based industries is hampered. Many of the industrial units import rattans from India. It has been estimated that more than 90% of total requirement of cane is brought in from India. But the Indian Government has officially banned export of raw rattan which has forced the middlemen to buy them in black markets in the states of Bihar and West Bengal. This raises the price of rattan products in Nepal to more than 100% of that prevailing across the boarder in India. Indian and Nepal custom offices charge custom and other fees on the raw material. The total charges float around 15% of the price of the transported products. Government support is practically nil to the otherwise rapidly growing rattan based cottage industry in Nepal (Dixit, 1998). There are about 50 rattan processors recorded in Nepal. All are small-scale processors employing 2-8 employees each (Paudel, 1997).

4.3 Cultivation

People in terai region cultivate rattan on marginal lands. Other than this there is no large-scale plantation.

5. BANGLADESH

Bangladesh, situated in the northeastern fringe of the subcontinent of South Asia has an area of 143,999 km^2. The land is mostly flood plain except for some hilly areas along the north and eastern boundaries and upland terraces on the central and northwest region. The area under state or public forest is 14% of the total land area. Of this about 9% is managed by the Forest Directorate and the remaining is under the control of district administration (Ara, 1997a).

5.1 Rattan resource

Rattans are one of the most important natural resources of Bangladesh forests and homesteads. Some indigenous people use young leaves, roots and shoot tips of rattans as medicines and vegetable (Ara, 1997b). Eleven species are reported under 2 genera, *Calamus* and *Daemonorops*. All species are found naturally in the forest, while *C. tenuis* also grows at the edge of water and marshy places in village groves.

Rattans occur in the northeastern hill forests of Chittagong, Cox's Bazar, Hill tracts and Sylhet districts of Bangladesh. In sal forests there are no rattans. Even though two species, *C. tenuis* and *D. jenkinsiana,* had been reported from northern parts of the mangrove forests of Sundarbans, at present no rattan is present in the mangroves (Ara, 1997a). *C. tenuis* occurs along the edge of the littoral forest towards the landside (Alam, 1990).

Only very little scientific work has been done on the rattans of Bangladesh. Consequently, only very little information is available on the silviculture, distribution and abundance (Ara, 1997a,b). Phenological data are available for six species (Ara, 1997a).

5.2 Resource management

Even though small-scale trials of some species are being raised by the Forest Department and other agencies, no commercial rattan cultivation exists, apart from *C. tenuis.*

5.3 Rattan processing and product development

The Forest Department and small-scale entrepreneurs carry out the collection and processing of NWFPs. Rattan collection is done by local people on payment of royalties to the Forest Department (Patric *et al.,* 1994).

5.4 Socio-economics

Usually rattans are harvested when money is needed. In the present harvesting systems wastage is high. After harvest the rattans are sold in the village markets. Skilled labour is not available for harvesting that during collection the clumps are damaged affecting the new shoot production.

Rattan is one of the materials of cottage industries and this resource adds considerable amount of revenue to rural households (Mohiuddin *et al,* 1986). But this natural resource is getting depleted at an alarming rate. Alam (1990) reports that most of the rattan supplies to the industries is smuggled from Assam and Myanmar. Annual average harvest from the forests during 1981-87 was 3,525,500 feet. At present most of the rattan used by the industries is imported. Ara (1997b) reports that cane based industries are being closed due to lack of raw materials. Most of the workers in these industries are women and they work to earn money to meet their basic needs. Hence closing down of these industries will directly affect the socioeconomic condition of the people.

6. SRI LANKA

The island of Sri Lanka has a land area of about 6.5 million hectares. Topographically the country consists of a highland area in the South Central part of the island which rises to about 2500 m with lowland plains surrounding it. The major climatic zones can be recognized based on the rainfall pattern, the wet zone (over 2500 mm/years), intermediate zone (1800-2500 mm/year) and the dry zone (below 1800 mm/year). The natural forests in the wet zone are tropical rain forests particularly fragmented and depleted to 8% of the wet zone area (Tilakaratna, 1997).

Sri Lanka has a total population of 17.4 million and about 74% of the people still live in rural areas. About 30% of the rural population in most of the areas has some form of involvement in collection or utilisation of rattan.

The decline in the natural forest cover in Sri Lanka has accelerated in the past 150 years. The average annual rate of deforestation during the past few decades had been around 2% or 36,000 ha (Bandaratilake, 1998). Nearly all of the natural forests in the country are state owned and cover over 30% of the land. Primarily three institutions have the authority over these natural forest areas, the Forest Department, the Department of Wildlife Conservation (DWLC) and the Local administrative bodies. All the forest areas managed by DWLC and some areas managed by the Forest Department have been categorized as protected areas. Other forest areas are categorized as production forests. Rattan is one of the major NWFPs obtained from these forests. Rattans are extracted from these forests under permits issued by the Forest Department.

6.1 Rattan resource

Sri Lanka has only one genus *Calamus* with 10 species (De Zoysa & Vivekanandan, 1995). Their taxonomic documentation goes back to the early eighteenth century; *Calamus rotang* was described in P. Herman's Musaeum Zeylanicum in 1717. Thwaites (1864), Beccari (1892, 1908) and Trimen (1898) documented the remaining species.

Rattans exhibit a high degree of endemism. Seven of the ten species are considered to be unique to this country. The other three, *C. thwaitesii, C. pseudotenuis* and *C. rotang* are also found in South India. *C. rotang* is reported from Myanmar also.

Except C. *rivalis* and C. *rotang,* the rattans are restricted to the wet zone, especially on the South Western part of the island. They are seen in the mixed-dipterocarp rain forests and lower montane area. C. *thwaitesii* extends to the semi-evergreen forests of the intermediate zone. C. *rotang* is the only species found exclusively in the dry lowlands which have an annual rainfall of less than 1000 mm. Very high degree of genetic diversity of rattans is seen in the lowland rain forests (De Zoysa & Vivekanandan, 1994).

Rattan species of the wet and intermediate zones occur mainly in forest habitats. In the dry zone, these species mostly occur in riverine forests but sometimes in the dry zone also. C. *rotang* and C. *rivalis* form impenetrable thickets. C. *rivalis* is also found in sub- mangrove conditions close to the coast. All rattan species have a clustering growth habit.

Phenology is very poorly known. It seems that the robust, high climbing species that reach the canopy and the non-forest species of the lowlands have marked seasonality in flowering and fruiting. The rest of the species appear to flower and fruit sporadically throughout the year. Observations indicate that pollination is by bees (De Zoysa & Vivekanandan, 1994)

– **Present status**

Many of the Sri Lanka's protected forests are small in extent and isolated. In general around 50% of the protected areas are less than 1000 ha in size. Absolute conservation of representative areas in the lowland wet zone is necessary in order to arrest the extinction of threatened rattan species, namely C. *pachystemonus,* C. *radiatus* and C. *ovoideus.* Among the ten native rattan species, seven have restricted distribution to lowland rain forests and submontane forests, five are endangered and one species vulnerable. The representation of the lowland rain forests within the protected area system is inadequate. C. *pachystemonus* is not afforded any protection by the present protected area system.

6.2 Resource management

– *Ex situ* conservation

Rattans are mainly cultivated by the Forest Department. The cultivation by private farmers is only on a very small scale. The Forest Department established an arboretum in 1988. Botanical Gardens at Peradeniya, Hakgala and Heneratgoda also have some species. A total of 148 ha of rattans have been planted between 1985-1991 as species trials with the financial help of IDRC.

The main objective of the Forest Department's rattan cultivation is to extend the resource base outside its natural habitat. Large diameter rattans are the ones that are mainly planted. Rattans have been planted as an enrichment species in logged-over natural forests, under-planted in mixed forest plantations or under-planted in pine plantations. Spacing adopted was 6.5x6.5 for enrichment planting, 3x3 m to 5x5 m in mixed plantations, and 2.5x2.5 m for under-planting in pine plantations. A total area of 248 ha of enrichment planting and 146 ha of under planting was done during 1989–94.

– **Seed storage**

De Zoysa & Vivekandan (1995) report that the seeds can be kept alive for up to two years if they are burried about 15 cm deep in moist soil.

– **Silviculture**

Silvicultural techniques have been standardised for Sri Lanka (De Zoysa & Vivekanandan, 1995).

– **Tissue culture**

Institute of Fundamental Studies, Kandy and University of Peradeniya are conducting research in this field.

6.3 Rattan processing and product development

– **Harvesting**

Rattans are harvested under permits issued by the Forest Departments and Divisional Secretaries. Harvest of rattans under the permits issued cover only a minor part of the total quantity harvested. The major part of the supplies from the main rattan producing regions comes from illegal harvests.

– **Marketing**

According to the survey carried out by IRED (1988), the export earnings from rattan–based products in 1986 were US$ 80,000 and since then exports have been negligible. This trend has been attributed to the poor quality of the products.

Stock accumulation is rare, because the scale of production is too small. This is because of the lack of capital to purchase raw materials in substantial quantities. Therefore the craft workers are dependent on middleman, who buy their goods at a minimum price, while both middleman and the retailer have the advantage of accumulating stocks and dictating prices. Middlemen usually have a profit margin of 10-15%. Local retailers may keep a further 30-90% while the craft centres keep a 10% profit. Due to the shortage of raw materials, bamboo, plastic or cotton materials are replacing rattan.

– **Socio- economics**

Rattan craft is a traditional occupation in Sri Lanka and it spreads over 18 of the 24 districts. It is estimated that about 3000 people are directly engaged in rattan industries, and they earn at least a third of the family income through the craft. State is not able to provide jobs for trained craftsmen. Only a few workers are attached to state run craft centres. There is an estimated surplus of about 400 trained people (De Zoysa & Vivekanandan, 1994). The rattan industry also provides significant opportunities for indirect employment in harvesting, transport and supply.

7. MYANMAR

Myanmar with an area of 676,577 square kilometers is rich in culture, traditions and resources. It is endowed with one of the highest forest cover in the Asia-Pacific region. Fifty percent (50%) of its area is covered with forests. The climate is tropical with well-defined seasons.

Rattans are common associates in the deciduous and evergreen forests (Forest Department, 1991). Rattans in Myanmar have high commercial potential. Little has been done to survey the rattan growing areas in Myanmar. The Forest Research Institute, Myanmar has conducted studies on taxonomy, physiology, plantation techniques etc.

7.1 Rattan resource

Six genera and 31 species of rattans are reported from the country. They are *Calamus* with 23 species, *Korthalsia* with 3 species, *Daemonorops* with 1 species, *Plectocomia* with 2 species, *Myrialepis* with one species and *Plectocomiopsis* with one species. The genus *Calamus* is widespread all over Myanmar.

Figure 6: Rattans growing in degraded forests (Dransfield)

Rattans are widespread in Myanmar both at lower attitudes and in the hills up to 3000 ft in the evergreen forests (Uhtay Aung, 1997). Rainfall varies from 80-120 inches (2032 to 3048 mm). Very few rattans are found in dry areas. The following areas in Myanmar are abundant with rattans:
1. Kachin State.
2. Upper Chinduin Forest Reserves, West Katha Forest Reserves of the Sagaing Division.
3. Momeile Forest Reserves and Shweli River Valley in Shan State.
4. Ternnesserim Division.

8. Conclusions

An analysis of the situation in all the countries reveals that scarcity of the raw material is a major problem. Hence natural resource development is urgently needed.

Present *in situ* protection systems are not sufficient to meet the immediate demands for the raw material. Hence cultivation of the commercially important rattans is essential. In the context of rapid depletion of the natural resources, germplasm preservation requires urgent attention. Ecosystem conservation is the best method for genetic conservation for which the area where most of the genetic diversity is concentrated is to be found out. For this, genetic mapping of the various species and varieties of rattans is needed. Methods of *ex situ* conservation also should be developed.

When compared to Southeast Asian countries, the rattan products of South Asian countries are of low market value. Hence value addition by means of better processing and manufacturing techniques is needed.

The rattan sector is characterized by a variety of stakeholders with different needs and interests such as rattan cultivators, raw material collectors, manufacturers and traders. Hence there is an urgent need for awareness raising on the importance of the rattan sector to decision makers all levels and to examine and modify the national policies covering harvesting, utilization and marketing of the resource.

REFERENCES

Alam, M.K., 1990. Rattans of Bangladesh. Bulletin 7. Plant Taxonomy Series. Bangladesh Forest Research Institute 33pp.

Ara Roshan, 1997 a. Taxonomy and ecology of rattan in Bangladesh. In A.N. Rao & V.R. Rao, 1997. Rattan Taxonomy, ecology, silviculture, conservation, genetic improvement and Biotechnology, IPGRI –APO, Serdang, Malaysia.

Ara Roshan, 1997 b. Silviculture, improvement and conservation of rattans in Bangladesh. In: A.N. Rao & V.R. Rao (eds.). 1997 Rattan Taxonomy, ecology, silviculture, conservation, genetic improvement and Biotechnology IPGRI –APO, Serdang, Malaysia.

Bandaratillake, H.M., 1998. Rattan genetic resources in Sri Lanka. In Vivekanandan, K., A.N. Rao & V.R. Rao (Eds.) 1998. Bamboo and Rattan Genetic Resources in Certain Asian Countries. IPGRI – APO, Serdang, Malaysia.

Basu, S.K., 1992. Rattans (canes) in India. A monographic revision. Rattan Information Centre, Kepong, Kula Lumpur.

Beccari, O., 1892. Palmae in J.D. Hooker. The flora of British India. Vol. 6: 441-446.

Beccari, O., 1908. Asiatic palms. Lepidocaryeae. Part 1. The species of *Calamus. Ann. Roy. Bot. Gard. Calcutta.* 11: 1-518. pl. 1-11, 1-238.

Beccari, O., 1911. Asiatic palms. Lepidocaryeae. Part 2. The species of *Daemonorops. Ann. Roy. Bot. Gard. Calcutta.* 12(1): 1-237. pl. i-ii, 1-109.

Beccari, O., 1913. Asiatic palms. Lepidocaryeae. Part 1. The species of *Calamus.* Appendix. *Ann. Roy. Bot. Gard. Calcutta.* 11: 1-142. pl. 1-83.

Beccari, O., 1918. Asiatic palms. Lepidocaryeae. Part 3. The species genera *Ceratolobus,Calospatha, Plectocomia, Plectocomiopsis, Myrialepis, Zalaca. Pigafelta, Korthalsia, Metroxylon, Eugeissona. Ann. Roy. Bot. Gard. Calcutta.* 12(2): 1-231, pl. i-iv, 1-120.

Bhat, K.M.,1996. Grading rules for rattan: a survey of existing rules and proposal for standardisation. INBAR Working Paper No. 6. IDRC, Canada. 44p.

Bhat K.M., Mohammed Nasser and P.K.Tulasidas. 1993. Anatomy and identification of South Indian rattans (*Calamus* spp.) IAWA Journal 14(1): 63-76.

De Zoysa, N. and K. Vivekanandan, 1995. Rattans of Sri Lanka – An illustrated field guide, Sri Lanka Forest Department.

Dhamodaran, T.K. and K.M Bhat. 1992. Quality improvement of cane (rattan) products. KFRI Info. Bull. No.11.

Dixit, P.M., 1998. Identification, validation and development of conservation and management methods of rattan in Nepal. Project proposal submitted to INBAR.

FAO, 1995. Forest Resource Assessment 1990. Global synthesis, FAO Forestry paper 124. Rome.

Forest Department, Myanmar, 1991. Forest Resources of Myanmar, Conservation & Management. Yangon.

Haridasan, K., 1997. The Silviculture Scenario of Canes in North East India with reference to Arunachal Pradesh. In Rao A.N. and V. R.Rao (Eds). Rattan Taxonomy, Ecology Silviculture, Conservation, Genetic improvement and Biotechnology. IPGRI- APO. Serdang, Malaysia.

IRED, 1988. Master plan for handicraft development in Sri Lanka, vols. I-VI, prepared for the Ministry of Rural Industrial Development, Development Innovations Network. IRED, Colombo.

Lakshmana, A.C., 1993. Rattans of south India. Evergreen publishers, Bangalore.

Li Yide, W. Zang, G. Yin and H.C. Xu, 2000. Studies on rattan communities. In Xu *et al.,* 2000. Research on rattans in China. IPGRI – APO, Serdang, Malaysia.

Mohanan, C, 1993. Protective measures against fungal staining in rattans. In: S.Chand Basha and K.M.Bhat (eds.) Rattan Management and Utilisation. KFRI, India and IDRC, Canada, pp 281-293.

Mohanan, C. and P. K. Muraleedharan, 1988. Rattan resources in the sacred grooves of Kerala, India. RIC Bulletin 7(4): 4-5

Mohiuddin, M., M.H. Rashid and M.A. Rahman, 1986. Seed germination and optimal time of transfer of seedlings of *Calamus* spp. from seed bed to polythene bag. Bano Biggyan Patrika 15(1&2): 21-24.

Muralidharan, E.M., 1994. Tissue culture of Indian rattan species. Paper presented at the 2nd Asia-Pacific Conference on Agricultural Biotechnology. March 6-10, 1994, Madras. Abstract No. 107.

Muraleedharan, P.K. and S. Anjana, 1994. Rattan based industry in Kerala: Raw material supply and marketing. J. Non-Timber For. Prod. 1(1&2): 83-88.

Muraleedharan, P.K. and V. Anitha, 1999. Some economic aspects of Harvesting, Processing and Marketing of cane products in Kerala, KFRI Scientific Paper 870. Proceedings on National Workshop on Rattans (Canes) held on 4th and 5th February 1999 in Bangalore: 79-88.

Muraleedharan P.K; B. Jayasankar and P. Rugmini., 1996a. Some economic aspects of cane harvesting in Kerala .J. Non-Timber For. Prod. 3(3&4): 202-207.

Muraleedharan P.K; B. Jayasankar. and P. Rugmini, 1996b. Economics of cane processing in South India. J. Trop. For. 12(3): 142-150.

Padmanabhan, D. and Krishnan, P., 1989. Rattan research and tissue culture in South India. *In*: A.N. Rao and A.M. Yusoff (Eds.). Proceedings of the Seminar on Tissue Culture of Forest Species. FRI, Malaysia and IDRC, Singapore. pp. 50-62

Padmanabhan, D. and Sudhersan, C., 1989. Laminoids in leaf cultures of a rattan palm. *In*: A.N. Rao and Isara Vongkaluang (Eds.). Recent Research on Rattans. Faculty of Forestry, Kasetsart University, Thailand and IDRC, Canada. pp. 148-151.

Padmanaban, D and Illangovan, R., 1989. Studies on embryo culture in *C. rotang.* RIC Bull 8(1/4): 1,6-9.

Padmanaban, D and Illangovan, R. 1994. Surgical induction of multiple shoots in embryo cultures of *Calamus gamblei.* Becc. RIC Bulletin 12(1/2): 8-12.

Patric, B.D., U. Ward and M. Kashio 1994. Non-wood Forest Products in Asia. FAO, Bangkok.

Paudel, S.K. 1997. Rattans in Nepal. In Rao, A.N.and V.R. Rao, 1997. Rattan- Taxonomy, Ecology, Silviculture, Conservation, Genetic improvement and Biotechnology. IPGRI-APO Serdang, Malaysia.

Pei, S.J. & Chen, S.Y. 1991. *Flora Reipublicae Popularis Sinicae.* Tomus 13(1). Science Press, Beijiung. 172 pp.

Renuka, C., 1991. How to establish a cane plantation. KFRI Info. Bull. No.10.

Renuka, C., 1992. Rattans of Western Ghats- a taxonomic manual. KFRI, Peechi. 60 p.

Renuka, C., 1995. A manual on the rattans of Andaman & Nicobar islands. KFRI, Peechi. 72 p.

Renuka, C., 1999. Indian Rattan distribution – An update. The Indian Forests 125 (6): 591-598.

Renuka, C., 2000. A field key for the rattans of Kerala. Kerala Forest Institute, Peechi.

Seethalakshmi, K.K., 1993. Propagation of clustering rattans using suckers. In: S. Chand Basha and K.M. Bhat (eds.) Rattan Management and Utilisation KFRI, India and IDRC, Canada, pp. 142-147.

Sunny Thomas, K. Haridasan and S.K. Borthakur, 1998. Floristic study on rattans and its relevance in forestry of Arunachal Pradesh. Arunachal Forest News. 16 (1&2): 19-24.

Thwaites, G.H.K., 1864. Enumeratio plantarum zeylaniae. 303 & 431. Dulau, London.

Tilakaratna, D., 1997. Taxonomy & Ecology of rattan in Sri Lanka. In: A.N. Rao & V.R. Rao (eds.) Rattan – Taxonomy, Ecology, Silviculture, Conservation, Genetic Improvement and Biotechnology. IPGRI – APO, Serdang, Malaysia.

Trimen, H., 1898. A handbook to the flora of Ceylon. IV. 329-335. Dulan, London

U Htay Aung, 1997. Rattans in Nepal. In Rao A.N. and V.R. Rao (eds.) 1997. Rattan- Taxonomy, Ecology, Silviculture, Conservation, Genetic improvement and Biotechnology. IPGRI- APO. Serdang, Malayasia.

Valsala, K. and Muralidharan, E.M., 1998. Plant regeneration from *in vitro* cultures of rattan (Calamus). In: Damodharan, A.D. (Ed.), Proceedings of 10th Kerala Science Congress, Kozhikode, January 1998. pp. 161-163

Valsala, K. and Muralidharan, E.M., 1999. In *vitro* regeneration in three species of Rattan (*Calamus* spp.), In: Plant Tissue Culture and Biotechnology- Emerging Trends. Kavi Kishor P.B. (Ed.), Universities Press. p. 118-122.

Xu, H.C; A.N. Rao; B.S. Zeng and G.T. Yin (eds.), 2000. Research on Rattans in China – IPGRI – APO, Serdang, Malaysia.

Xu, H.C; B.S. Zeng and G.T. Yin, 1999. Herbarium survey and rattan conservation in China. In Rao.A.N. and V.R. Rao (eds). Bamboo and Rattan Genetic Resources and Use. IPGRI, Malaysia.

Yin, G; H.C. Xu; Z. Zhou and B.S. Zeng, 2000. Biomass accumulation and nutrient recycle in rattan plantations. In Xu *et al* (eds.) 2000. Research on Rattans in China. IPGRI –APO, Serdang, Malaysia.

Yin, G and H.C. Xu, 2000. Suitable storage conditions for seeds of *Daemonorops margaritae* In Xu *et al* 2000. Research on Rattans in China. IPGRI –APO, Serdang, Malaysia.

Yekantappa, K; K.M Bhat and T.K. Dhamodaran, 1990. Rattan (cane) Processing techniques in India: A case study of oil curing. RIC Bull. 9(2): 15-21.

Zeng, B.S. and G.Yin, 1997. Silviculture, genetic improvement, conservation, utilisation and socio-economics of rattan in China. *In* Rao A.N. and V.R. Rao (eds.). Rattan-Taxonomy, Ecology, Silviculture, Conservation, Genetic improvement and Biotechnology. IPGRI – Serdang, Malaysia.

Zeng, B.S.; H.C. Xu; Y. Liu; G. Yin and Z. Qiu, 2000. Tissue culture for mass propagation and conservation of rattans. *In* Xy, H.C. *et al* (eds.) Research on Rattans in China. IPGRI – APO. Serdang, Malaysia.

Zeng, B.S.; Y. Liu; Z. Qiu; H.C. Xu and G. Yin, 1999. Evaluation of *ex situ* conservation status of rattan germplasm in China. *In* Rao A.N. and V.R. Rao (eds.) Bamboo and Rattan Genetic Resources and Use. IPGRI-APO. Serdang, Malaysia.

THE STATUS OF THE RATTAN SECTORS
IN LAO PEOPLE'S DEMOCRATIC REPUBLIC, VIET NAM AND CAMBODIA
– WITH AN EMPHASIS ON CANE SUPPLY

Tom Evans

Abstract

The rattan cane sectors in Viet Nam, Lao People's Democratic Republic and Cambodia are thought to be developing along similar paths, but they have reached different stages. In Viet Nam wild stocks are probably almost exhausted and plantation development is under way. In the Lao People's Democratic Republic, wild stocks are substantial but declining and plantation trials are just beginning (with plantations for edible shoot production forming a dynamically growing subsector). In Cambodia the limited information available suggests that overharvesting of wild stocks is well under way but little if any work has yet been done on plantations.

Low international cane prices are a key constraint to investment in either plantations or sustainable wild harvesting. Many other deep-rooted socio-economic constraints also combine to make the prospects for large-scale sustainable harvesting poor in all three countries, but especially in Viet Nam with its high population density and degraded forests. It may be possible to foster sustainable wild harvesting at certain sites to boost rural livelihoods or to support conservation, but it is most unlikely to satisfy market demands.

Some suggestions for interventions to support the rattan cane sector are made. A region-wide taxonomic revision has just been completed, focused on the Lao People's Democratic Republic, but further work is needed, especially in herbaria in Viet Nam, in the field in Cambodia and on information exchange between countries. To set priorities for other interventions in Cambodia an initial survey of requirements and capacity is needed. For Viet Nam a similar survey is needed, but the focus is almost certain to be on aspects of plantation development through agencies already active in this field in the country.

For the Lao People's Democratic Republic, it is already possible to discuss options in some detail. The most promising field is the edible shoot sector and some existing programmes have begun to work in this area. Given the low present economic incentives for intensified cane production, the best immediate strategy in that sector may be a wait-and-see one focused on capacity-building, small-scale trials of planting and management in the wild, direct support for protected areas and continued dialogue with the government on policy incentives (especially tenure and reform of trade regulations).

1. Background

1.1 Geographical background

The three states often collectively called Indo-China differ greatly, with Viet Nam being particularly distinct from the other two, Cambodia and the Lao People's Democratic Republic. Annex 1 presents some key statistics.

The Lao People's Democratic Republic and Cambodia have sparse populations, a high proportion of forest and remain amongst the least developed countries in the world. Viet Nam, by contrast, is slightly better developed, very densely populated, has little forest remaining in most regions, and includes two large cities where major industrialization is occurring. Until recently it was on the way to becoming a booming "tiger economy" but growth and foreign investment are currently at a low ebb (Anon, 2000a).

The Lao People's Democratic Republic and Viet Nam are strictly controlled one-party states whereas Cambodia is a kingdom with a fledgling multi-party democracy and a very weak regulatory environment (FAO, 1997a; Global Witness, 1998). Viet Nam and Cambodia have coastline and seaports but the Lao People's Democratic Republic is hindered in its trade options because it is land-locked. All three have tropical monsoonal climates with a prolonged dry season, especially in the lowlands.

1.2 One plant, two commodities

Globally rattan is seen principally as a cane-producing plant. Nonetheless in the Lao People's Democratic Republic and northeast Thailand rattans also supply great quantities of edible shoot tips. These are consumed locally or exported to Southeast Asian communities in France, United States and elsewhere. Although this paper concentrates on cane, the shoot production subsector is discussed where relevant.

1.3 Taxonomic overview

Good taxonomy is crucial since the very many Asian species differ in quality, abundance, growth rates and many other aspects. Without a common, shared system of species names it is almost impossible to discuss the status of the resource or the means to manage it.

Much of the existing knowledge on rattans, at least in the Lao People's Democratic Republic, is held by district and provincial foresters and organized according to the local names of the species as used by villagers. Recent research has shown that, whilst these local names are often quite consistently applied within individual villages, they are virtually useless in comparing one area with another (Evans *et al.*, in press (a)). This is because the same species can have different local names in different places, the same names are often applied to different species in different areas and pairs of species can even swap their local names from one area to the next. This problem has also been stressed by workers elsewhere (Dransfield and Manokaran, 1993).

The only way to establish shared, agreed names is by collecting and comparing herbarium specimens. These can then form the basis of published opinions and remain available for re-examination by other researchers. Field guides are valuable for day-to-day use but ultimately one often needs to return to the herbarium to confirm the identification of a given plant.

The weak taxonomic basis in Indo-China has slowed the scientific development of the sector. No revision or guide has been written since the rather poor *Flora of French Indo-China* account (Gagnepain and Conrard, 1937). For the Lao People's Democratic Republic in particular knowledge was almost nil – only about five historical specimens existed and the 50 or so specimens collected as part of an IDRC/INBAR-funded project from 1992 to 1997 had not been critically named, due to lack of access to other herbaria (Ketphanh and Sengkhonyang, 1997).

During the period 1997-2000, the Department of the Environment of the United Kingdom Government funded a rattan research project in the Lao People's Democratic Republic through the Darwin Initiative for the Survival of Species. The project included a taxonomic component with a great deal of new fieldwork. About 250 herbarium specimens are now available from the Lao People's Democratic Republic and over 500 others have been examined from the surrounding areas (Viet Nam, Cambodia, non-Peninsular Thailand and southern Yunnan, China).

The results of this study will shortly be published in two formats – a formal paper detailing the taxonomic revisions and a field guide suitable for non-specialists (Evans *et al.* in press (a)). It is hoped that these will catalyze better sharing of information between foresters and researchers within and beyond the region. A small rattan herbarium has been established in the Lao People's Democratic Republic and the Lao Government has wisely deposited many duplicates at Kew. Together with plans for an exchange programme within the region, this should ensure that the specimens can easily be seen by active researchers around the world.

In total 50 species are now recognized from the studied region, of which 44 have been found in the Lao People's Democratic Republic, Viet Nam and Cambodia. This includes eight previously undescribed species and one subspecies (Dransfield, 2000; Evans, in press; Evans *et al.* in press, (a) and (b)). It also includes 16 species names and 12 variety names which are now known to be synonyms. The net result has been a simplification and clarification of the system of names in the region. Some species doubtless remain to be discovered.

Of these 43 species, 37 climb and so produce cane. At least 20 have canes of moderate or high quality. Most others have very short or brittle canes but there are still some species lacking information. A few species are predominant in the trade at present, because they are both of high quality and abundant. Other high-quality species are much scarcer and so appear insignificant at present; however, some may gain greater importance if plantations become widespread.

The situation is now discussed in more detail for each country. Resource-side and processing-side issues are discussed separately, with an emphasis on the former since that is where the greatest weaknesses are, in Indo-China and throughout Asia (Belcher, 1999). New information is mainly presented for the Lao People's Democratic Republic, where the author has extensive experience; a review of the literature was made for Viet Nam and Cambodia.

2. THE LAO PEOPLE'S DEMOCRATIC REPUBLIC

2.1 Resource-side issues

2.1.1. The significance of forest products to the economy

The Lao People's Democratic Republic has a poor, predominantly rural population which relies on wild-harvested resources for an important part of their diet (especially after bad harvests) and cash income (Foppes and Ketphanh, 1997 and 2000). Timber harvesting contributed 35–40 percent of export earnings in 1997 (FAO, 1998) but this has dropped off sharply since the Asian economic crisis of 1997, when timber prices collapsed. The trade in other forest products is also substantial (about 2.5% of total export earnings in 1996) and very diverse (Foppes and Ketphanh, 1997). Recent estimates suggest that local subsistence uses of NTFPs, taking place outside the cash economy, may be equivalent to a very significant 20 percent or more of Gross Domestic Product (Foppes and Ketphanh, 2000).

2.1.2. Wild sources of cane

Cane production in the Lao People's Democratic Republic is entirely from wild stocks. Forest cover was estimated at 110 000 km^2 in 1989 (FAO, 1998) but may now be closer to 95 000 km^2 (Duckworth *et al.*, 1999). Deforestation is rapid, but there are no recent published figures.

Two broad forest types probably hold the main commercial populations of rattans: evergreen or semi-evergreen lowland forests dominated by dipterocarps and evergreen hill forests dominated by Fagaceae and Lauraceae (*pers. obs.*). The total area of forest suitable for rattans is not clearly known, but Berkmüller *et al.* (1995a) estimated the total area of well-stocked, closed canopy forest as about 22 000 km^2 in 1989. The total is undoubtedly much less now. Rattan stocks within these forests are known only in broad, qualitative terms, based mainly on the observations of professional foresters, since no formal inventories have been made. They are generally not included in timber inventories and no formal monitoring is undertaken.

Forest plantations are not yet abundant in the Lao People's Democratic Republic [< 400 km² in 1997, most newly established (FAO, 1998)] and produce almost no rattan cane since wild regeneration is slight and deliberate planting does not yet occur.

2.1.3. Key commercial species

Species of known or suspected commercial importance in the Lao People's Democratic Republic are listed in Annex 2. The most important large-diameter species is *Calamus poilanei* and the most important small-diameter species include *C. palustris, C. gracilis, C. tetradactylus* and the newly described *C. solitarius* (Evans *et al.*, in press (b)). The medium-large *C. platyacanthus* (especially important in Viet Nam) also occurs. The list is surely incomplete: knowledge is best for the central Lao People's Democratic Republic and the southern half of the north, and fieldwork has been much less extensive in other parts of the country. Several of these species have been identified by INBAR as high-priority species for further research at an international level (Williams and Rao 1994, Anon, 1997).

2.1.4. Data on abundance and depletion

There is very little resource information on which to base management decisions at a national level. The qualitative local knowledge held by forestry officers and villagers is not available in a format suitable for incorporation into a national-level review.

During the Darwin Initiative project, preliminary studies have been conducted on the growth rates of *C. viminalis* and *C. solitarius* but the results of these studies are not yet available for publication. As far as the author is aware, quantitative information is only available for *C. poilanei*, and even then it is very fragmentary.

C. poilanei is the country's elite large-diameter cane. It is single-stemmed and thus regenerates only from seed after cutting. This gives it a poor regeneration capacity, like *C. manan* from Malaysia (Dransfield and Manokaran, 1993). Current evidence suggests that heavy harvesting is putting the species at high risk of commercial extinction in the Lao People's Democratic Republic in the foreseeable future. Further information supporting this conclusion is presented in Annex 3.

The status for most other commercial species is unknown. There are extensive forest areas where small-diameter canes have been seriously overharvested. During a recent socio-economic survey, traders and manufacturers widely reported that lack of raw materials was limiting their businesses and some factories had already closed for this reason (Sengdala *et al.*, 1997), although this was partly due to administrative difficulties with permits and poor road access to rattan-bearing forests. Nonetheless, the national picture is likely to be rather better than for *C. poilanei* if only because other species are less valuable (and so less sought after), had higher initial densities and are mostly clustering (i.e. resprout after cutting). They thus persist in harvested areas and have the potential to regenerate more rapidly when pressure eases. One small-diameter species (*C. solitarius*) is solitary-stemmed and, although more abundant than *C. polianei* even in areas where both have been harvested, it should give some cause for concern. The high-altitude species *C. acanthospathus* apparently tends to produce only one or two stems; this may make it especially vulnerable to overharvesting in the Lao People's Democratic Republic.

Three Lao species have been listed as being at global risk of extinction (Annex 3).

2.1.5. Management regimes

Land tenure in the Lao People's Democratic Republic is being restructured. Partial control of forest land is being allocated by the state to some communities but this programme is new and still evolving. The present *de facto* situation is that most forest areas (in particular those areas so far from villages that they retain rattan stocks) are not under the control of any one individual or community. For most non-timber resources (fish, game, plants, grazing) in most places the customary land-use regime in remote areas is open access harvesting, with no community ownership of particular areas or resources. This has been fostered by low population pressure, huge forest areas and the cultural preference for conflict avoidance. By encouraging unregulated competition between users, open access is one factor which encourages the rapid overharvesting of rattans. Many other factors also act in the same direction: they are discussed further in section 2.3 below.

There are government regulations on rattan harvesting but it is not clear that they are intended to preserve stocks and they can sometimes accelerate declines. For example, the harvesting that led to a collapse of *C. poilanei* stocks around Sayphou Phaphet (Annex 4) was done to satisfy official, legal quotas given to a legitimate trader by the provincial authorities. An accurate analysis of the effect of policy and regulations on rattan stocks is not possible because those regulations are rarely publicly available (Enfield *et al.*, 1998). Total provincial quotas are granted by the central government and then each province apportions its allowance to a number of approved manufacturers (Sengdala *et al.*, 1997; Enfield *et al.*, 1998). They subcontract traders who employ villagers to cut the rattan and bring it to a roadhead. The quota specifies a volume and a general source area and is valid for one year. The process of deciding quotas is very unclear to most participants in the supply and marketing chains, and is not open to independent scrutiny (Enfield *et al.*, 1998). Although some regulations mention the need to inventory the wild stocks and assess the effect on future yields at the site this is not thought to be enforced other than by reliance on vague "common knowledge" that certain areas have many rattans (Enfield *et al.*, 1998).

Individuals or communities cannot sell large quantities (i.e. truckloads) of rattans unless approached by a quota-holding trader. In general this is a restraining influence. However, when a quota-holder does visit, the incentive for all parties is to harvest the greatest volume of rattans possible during the brief window of opportunity. The result is the stripping of commercial rattan species from an area of forest, removing the possibility of any worthwhile harvest in the following years, let alone a sustainable one.

A significant feature of the trade is that the rural collectors are often paid very low rates for their cane, as is found throughout Asia (Belcher, 1999). This is partly because they have little access to price information, partly because one trader has a monopoly in any particular village and year and partly because rural people are in a weak position to argue with the politically and financially strong traders. Rattan collectors are often extremely poor and may ask to be paid directly in rice, rather than cash.

The export of unprocessed or "initially processed" cane is forbidden (Prime Minister's Order 14/psl, September 1990, cited and reproduced by Enfield *et al.*, 1998). Large recent exports (see Annex 4) may avoid this regulation by being considered "semi-finished".

The above rules refer to non-conservation forests, where the Lao Government appears to be acting on an overriding need to generate income in the short term to support national development despite the damage to future productivity. The Lao People's Democratic Republic also has a large extent of conservation forests (the Lao term translates simply as "national protected forest" but the preferred English term is National Biodiversity Conservation Area or NBCA). These 18 NBCAs were established in 1993, range from 500 km^2 to over 3500 km^2 and cover about 10 percent of the country (25 000 km^2), including large areas of all the major forest types (Berkmüller *et al.*, 1995a). The law prohibits commercial harvesting in NBCAs. However, three factors reduce their value for preserving species and genetic diversity of rattans:

- some reserves had been heavily logged, harvested or impacted by shifting cultivation before being declared [e.g. Dong Hua Sao, Phou Khaokhoay, Nam Et/Phou Loeuy (Berkmüller *et al.*, 1995b)] and thus are unlikely to support strong commercial rattan populations;
- large quotas are still sometimes issued for extraction within protected areas [e.g. Nakay Nam-heun NBCA in Nakay District in 1996 (Sengdala *et al.*, 1997; C. Marsh and J. Baker, *pers. comm.*)];

- official support for protected areas remains somewhat ambivalent at all levels so that they face great difficulties in competing against powerful commercial interests, or even in obtaining operating funds and skilled staff.

There has been no review of how many rattan species occur within the existing protected areas network or how well they are protected. Data are probably still too poor to allow this. Nonetheless, judging from experiences with commercially valuable animal species (Duckworth *et al.*, 1999) the system is probably not able to preserve viable populations of those species it does contain without greatly increased political and practical support.

The above discussion refers to large-scale extraction, in particular of larger diameter canes. Another large but unquantified part of the total harvest is done by many rural inhabitants who make frequent trips to collect and sell small quantities (a few kilos to tens of kilos), especially of smaller diameter canes. Onward transport is often by public transport or in trucks carrying other goods. These small shipments are often destined for handicraft factories in the larger towns, although some may eventually be exported. Regulations concerning this trade are rather unclear (Enfield *et al.*, 1998), and they may well vary from province to province. Technically it is illegal to harvest materials from NBCAs for sale and all commercial transactions of rattan should be approved by quotas. However, there is also a contradictory presumption in law that villagers, who have customarily done so, will be allowed to continue harvesting for household subsistence purposes, which is widely interpreted to include small scale-trade (Government of Lao PDR, 1996; Enfield *et al.*, 1998). This makes the diffuse, small-scale trade in rattans a *de facto* legal trade, although confiscation may occur e.g. if the trade is conducted in a too obvious way. The net impact on rattan populations and productivity is unclear, but is probably significant and deserves further study. The trade is probably also important to the livelihoods of the people involved.

The trade in edible rattan shoots from wild plants is large, unquantified and essentially unregulated. *Daemonorops jenkinsianus* thrives in the north in areas of shifting cultivation and appears to be the main source of shoots in the markets there. Its profusely clustering clumps survive fire, deforestation and repeated shoot removal very well. The cane of this species is not highly sought after, so trade in its shoots has little effect on overall commercial cane production. However, in some places valuable cane-producing species are targeted (e.g. *C. wailong* in Bokeo Province or *C. poilanei* in Bolikhamxay Province) and this trade is of greater concern.

2.1.6. Domestication of rattans in the Lao People's Democratic Republic

Small-scale nursery trials have been made for six or seven species with commercial potential, and a small germplasm collection established. Only one or two very small trials have begun of plantations for cane production, but one species (*Calamus tenuis*) has already become a major commercial success in plantations for edible shoot production (Sengdala and Evans, 1998; Evans and Sengdala, 1999). Many fields begin producing saleable shoots only a year or so after planting and can then be harvested monthly for many years thereafter, offering a return competitive with rice production and preferring sites where regular flooding would damage most other crops. In the Lao People's Democratic Republic, the techniques were first developed in 1994, but there are now estimated to be over 100 ha planted by over 50 planters in at least five provinces. This new development was inspired by large-scale commercial planting in Thailand of three species (mainly *C. viminalis* with some *C. siamensis* and *C. tenuis*) which began in 1991 (Jarenrattawong, 1997; Evans and Sengdala, 1999).

2.1.7. Likely future trends in the supply of cane in the Lao People's Democratic Republic

It is likely that heavy extraction from wild stocks will continue, leading to the commercial extinction of many or most species. This will be accompanied by scattered, externally funded attempts at sustainable wild harvesting together with small-scale attempts at plantation establishment, with limited initial commercial success. The reasons for these predictions are outlined below.

Lao has a small domestic cane market and a much larger export market. The domestic market seems unlikely to grow much, given the small size of the national economy. In the huge export market, raw Lao cane will presumably continue to receive a low unit price until global stocks as a whole run low.

Plantations cannot presently compete in price with wild harvested cane. As discussed in section 2.1.4, legally harvestable wild stocks, especially of large-diameter species, are likely to become exhausted in the Lao People's Democratic Republic in the foreseeable future given current harvesting practices. The trend in production will probably then depend mainly on the price of cane in international trade.

If the price stays low over the next 10-20 years, as seems likely, there will probably remain little market incentive to develop plantations or to establish large-scale sustainable production from the wild. National production will thus be very low and the international trade will turn elsewhere for sources of cane. The domestic market will be unable to afford imported materials and will either switch to other materials (from large cane to bundled small canes, from rattan to bamboo or wood), obtain supplies from the small remaining wild stocks or stimulate a small, low-cost domestic plantation industry. The latter would be a valuable addition to the small national economy but of little significance at an international level.

If the world price improves, then possibilities for the sector are more positive and varied, since investment should increase.[5] A price rise is most likely to happen if Indonesia replaces the recently lifted restriction on exports of raw cane. Failing that, a general tightening of restrictions on harvesting in protected areas across the region, together with declining wild stocks, may reduce supply and increase the price, although large unprotected reserves probably still remain in Indonesia and Myanmar. The same effect would result if rattan production in the Lao People's Democratic Republic was subsidized by the state or development agencies.

What is the likely balance then between production from sustainably managed natural forests and from plantations? The general trend for Asian timber and NTFPs is towards plantations (FAO, 1997c; FAO, 1997d) and the present author's assessment of the situation is that this will be the case for rattans in the Lao People's Democratic Republic since they experience fewer economic and social constraints (Belcher, 1999; section 2.3.1 below). However, it may be possible to establish sustainable wild harvests at certain sites given strong intervention at a number of levels. It is an open question whether the latter course would be beneficial enough to merit the costs.

2.2 Processing-side issues

There are probably fewer than ten major rattan factories in the Lao People's Democratic Republic, none of them employing more than a few tens of people. Details can be obtained from the Lao Forestry Information Service Section (Annex 5). They handle much of the raw cane and then export it "semi-finished" to Thailand and Viet Nam. They also produce finished goods which are mainly sold within the Lao People's Democratic Republic and are rarely of export quality. Advanced processing techniques are not yet widely used; for example, drying often relies on sun-drying or ovens and no fumigation is carried out (Sengdala *et al.*, 1997). Some villages specialize in small-item rattan handicrafts which are mainly sold within the Lao People's Democratic Republic, including the lucrative and growing tourist market. There is also extensive domestic use in rural areas of home-made or locally traded items; this usage has little direct cash value but is important within the poor communities practising it.

In addition to the growing difficulties in finding wild stocks, a variety of other constraints have been discovered during three recent socio-economic studies.

---

[5]Although there is a risk that increased prices will simply cause rattan to be replaced by affordable substitutes like wood, bamboo or plastic. This trend has already been observed in the Philippines, Indonesia and China (Belcher, 1999).

Traders note a number of difficulties in procuring cane even in areas where accessible stocks remain. These include the erratic and non-transparent nature of the quota system, physical difficulties in pulling down cane trapped in the canopy (which is thus often wasted) and spoilage caused by the immediate lack of post-harvest treatments. Providing credit to villagers who cannot then find enough cane to repay the debt is also a problem in some areas. The poor road network hinders collection of harvested cane.

Traders also have to deal with a variable and unpredictable number of supplementary taxes as they pass the numerous district and provincial trade checkpoints. These can add significantly to the cost of the raw material.

Manufacturers report that they lack the operation capital and bargaining power to obtain a high price for their products. The low levels of technology, simplistic designs and poor finish reduce the ability of Lao firms to reach export markets.

The last survey based on interviews with manufacturers was almost five years ago and many economic changes have occurred in the Lao People's Democratic Republic since then, so the above information may be somewhat outdated.

2.3 Opportunities to support the rattan sector in The Lao People's Democratic Republic

Well-targeted investment in the Lao cane sector by outside bodies might have development and conservation benefits. Table 1 outlines some potential beneficiaries from support in each subsector. The three cane subsectors are discussed in more detail below, followed by comments on the edible rattan shoot subsector.

Table 1. Potential beneficiaries and problems caused by future development in three parts of the Lao rattan industry

Subsector *Beneficiaries*	Sustainable harvesting	Plantation development	Processing and marketing chain
Rural communities	In longer term	In some cases	In some cases
Traders/middlemen	?	✗	?
Factories	In longer term	✓	✓
National exports	In longer term	✓	✓
Biodiversity conservation	✓	Indirect (need to preserve germplasm sources)	✗
Overseas buyers (e.g. factories)	In longer term	✓	Neutral? (transfers market share to the Lao People's Democratic Republic)
Potential problems	Reduced production in short term	Natural forest no longer needed to supply cane	Increased demand speeds destruction of wild stocks. Increased capacity useless if stocks fail.
	Increased costs[1] and technical difficulty	Rural communities could lose a source of income	
	Significant risk of failure	Vulnerable to changes in trade price	

[1]E.g., Evans and Viengkham (in press) highlight the impractically high costs that can be involved in statistically meaningful inventory and monitoring of wild rattan populations.

The choice of intervention must depend partly on the objectives of the funding agency (e.g. poverty alleviation, boosting national income, biodiversity conservation).

2.3.1. Sustainable harvesting

If trade prices increased this would place increased pressure on the wild resources but would also offer the possibility, currently lacking, of livelihoods well above the poverty line for sustainable harvesters. Given such a price rise, it would also require a strong political will and prompt technical intervention to establish good management. It is perhaps unlikely that this would happen before accelerated harvesting had destroyed the resource base.

No functioning model of a planned extractive reserve for timber or other forest products yet exists in the Lao People's Democratic Republic. This is despite strong enthusiasm in the conservation and aid communities for the concept, which seems to promise simultaneous biodiversity conservation and improved rural livelihoods.

Two pilot Natural Forest Management schemes exist for timber but have encountered substantial political and social difficulties, in part because of the sums of money involved. One activity of the Lao IUCN-NTFP project is a pilot to improve management of village bamboo stands in the northern Lao People's Democratic Republic for shoot production. Early successes have been mainly in the establishment of marketing groups and the acid test, when harvests need to be reduced by the community to ensure sustainability, has not yet been faced. The project has also begun to tackle unregulated malva-nut and rattan extraction at a few villages in the south, but work is at an early stage. The one shining example of community management of a wild Lao resource involves fisheries in southern Champassak (Baird and Singsouvan, 1996; Foppes and Ketphanh, 2000), but its lessons have yet to be applied to the forestry setting, which has important qualitative differences.

Agencies involved in sustainable NTFP management in the Lao People's Democratic Republic are listed in Annex 5. They have identified a great number of local and national constraints and begun addressing some of them at the technical, policy and village levels. Some of these obstacles are listed in Annex 6. They fall under three broad headings:

Social constraints
- Scarcity of cultural precedents for managing the forest;
- Shortage of trained staff or operating budgets in relevant government agencies;
- Imbalances in the political power of different actors.

Policy constraints
- Open-access to resources;
- Short, erratic quotas permitting (even encouraging) overharvesting;
- Complex, obscure regulatory framework in the forestry sector;
- Government imperative to maintain its short-term income to fund development;
- Economic and biological constraints;
- Increasing market access through road expansion and macro-economic changes;
- Intense rural poverty driving rural people to overharvest for short-term survival;
- Physical difficulty of policing remote, dispersed stocks;
- Long travel times to harvest remaining stocks;
- Low bargaining power of harvesters;
- Uncertain legal status of diffuse trade in small canes, preventing investment or regulation;
- Low unit value of wild cane and suspected low annual productivity per hectare;
- Substantial costs of planning, inventory and monitoring;
- Lack of technical knowledge about the ecology of commercial species.

The prospect of overcoming all of these constraints seems quite poor in the majority of cases, but none of them is technically insurmountable if the likely benefits are considered worthwhile. Specific constraints have been overcome in specific situations in an increasing number of cases (Foppes and

Ketphanh, 2000). If widespread progress in this field is to be achieved, three broad approaches are required (see Annex 6 for more detail):

1. *International level.* Policy changes to boost prices. Regulation of world trade (e.g. CITES).
2. *National level.* Enhanced political commitment to sustainable harvesting (e.g. resolution of tenure issues, improved quota system, tightened protection for NBCAs, and regulation of the diffuse trade in small-diameter canes).
3. *Local level.* Establishment of biologically and economically sound pilot schemes.

In the Lao People's Democratic Republic, it may be most practical to await an increase in world prices to stimulate the planting and management of rattan for cane production. Meanwhile, the ground could be prepared through capacity building (in biological and participatory social research), small-scale planting and management trials, protection of key genetic stocks in NBCAs and an ongoing dialogue with the Government on the regulation of quotas and land use.

2.3.2. Plantations

The reason rattan plantations will probably be favoured in the long run is that they escape most of the above difficulties. They nonetheless have some significant constraints which have prevented any substantial investment to date (Table 2). Low market prices and slow returns are also preventing private investment in Yunnan (Chen Sanyang, *pers. comm.*) and Thailand (various researchers, *pers. comm.*). There is some scope for further external support to enable the sector to overcome these.

Table 2. Current constraints to the development of rattan cane plantations in the Lao People's Democratic Republic

Constraint	Current trend	Possible solutions
Weak market price	Not improving	International interventions? Wait-and-see approach Reduce costs through improved techniques (especially fast-maturing stock) National subsidies
Limited research capacity	Improving	Further capacity-building support
Limited published research on species found in Lao PDR	Gradually improving	Expanded research programme Increased publication of results from neighbouring countries Study visits to those countries
Poor establishment and growth rates for the elite *C. poilanei*	No progress	Focused research Test alternative species (especially clustering ones)
Lack of a governmental extension agency	Proposals in existence	Feed into extension programmes of other organizations or fund rattan-specific programmes.

Furthermore, the regulatory and economic environment for establishing new businesses is not particularly attractive in the Lao People's Democratic Republic (e.g. slow approvals, high inflation) and this may be expected to hinder investment, especially for a venture which requires long-term confidence.

2.3.3. Processing and trade

Requirements for this subsector in the Lao People's Democratic Republic are not well understood. The first need is probably for an updated socio-economic survey to identify constraints and current priorities. Likely priorities include the following:

- Reform of the regulatory system at central, provincial and district government level (building on the findings of Enfield *et al.*, 1998).
- Introduction of better treatment methods for cane immediately post-harvest to boost both the price villagers receive and the quality of exported cane.
- Introduction of new manufacturing technologies, if the extra costs can be supported by the weak domestic market.
- Assisting Lao companies to explore higher value export markets.

Belcher (1999) suggests that support for this sector might boost demand and so encourage intensified investment in production of raw cane; given current socio-economic conditions in the Lao People's Democratic Republic, it seems equally likely that this would just accelerate overharvesting, causing a brief boom followed by a bust.

2.3.4. Edible shoot production

The outlook for expanding edible shoot production is much better than for cane. There is a large domestic market and the Lao People's Democratic Republic only competes with Thailand in supplying the substantial export market. Furthermore, planting is spreading rapidly without needing special policy support because, unlike cane, shoot-producing plantations of *C. tenuis* offer a rapid and proven return on the open market. Shoot-producing species are likely to be the focus of activity in the Lao People's Democratic Republic over the next few years. The following areas (in no order of priority) would benefit from increased investment:

1. Resolution of the remaining taxonomic difficulties and searches for additional Lao species.
2. Trials of species suitable for a wider range of environments.
3. Management techniques for maximizing the yield from established *C. tenuis* plantations.
4. Protection of genetic resources by improved protection of known seed sources and of NBCAs in general.
5. Provenance improvement, initially for *C. tenuis*.
6. Extension of techniques to those rural areas where the effects on poverty alleviation and stabilization of shifting cultivation might be greatest.
7. Processing techniques for export, in particular canning (which already occurs in Thailand but not the Lao People's Democratic Republic).
8. Marketing, particularly for export to Thailand, United States and France.

A proposal to partly address aspects 1- 4 has been drawn up by the Lao Forestry Research Centre, Oxford University and Royal Botanic Gardens Kew and funds are currently being sought. Increased external support from major donors is vital for general NBCA policy improvements and site management, especially given the recent cessation of several major protected area projects. Rattans are just one facet of the biodiversity under threat within the NBCA system (e.g. Thewlis *et al.*, 1998; Duckworth *et al.*, 1999). Aspect 6 is likely to be incorporated into existing multi-sectoral development projects as planting stock and proven techniques become available.

This subsector offers some spin-off benefits for the cane sector. The plantations themselves have little potential for conversion to cane production in the future since they are grown in open sun with no available climbing supports. However, the abundance of cheap seedlings and the widespread expertise in growing these species will make cane plantations easier to establish if economic conditions become attractive in the future.

3. VIET NAM

3.1 Resource-side issues

3.1.1. The significance of forest products to the economy

Forestry represents only a small fraction of Vietnamese GDP, although a fairly large absolute value. The direct value of all timber and non-timber harvesting was estimated at US$ 132 million or 0.65 percent of GDP in 1995, and additional wood and forest products processing industries were considered poorly developed (FAO, 1997b). Furthermore, the farm and forestry sector was growing only a third as fast as the other two major sectors of the economy (non-farm/forest industries and service industries), so the present share is probably even smaller. This difference from the Lao People's Democratic Republic and Cambodia is due to the greater industrialization of Viet Nam and also to the relatively small and heavily exploited forest estate. Nonetheless, many people, particularly ethnic minority groups in upland areas, remain quite dependent on harvesting forest resources (de Beer *et al.*, 2000).

3.1.2. Wild sources of cane

Forest cover is still moderately extensive in Viet Nam, with an official figure of 83 000 km^2 of natural forest in 1995 and 10 000 km^2 of plantations. Although the total cover is thought to be increasing there are substantial ongoing losses of natural forest (perhaps 1 000 km^2/a, see Annex 1) and gains in the plantation sector. Most natural forests have been severely degraded by logging, shifting cultivation and the effects of the Viet Nam war. As with the Lao People's Democratic Republic, the principal sources of wild cane, at least historically, would have been in the evergreen/semi-evergreen lowland forests and the evergreen hill forests. Each part of the country supports several commercial species.

3.1.3. Key commercial species

No recent published primary taxonomic research is available from Viet Nam. Vu Van Dung and Le Huy Guang (1996), which is primarily a report on cultivation research, includes an appendix with an updated species list, ecological and distributional notes. A summary of these data with minor additions was given by Ngo Thi Min Duyen and Nguyen Truong Thanh (1997). Vu Van Dung and Le Huy Guang (1996) make no reference to herbarium specimens and, despite the first-hand descriptions they give, it seems preferable to treat the accuracy of the names used as unknown until there has been an opportunity for Vietnamese botanists to exchange voucher specimens with other herbaria and to examine the crucial types. With this caveat, Vu Van Dung and Le Huy Guang's names are used in this review. Some likely modern synonyms are given in brackets.

Important commercial species are listed in Annex 7. The main large canes are *C. poilanei*, *C. rudentum* and the slightly smaller *C. platyacanthus*. Important smaller canes include *C. tetradactylus* and *C. tonkinensis* (= *C. walkeri*?). There is considerable overlap with the list for the Lao People's Democratic Republic. There are doubtless other species of usable quality amongst the many rarer species known or expected to occur in Viet Nam.

3.1.4. Data on abundance and depletion

No published quantitative data were traced on the abundance or population dynamics of wild stocks. Rake *et al.* (1993) present official figures for the extraction of large-diameter canes in two northern provinces which total approximately 34 million linear metres (approximately 7.5 million 4.5 m canes or 14 400 t) for the seven-year period 1986-1992. This was stated to be 10 percent of the national production, which can thus be very coarsely estimated at 340 million linear metres, 75 million canes or 144 000 t during that period. Of this, 90 percent was destined for export, and three-quarters of that was exported raw. The authors believed this referred solely to one large-diameter species although the name they give, *C. rudentum*, seems unlikely to be correct and it is odd that there is no mention of trade in small-diameter canes.

Vu Van Dung and Le Huy Guang (1996) repeatedly mention the very low remaining stocks, caused partly by the heavy recent export trade. Stocks seem likely to be in a much worse state than those of the Lao People's Democratic Republic and thus be close to exhaustion. This is supported by the fact that large quantities of rattan are now bought from the Lao People's Democratic Republic by Vietnamese traders (*pers. obs.*; Belcher, 1999; de Beer *et al.*, 2000) who presumably find it difficult to obtain supplies at home. A similar situation has been observed for many other tradable forest resources, including timber (Anon, 2000b), eaglewood (Broad, 1995) and a great variety of wildlife species (Duckworth *et al.*, 1999). Access to cheap Lao rattan stocks may be an important element in the Vietnamese domestic processing industry and export trade, and is facilitated by the close political relationship between the two countries.

Several Vietnamese species are believed to be at global risk of extinction (Annex 3).

3.1.5. Management regimes

A quota system is believed to control wild cane harvesting in Viet Nam, but details were not available during the writing of this paper.

There has been a relatively long programme of land allocation to individuals in Viet Nam (FAO, 1997c), with one of its aims being to enhance productivity through the establishment of clear tenure rights. No data could be traced for this paper on the effect this is having on rattan harvesting practices.

There are approximately ten national parks and 53 nature reserves. Together with cultural sites these are called "special-use forests" and total 6 600 km^2 (Nguyen Truong Thanh and Ngo Thi Min Duyen, 1997). Current regulations forbid the harvesting of NTFPs except in buffer zones, although the reality is that these rules are not often applied successfully. New reserves are being added and there is a policy target that 12 000 km^2 should come under "special-use" by 2005 (FAO, 1997b), with several times that area given over to protection of watersheds and erosion prevention in coastal areas. These protected areas have the potential to conserve genetic stocks of many rattan species, but their success in doing so has not been assessed. De Beer *et al.* (2000) list many "conservation and development" projects, often in buffer zone areas, which involve an element of NTFP development (sometimes including rattan), but state that few have invested sufficient expertise or time to achieve substantial results.

3.1.6. Domestication of rattans in Viet Nam

Vu Van Dung and Le Huy Guang (1996) report that smallholder cultivation of the small-diameter clustering species *C. tetradactylus* in North Viet Nam began over 100 years ago and is thus one of the longest established rattan cultivation systems in the world. From beginnings in the Red River Delta (which has long been severely deforested), the practice has spread to almost every northern province and, since unification in 1975, it has also spread to much of the south. An estimated 1 500–2 000 t/a of cane is produced annually by these smallholders. *C. amarus* (= *C. tenuis*?) has also long been cultivated, albeit on a smaller scale, and recent research efforts have led to the planting of *C. platyacanthus*, *C. rudentum*, *C. tonkinensis* (= *C. walkeri*?) and *C. poilanei*, much of it in tree plantations. In a paragraph on plantations Nguyen Truong Thanh and Ngo Thi Min Duyen (1997) cite a figure of 1.2 billion rattan clumps and a (very large) rattan growing area of 6 250 km^2 but the source and meaning of these figures are not clear since they later state that "the total area of rattans in plantations and spare plantings is about 25 000 ha (250 km^2), including (?plus) 60 000 ha (600 km^2) growing in natural forest."

There appears to be no mention in the literature of widespread consumption of rattan shoots or cultivation for edible shoot production.

3.1.7. Likely future trends in the supply of cane in Viet Nam

Vu Van Dung and Le Huy Guang (1996) estimated the demand for raw cane at 10 000 t/a for large-diameter cane (approximately 5 million 4.5 m sticks) and 15 000 t/a for small-diameter cane. This was mainly for the export trade, which Nguyen Truong Thanh and Ngo Thi Min Duyen (1997) put at US$ 4–37 million annually during 1993-1995. Recorded exports were 35 percent raw cane, 40 percent semi-finished products and 25 percent finished products, although it is not clear whether this refers to value or volume. If value, then clearly the vast majority of the volume is for export. There is also a substantial domestic market, especially in the larger towns and cities. Existing plantations satisfy only a fraction of this demand; wild stocks must provide the balance, an unknown but significant proportion of which is sourced from the Lao People's Democratic Republic and presumably also Cambodia.

Sustainable harvesting from wild populations seems even less likely to be an important large-scale source of cane in Viet Nam than in the Lao People's Democratic Republic, given the low expected yields per hectare, the much higher pressure on the remaining forests, the lower remaining stocks and the focus of existing research on plantations. Nonetheless, there do remain protected areas where sustainable harvesting of rattan and other NTFPs may be a useful tool in supporting forest conservation, and various scattered efforts exist and will continue in this direction (de Beer *et al.*, 2000). However, it seems likely that the main thrust will be toward increasing the amount of planted rattan, and that this will become the predominant source of cane for domestic processing and export.

There is a very ambitious stated goal of 800 km^2 of cane plantations by 2000 (?possibly an error for 2005), producing an enormous 150 000 t dry weight, mainly for export. This implies an impractically high average production of about 2 t/ha/a or 1 000, 4.5 m sticks/ha/a. The goal was to be achieved by a combination of allocating forest lands to people, social forestry programmes, investing more in state-owned forest enterprises, attracting non-state investment and offering favourable tax and credit opportunities (FAO, 1997b).

3.2 Processing-side issues

No substantial information was found on the rattan processing sector in Viet Nam, other than that it is quite large and more technically advanced than elsewhere in Indo-China. Exports of semi-finished and finished products exceed US$ 20 million and there is a substantial domestic market. Several companies advertise finished and semi-finished products online, which is not the case for the smaller, low-tech factories in the Lao People's Democratic Republic and Cambodia.

3.3 Opportunities to support the rattan sector in Viet Nam

All aspects of the Vietnamese rattan sector would probably benefit from work to bring the taxonomic system there into harmony with other countries. Further field work would undoubtedly add species to the country and provincial lists and perhaps also undescribed taxa, but should be preceded by a review of existing collections held in Viet Nam. This requires specimen exchanges and visits by botanists to herbaria holding types, especially Kew, Paris and the southern Chinese herbaria, followed by publication in international journals.

Sustainable harvesting of wild rattan populations is apparently not a major active concern in Viet Nam. No mention is made of it by Vu Van Dung and Le Huy Guang (1996) or Nguyen Truong Thanh and Ngo Thi Min Duyen (1997) and the IUCN-NTFP Project in Viet Nam is not promoting any such activities (J. Foppes, *pers. comm.* 1999). Support for this aspect might be best integrated with multi-sectoral projects working on protected area management or upland development, as has been the case in the past. Successful pilots need to be established before any large-scale effort can be considered. Many of the constraints are likely to be similar to those in the Lao People's Democratic Republic (section 2.3.1).

Support for the plantation sector would be welcomed. Nguyen Truong Thanh and Ngo Thi Min Duyen (1997) outline some areas where further research is needed, as follows:

- provenance improvement;
- advanced technology for propagation and seed source improvement;
- research on silvicultural techniques;
- extension of techniques to increase the area of plantations;
- improved capital investment in plantations and processing.

There may be other aspects which also require support. A great number of outside organizations are involved in the NTFP sector in Viet Nam (de Beer *et al.*, 2000) and new inputs, if appropriate, would have to be carefully coordinated with these. The relative merits of planting under tree crops in degraded natural forests or on smallholdings need to be considered depending on the goal of the support. There is a growing interest in small-scale cultivation of NTFPs (including rattan) as a tool in the development of impoverished highland communities (de Beet *et al.*, 2000).

It would be valuable to explore the factors which have allowed the plantation sector to grow so rapidly in Viet Nam when it has been so sluggish in equally overharvested regions such as Thailand and Yunnan. This would be especially helpful to guide policy in the Lao People's Democratic Republic and Cambodia, where the process has yet to begin. It is possible that much of the recent expansion in Viet Nam has been driven by state subsidies, as is the case for timber plantations; the profitability of rattan plantations has yet to be reported upon.

Consideration of support for the rattan-processing sector would have to be preceded by an up-to-date review of issues and requirements based on new surveys.

4. CAMBODIA

No national level information is available for the rattan sector in Cambodia. It is believed to be dominated by the extraction of raw cane for export. The country is not a member of INBAR and has not contributed participants to recent international rattan meetings (Rao and Rao, 1997; Bacilieri and Appanah, 1999).

4.1 Forest cover, commercial species and cane stocks

Satellite data put 1993 forest cover at 63 percent, but this uses a very broad definition of forest (FAO, 1997a). Only 18.2 percent of it covered evergreen, "mixed" and flooded forests, which are likely to include the main rattan habitats; the remainder involved deciduous forests, shrub land, scattered woodland and plantations. Dependency on forest resources is high among rural people, especially in the hillier northeastern parts (Bann, 1997; FAO, 1997a), and forestry is a major source of government income (FAO, 1997a).

Very little taxonomic work has been conducted in Cambodia, almost all of it during the French colonial period. There are records supported by specimens of only 11 species from Cambodia using current taxonomy (Evans *et al.*, in press (a)), but the total is likely to be at least twice this. Commercial species probably include two large-diameter species (*C. rudentum* and, if it occurs, *C. poilanei*) and four or more small-diameter species (*C. tetradactylus*, *C. tenuis*, *C. viminalis* and *C. palustris*). One species known from Cambodia is listed as globally threatened (Annex 3), as is *C. poilanei*.

Feil (1998) surveyed rattans and other NTFPs in the 1 400 km² Bokor National Park. He found low densities of large-diameter rattans (standing stocks of 0 m, 160 m and 350 m of harvestable cane/ha along three transects) and noted that even the remotest areas he visited seemed to have been depleted of cane. Since one of the three species he counted, '*Phdav preas*', is now known to be a *Korthalsia* with low cane quality (vouchered by specimen *Feil et al.* 03 at AAU), the stock of valuable species is probably still lower. Export records from the southern part of the park led him to estimate that

120 000 sticks 6-8 m long (and thus about 840 000 linear metres) had been legally removed from that area in 1998. This is a very high level of extraction in relation to the standing stock and seems unlikely to be maintained for long. Preliminary calculations suggested a sustainable maximum of only 12 000 sticks per annum.

Bann (1997) included rattan in a comparison of the economic benefits of customary land use and commercial logging in Ratanakiri Province, northeast Cambodia, an area where large-scale extraction was not reported. She found an average of 266 canes/ha but unfortunately her inventory results give no indication of the species other than locally used names, nor of the proportion of large-diameter or saleable canes. She estimated the sustainable annual offtake of rattans would be worth only about US$ 5/ha/a, given a ten-year harvest cycle and not taking into account harvesting costs (which were currently negligible due to low alternative employment opportunities). Overall, though, she estimated that the economic potential of NTFPs in the forest exceeded that of timber production.

Exploitation of Cambodian timber during the 1990s has been very rapid and in practice virtually unregulated, even though a total logging ban was announced in 1995 (FAO, 1997a). Known concessions at times exceeded some estimates of the total forested area of the country and extraction rates have been vastly in excess of sustainable levels (Global Witness, 1998). During 2000 far-reaching reforms have been implemented but their efficacy is not yet clear (Global Witness, 2000). Against this background the recent pressure on rattan also seems likely to have been high and unregulated. Narith (FAO, 1997a) states simply that rattans "have been harvested throughout the country". Enfield *et al.* (1998) found that shipments of Cambodian cane had been imported to the Lao People's Democratic Republic by factories there unable to obtain permits to harvest in-country.

There are no reports of rattan plantations in Cambodia, although the extensive areas of seasonally flooded forest perhaps offer high productive potential if suitable species can be found. Both *C. tenuis* and *C. godefroyi* deserve consideration in this habitat, and trials of *C. trachycoleus* may also be worthwhile.

4.2 Opportunities for assistance to the rattan sector in Cambodia

Current restructuring in the timber sector may have an influence on opportunities for sustainable rattan harvesting in the extensive natural forests. However, many of the constraints mentioned above for the Lao People's Democratic Republic will also apply to Cambodia and the low returns predicted by Bann (1997) suggest that sustainable rattan harvesting may fail to stand alone as a livelihood. Plantation development is a possibility but research capacity is likely to be quite low at present.

If external assistance is to be considered for Cambodia, the first requirement is a broad survey of needs and opportunities, in collaboration with the relevant government agencies. This should include basic taxonomic and distributional information, the status and harvesting patterns of wild stocks, existing processing capacity, inclusion of rattan in existing development and conservation activities and available research/extension capacity in relevant agencies. A list of possible activities could then be developed for consideration.

5. Conclusions

The rattan cane sectors in Viet Nam, the Lao People's Democratic Republic and Cambodia are thought to be developing along similar paths, but they have reached different stages. In Viet Nam wild stocks are probably almost exhausted and plantation development is under way. In the Lao People's Democratic Republic, wild stocks are substantial but declining and plantation trials are just beginning (with plantations for edible shoot production forming a dynamically growing subsector). In Cambodia the limited information available suggests that overharvesting of wild stocks is well under way but little if any work has yet been done on plantations.

Figure 7: Rattan mats (Belcher)

Sustainable harvesting of wild rattans remains poorly researched but evidence to date suggests that it is likely to offer relatively low annual returns per hectare and to face many other deep-rooted socio-economic difficulties. Whilst it may be possible to foster it in certain sites this will be an uphill struggle under current conditions, which include a pervasive lack of government action to curb overharvesting.

Widespread enthusiasm for commercial plantation development is unlikely to come until cheap wild stocks in a country are very low or unavailable due to protective measures. Even then, plantations will probably struggle to compete with cheap wild-cut supplies in trade from elsewhere in Asia except when they receive government subsidies and/or have access to a substantial domestic market. There is a risk that increased costs through intensified production of any kind may cause the replacement of rattan by cheaper substitutes such as wood or bamboo.

Suggestions for interventions to support the rattan cane sector are made in sections 2.3, 3.2 and 4.2. A recent taxonomic revision has been completed, focused on the Lao People's Democratic Republic, but further work is needed, especially in herbaria in Viet Nam, in the field in Cambodia, on information exchange between countries and on the adequacy of existing protected areas. To set priorities for other interventions in Cambodia an initial survey of requirements and capacity is needed. For Viet Nam a similar survey is needed, but the focus is almost certain to be on aspects of plantation development through agencies already active in this field in the country.

For the Lao People's Democratic Republic, it is already possible to discuss options in some detail. The most promising field is the edible shoot sector and some existing programmes have begun to work in this area. For the cane sector, the choice between focusing on wild stocks, plantations or the processing sector depends on the development and conservation objectives of the body offering assistance. If successful, plantation and factory development is likely to benefit mainly investors and urban workers whereas smallholder cultivation, micro-processing enterprises and perhaps management of wild stocks offer more prospect of assisting the rural poor. Biodiversity conservation may be best served by increasing support for protected areas in three ways: through direct funding, through subsidized sustainable harvesting schemes and also by publicizing the opportunities (e.g. for domestication) which will be lost if extinction occurs. The Lao People's Democratic Republic has much lower agricultural potential and lower population densities than its neighbours and thus offers more scope for maintaining very extensive protected areas, given sufficient external support.

Given the low present economic incentives for intensified cane production in the Lao People's Democratic Republic, the best immediate strategy may be a wait-and-see approach with a focus on capacity-building, small-scale planting trials, support for protected areas and continued dialogue with the government on policy constraints (especially tenure and reform of trade regulations).

Acknowledgements

Joost Foppes provided valuable information during the writing of this paper. He, Nick Brown and Laura Watson also commented on draft versions. I would like to thank the following people for their involvement in all stages of the Darwin Initiative Lao Rattan Research Project: Khamphone Sengdala, Oulathong V. Viengkham, Banxa Thammavong, Nick Brown, John Dransfield, Khamphay Manivong, Sounthone Ketphanh, Joost Foppes and the staff of the Forest Research Centre, the National Agriculture and Forestry Research Institute, the IUCN-NTFP Project and the NTFP Information Centre. Approval for the work was granted by the Ministry of Agriculture and Forestry, Lao PDR, and funding was provided by the Department of the Environment of the United Kingdom Government through the Darwin Initiative for the Survival of Species. A great many herbaria contributed to the taxonomic aspect of the project, especially Kew Gardens; the Institute of Ecology, Kunming; Xishuangbanna Tropical Botanical Gardens; Bangkok Herbarium, Bangkok; Forest Herbarium, Paris; and the personal collection held by Dr Issara Vongkaluang.

REFERENCES

Anon, 1997. Extended lists of priority rattan and bamboo species. *INBAR Newsletter* 4(3): 4-5.
Anon, 2000a. Goodnight Viet Nam. *The Economist* 8/1/00: 74-75.
Anon, 2000b Viet Nam importing huge amount of timber from Laos. *Nong Thon Ngay Nay Newspaper*, 5/6/2000, cited by enviro-vlc list server, 11/6/2000.
Anon, (2000c) Editorial. *Nhan Dan Newspaper*, Hanoi, 27/10/00, cited by the enviro-vlc list-server, 7/11/00.
Bacilieri, R. &Appanah, S., eds. 1999. *Rattan cultivation: achievements, problems and prospects.* CIRAD-Forêt, Montpellier, France and Forest Research Institute Malaysia, Kuala Lumpur.
Baird, I. & Singsouvan, S., 1996. The fish give us strength. *Watershed* 1(3): 30-32.
Bann, C., 1997. *An economic analysis of tropical forest land use options, Ratanakiri province, Cambodia.* Economy and Environment Program for Southeast Asia, Singapore.
Belcher, B., 1999. Constraints and opportunities in rattan production-to-consumption systems in Asia. *In R.* Bacilieri & S. Appanah, eds. *Rattan cultivation: achievements, problems and prospects.* CIRAD-Forêt, Montpellier, France and Forest Research Institute Malaysia, Kuala Lumpur. pp. 116-138.
Berkmüller, K., Evans, T., Timmins, R. & Vongphet, V., 1995a. Recent advances in nature conservation in the Lao PDR. *Oryx* 29: 253-260.
Berkmüller, K., Southammakoth, S. & Vongphet, V., 1995b. *Protected area system planning and management in Lao PDR: Status Report to mid-1995.* Vientiane, Lao-Swedish Forestry Cooperation Programme.
Bogh, A., 1997. *A comparative study of the demography of three species of* Calamus *(Arecaceae) in southern Thailand.* Manuscript submitted in partial fulfilment of Ph.D. Degree. Risskov, Denmark, University of Aarhus.
Broad, S., 1995. Agarwood harvesting in Viet Nam. *TRAFFIC Bulletin* 15(2): 96.
de Beer, J., Ha Chu Chu & Tran Quoc Tuy, 2000. *Non-timber forest products sub-sector analysis, Viet Nam.* Hanoi, IUCN-Viet Nam and the NTFP Research Centre.

Dove, M.R., 1993. A revisionist view of tropical deforestation and development. *Environmental Conservation* 20(1): 17-24.

Dransfield, J. & Manokaran, N., eds. 1993. *Plant Resources of South-East Asia 6. Rattans.* Wageningen, Pudoc Scientific Publishers.

Dransfield, J., 2000. *Calamus bousigonii. Kew Bulletin* 55: 711-716.

Duckworth, J.W., Salter, R.E. & Khounbouline, K., compilers. 1999. *Wildlife in Lao PDR: 1999 status report.* Vientiane, IUCN-The World Conservation Union/Wildlife Conservation Society/Centre for Protected Areas and Watershed Management.

Enfield, N.J., Ramangkhoun, B. & Vongkhamsao, V., 1998. *Case study on constraints in marketing of Non-Timber Forest Products in Champasak Province, Lao PDR.* IUCN-NTFP Project, Department of Forestry, Vientiane.

Evans, T., 2000. The rediscovery of *Calamus harmandii*, a rattan endemic to southern Laos. *Palms* 44(1): 29-33.

Evans, T., In press. A new species of *Calamus* from Thailand. *Thai Forest Bulletin.*

Evans, T., Sengdala, S., Viengkham, O., Thammavong, B. & Dransfield, J., In press (b). Four new species of *Calamus* (Arecaceae: Calamoideae) from Laos and Thailand. *Kew Bulletin.*

Evans, T.D. & Sengdala, K., 1999. From Non-Timber Forest Product to cash crop - the recent spread of rattan cultivation for edible shoot production in the Lao PDR. *Lao Journal of Agriculture and Forestry* 1999(2): 40-47.

Evans, T.D. &Viengkham, O.V. In press. Inventory time-cost and statistical power: a case study of a Lao rattan. *Forest Ecology and Management.*

Evans, T.D., Sengdala, K., Thammavong, B. & Viengkham, O.V. In press (a). *A field guide to the rattans of Lao PDR.* Kew, United Kingdom, Royal Botanic Gardens.

FAO. 1997a. *Aspects of forestry in Cambodia,* by H. Narith. Asia-Pacific Forest Sector Outlook Study Working Paper No. 18. Bangkok.

FAO. 1997b. *Country Report – Viet Nam,* by Nguyen Tuong Van. Asia-Pacific Forest Sector Outlook Study Working Paper No. 31. Bangkok.

FAO. 1997c. *Non-wood forest products outlook study for Asia and the Pacific: towards 2010,* by A.J. Mittelman, C.K. Lai, N. Byron, G. Michon & E. Katz. Asia-Pacific Forest Sector Outlook Study Working Paper No. 28. Bangkok.

FAO. 1997d. *People and forests in Asia and the Pacific: Situation and prospects,* by R.J. Fisher, S. Srimongkonthip & C. Veer. Asia-Pacific Forest Sector Outlook Study Working Paper No. Bangkok.

FAO. 1997e. *Technology scenarios in the Asia-Pacific forestry sector,* by T. Enters. APFSOS Working Paper No. 25. Bangkok.

FAO. 1998 *Summary of the country outlook: Lao PDR,* by K. Kingsada. Asia-Pacific Forest Sector Outlook Study Working Paper No. 38. Bangkok.

Feil, J. P., 1998. *A preliminary valuation and assessment of sustainable limits for harvest of forest products in Bokor National Park, Cambodia.* Phnom Penh, European Commission Support Programme to the Environmental Sector in Cambodia/Ministry of Environment.

Foppes, J. & Ketphanh, K, 1997 *The use of non-timber forest products in Lao PDR.* Paper presented at the Workshop on Sustainable Management of Non-Wood Forest Products, IDEAL, UPM, Serdang, Selangor, Malaysia.

Foppes, J. & Ketphanh, S., 2000. *Forest extraction or cultivation? Local solutions from Lao PDR.* Paper presented at the workshop on the evolution and sustainability of 'intermediate systems' of forest management, FOREASIA, 28 June-1 July 2000, Lofoten, Norway.

Gagnepain, F. & Conrard, J., 1937. Palmiers. *In* H. Lecomte, ed. *Flore Générale de l'Indochine* Vol. 6 (parts 7 & 8). Paris, Massie and Co. pp. 946-1056.

Global Witness. 1998 *Going places. Cambodia's future on the move.* Accessed at URL http://www.oneworld.org/globalwitness/reports/GoingPlaces/ on 8/11/00.

Global Witness. 2000. *Chainsaws speak louder than words.* Accessed at URL http://www.oneworld.org/globalwitness/reports/chainsaws/ on 8/11/00.

Government of Lao PDR. 1996. *The Forestry Law.* Dirksen Flipse Doran and Le, Vientiane. [unofficial translation for the Ministry of Justice, Vientiane].

Jarenrattawong, J., 1997. Rattans in Thailand. *In* A.N. Rao & V.R. Rao, eds. *Rattan – Taxonomy, ecology, silviculture, conservation, genetic improvement and biotechnology.* Serdang, Malaysia, International Plant Genetic Resources Institute. pp. 115-116

Ketphanh, S. & Sengkhonyang, B., 1997. Rattan in Laos. *In* A.N. Rao & V.R. Rao, eds. *Rattan – Taxonomy, ecology, silviculture, conservation, genetic improvement and biotechnology.* Serdang, Malaysia, International Plant Genetic Resources Institute. pp. 77-92

Ngo Thi Min Duyen & Nguyen Truong Thanh, 1997. Taxonomy and ecology of rattan in Viet Nam. *In* A.N. Rao & V.R. Rao, eds. *Rattan – Taxonomy, ecology, silviculture, conservation, genetic improvement and biotechnology.* Serdang, Malaysia, International Plant Genetic Resources Institute. pp. 117-120

Nguyen Truong Thanh & Ngo Thi Min Duyen, 1997. Silviculture and utilisation of rattans in Viet Nam. *In* A.N. Rao & V.R. Rao, eds. *Rattan – Taxonomy, ecology, silviculture, conservation, genetic improvement and biotechnology.* Serdang, Malaysia, International Plant Genetic Resources Institute. pp. 77-92.

Rake, C., Meyer, G., Luong Van Tien, Nguyen Quang Long, Vu Dinh Quang, Nguyen Viet Tuc & Do Cong An., 1993. *Markets of important products, non-timber forest products and agricultural products in the provinces of Hoa Binh, Son La and Lai Chau in the North West of Viet Nam.* Social Forestry Development Project. Hanoi, Ministry of Forestry/GTZ.

Rao, A.N. & Rao, V.R., eds. 1997. *Rattan – Taxonomy, ecology, silviculture, conservation, genetic improvement and biotechnology.* Serdang, Malaysia, International Plant Genetic Resources Institute.

Sengdala, K. & Evans, T., 1999. Rattan cultivation in Lao PDR: Achievements, problems and prospects. *In* R. Bacilieri & S. Appanah, eds. *Rattan cultivation: Achievements, problems and prospects.* CIRAD-Forêt, Montpellier, France and Forest Research Institute Malaysia, Kuala Lumpur. pp. 210-216

Sengdala, K., Sengkhamyong, B. & Vongkhamsao, V., 1997. *Socio-economic study of rattan in Lao PDR.*, Vientiane, International Network for Bamboo and Rattan/Department of Forestry.

Thewlis, R.M., Duckworth, J.W. Evans, T.D. & Timmins, R.J., 1998. The conservation status of birds in Laos: A review of key species. *Bird Conserv. Internat.* 8 (Suppl.) 1-159.

Turner, B., ed. 2000.) *The statesman's yearbook 2000.* London, Macmillan.

Vu Van Dung and Le Huy Guang. 1996. [*Results of attempts to grow, plant and develop rattans.*] Hanoi, Department of Agriculture. [In Vietnamese, unofficial translation by O.V. Viengkham. Vientiane, Darwin Initiative Lao Rattan Research Project.]

Williams, J.T. & Rao, V.R., eds. 1994. *Priority species of bamboo and rattan.* New Delhi, International Network for Bamboo and Rattan/International Plant Genetic Resources Institute.

Annex 1

Key statistics for the Lao People's Democratic Republic, Cambodia and Viet Nam

	Lao PDR	Cambodia	Viet Nam
Population (estimate for 2000)[1]	5.69 m	11.21 m	80.55 m
Land area (km^2)[1]	236 800	181,\035	331 690
Population density (people/km^2)[2]	24	62	243
Rural population (%) (1995)[1]	c.80	c.80	c.80
Population growth rate (% pa) up to 1995 [1]	3.0	2.8	2.4
Adult literacy (%)[1]	57	66	93.7
Life expectancy (years)[1]	53.6	49.9	65.1
Human Development Index* (/and global position) [1]	0.465/136	0.422/140	0.560/122
Natural forest cover (%)	c. 47 (1989)[3] c. 40 (1996)[4]	38 (1993)[5]	25 (1995)[6]
Natural forest cover (km^2)[2]	95 000	>90 000	83 000
Estimated rate of loss of natural forest (km^2/year)	c.2000[2]	unclear	c.1000[7]

Sources: [1]Turner (2000)
[2]calculated from figures given here
[3]Kingsada (1998)
[4]Duckworth *et al.,* 1999
[5]Varith, 1997, excluding scrub, wooded grasslands, plantations and orchards
[6]Nguyen Tuong Van, 1997, may include extensive open formations
[7]Anon, 2000c

* For comparison, Malaysia scores 0.834/60. The maximum score is 1.000 and the lowest position 174.

Annex 2

Rattan species of known or suspected commercial importance in the Lao People's Democratic Republic

Species	Diameter	Known distribution[1]	Status and regeneration habit (all species clustering unless stated)
Calamus poilanei	Large	C and N, 300–1 300 m	Very heavily harvested throughout. Declining. Solitary stemmed.
C. nambariensis	Medium	Mainly N and parts of C, 1 400–1 800	Status unknown. Records remain unconfirmed.
C. platyacanthus	Medium	Mainly N and parts of C, 750–900 m	Status unknown. Records remain unconfirmed.
C. wailong	Medium	Mainly N and parts of C, 350–600 m	Status unknown.
C. viminalis	Medium	Thrt, 100–600 m	Populations in scrub secure, but cane stocks low. Forest populations harvested in N.
C. gracilis	Small	C, southern half of N, 300–750 m	Heavily harvested, Status unknown.
C. solitarius	Small	C, southern half of N, 200–600 m	Heavily harvested. Probably declining. Solitary stemmed.
C. palustris var *cochinchinensis*	Small	Probably thrt, rarer in far N, 100–650 m	Heavily harvested. Status unknown.
C. tetradactylus	Small	Mainly S, 100–600 m.	Status unknown.
C. acanthospathus	Small	N, 1800 m	Status unknown.
C. tenuis	Small	Southern half of N, 200–300 m	Heavily used around Vientiane. Survives in scrub, but localized.

[1]   From Evans *et al.*, in press (a)
N   = North (North of Route 8)
C   = Central
S   = South (South of Xe Banghiang)
thrt   = throughout (following Duckworth *et al.,* 1999)

Annex 3

Globally threatened rattans in Indo-China

Species	Comments	Viet Nam	Cambodia	Lao PDR
D. sp "longispatha"	Taxonomy under revision (J. Dransfield, *pers. comm.*)	+		
Calamus ceratophorus		+		
C. dioicus		+		
C. dongnaiensis		+		
C. godefroyi	Also historically in Thailand		+	+
C. harmandii				+
C. poilanei	Also recorded in Thailand	+		+
C. scutellaris	Now a synonym of the Chinese *C. thysanolepis*, which is not listed as threatened.	+		
C. tonkinensis	Now a synonym of the Chinese *C. walkeri*, which is not listed as threatened.	+		

Source: Walter and Gillett, 1998. Distributions from Evans *et al.*, in press.

These categorizations have not yet been fully reviewed in the light of recent new data. Information is still scanty from Viet Nam, where most of the listed species occur. The first six species listed are still known from very few sites, and are thus at risk from habitat loss, even when they are not targeted by harvesters. For example, *C. harmandii* is restricted to the Lao People's Democratic Republic, where it is currently known from only one site which is at risk from deforestation (Evans, 2000). *C. poilanei* is now known to be widespread (Evans *et al.*, in press a); nonetheless it is under severe commercial pressure everywhere and has poor prospects of regeneration, so it should be retained on the red list for the time being.

Species not listed here may also be at risk and an updated review of these is also needed. *C. solitarius* is of some concern, although it cannot yet be considered threatened. The isolated Indochinese population of *C. kingianus* is presently only known from the Nakay Plateau, the Lao People's Democratic Republic (Evans *et al.*, in press (a)), where a planned hydropower project may threaten its survival. This species also grows in northeast India.

Annex 4

The status of *Calamus poilanei* in the Lao People's Democratic Republic

C. poilanei is the elite large-diameter cane for domestic handicrafts and export in the Lao People's Democratic Republic. It is widespread and was apparently once common or even abundant. In very remote areas, stems 100-150 m long can still be found quite easily. However, most areas visited by the Darwin project team during 1997-2000 had already experienced very heavy harvesting of this species, confining it to the remotest and least accessible areas at the heart of major forest blocks in mountainous areas. Even these are likely to be harvested if quotas are issued for them. Current evidence suggests that the species is at high risk of commercial extinction in the Lao People's Democratic Republic in the foreseeable future.

Sengdala *et al.* (1997) presented data on quotas listed by the Department of Forestry annual reports (Table 3). Almost all of the large-diameter cane listed is thought to be *C. poilanei*. In 1996/97, one quota holder alone sought to cut 400 000 sticks of large-diameter cane.

Table 3. Total reported harvest quotas for "large-diameter poles" in Lao PDR

Year	Number of concessionaires	Total allowable cut (4.5 m sticks)
1993-1994	4	261 000
1994-1995	16	1 129 000
1995-1996	15	649 000

National export figures compiled from official provincial statistics and supplied by the Lao IUCN-NTFP project are shown in Table 4. The table shows moderately large quantities, with a substantial increase in 1998. Internal inconsistencies in the data sources, major fluctuations from year to year in individual provinces and experiences from the timber sector suggest that these recorded figures represent only a part of the total export (J. Foppes, *pers. comm.*). Note, for example, the striking difference between the low reported exports for 1995 and the very high permitted cut for both 1994/95 and 1995/96.

Table 4. Recorded exports of large-diameter rattan cane from Lao PDR 1995-1998

Year	1995	1996	1997	1998
Recorded exports (4.5 m sticks)	45 500	91 995	93 355	367 196
Calculated total length (m)	205 000	414 000	420 000	1 652 000
Recorded value (US$)	17 961	37 095	36 564	117 503

Data on growth rates come from a single minor study (Darwin Initiative Lao Rattan Research Team, unpublished data). The growth of a small non-random sample of *C. poilanei* stems in natural forest at 550 m in central the Lao People's Democratic Republic was followed over a period of 13 months.

There was a wide range of growth increments from 15 cm to 300 cm with most over 100 cm and several over 200 cm. Growth was apparently least in deep shade. These rates are broadly comparable with rates for *C. manan* from Malaysia (Dransfield and Manokaran, 1993) and suggest the possibility that economically viable growth rates might be obtained in some situations.

The only known quantitative survey of density was conducted along a short transect in evergreen forest at 600 m on the Sayphou Phaphet massif in central the Lao People's Democratic Republic in 1999. The survey covered randomly sampled plots within a relatively rattan-rich 12 ha block deep in a very large unlogged forest block and found a density of 20.0 aerial stems/ha (95% confidence limits 8.7-31.3 stems/ha) (Evans and Viengkham, in press), or approximately 550 m of harvestable cane/ha (Darwin Initiative Lao Rattan Research Team, unpublished data). *C. poilanei* had been partially harvested in this general area during 1992-1996 and the original density was probably somewhat higher. Wider exploration showed that, in even more remote areas of the forest, similar densities could be found, although the distribution was somewhat patchy. In less remote, more accessible forest around the feet of the massif (but still 5-10 km from the nearest village or road) the *C. poilanei* population, reportedly once strong, had collapsed since harvesting began in 1992/93. Only very occasional adult stems could be found despite prolonged searching, and seedlings were also scarce. This provides striking evidence of the speed with which commercial extinction of a rattan can occur over a large tract of essentially pristine forest. It parallels the well-documented effect hunting has had in vastly reducing the populations of most quarry species of bird and mammal in the Lao People's Democratic Republic, despite the great extent of suitable habitat that remains (Thewlis *et al.*, 1998; Duckworth *et al.*, 1999).

Three factors severely hinder the regeneration of *C. poilanei* at this site. Firstly it is a solitary-stemmed species so that stumps do not resprout and regeneration has to occur from seed. Secondly, the original harvest was so thorough over large areas that no adults remained as seed sources. Since rattans show little seed dormancy, there will be few new recruits to the seedling bank until, eventually, a new generation of adults has developed from the existing seedlings. Thirdly, the practice of harvesting edible shoot tips for trade is expanding (at least at this site). This is leading to the deaths of many of the remaining seed plants and also (since juveniles are harvestable long before they reach reproductive age) it is preventing many, perhaps all of the existing seedlings from reaching adulthood. The combined impact of these factors is likely to prevent a second harvest of *C. poilanei* from the same forests in the foreseeable future and might drive the species to extinction. A similar situation is suspected to prevail over wide areas of the country.

Table 5 summarizes observations on the status of *C. poilanei* province by province during 1992-2000.

Table 5. The status of *Calamus poilanei* in each province of the Lao People's Democratic Republic during 1992-2000

Province (arranged NW to SE)	Observations on status of *C. poilanei* populations
Luang Namtha	Believed to be naturally scarce or absent.
Phongsaly	Believed to be naturally scarce or absent.
Oudomxay	One probable record, population status unknown. Little pristine forest remains, populations unlikely to be strong.
Bokeo	Believed to be naturally scarce or absent.
Luang Phabang	Status unknown. Little pristine forest remains, populations unlikely to be strong.
Huaphanh	Believed to be naturally scarce or absent.
Xayaboury	Believed to be present. Heavy harvesting of a large-diameter species with appropriate local name reported.
Xiengkhuang	Unknown.
Xaysomboun Special Zone	Present, possibly still locally abundant. Difficult access and security problems may have hindered overharvesting.
Vientiane Province	Present, believed scarce as a result of past harvesting. Populations likely to remain in remote areas, but very low in much of Phou Khaokhoay NBCA.
Vientiane Municipality	Very scarce or absent, presumably as a result of deforestation and harvesting
Bolikhamxay	Locally common in some remote forest areas (e.g. Nam Kading NBCA*) but removed by harvesting from large tracts of forest in at least three districts. Very severely reduced in Nakai-Nam Theun NBCA due to local and trans-boundary harvesting.
Khammuane	Reported by District forestry officials to be almost gone from Nakay District due to heavy recent harvesting. Heavy harvesting of large-diameter canes in Hinboun District reported during 1999. Status in other districts unknown.
Savannakhet	Probably present, status unknown.
Saravane	Probably present, status unknown.
Sekong	Probably present, status unknown.
Champassak	Probably present, status unknown but probably very poor. Large-diameter canes reportedly almost all harvested already from this province (Enfield *et al.,*1998; Foppes and Ketphanh, 2000; J. Foppes, *pers. comm.*).
Attapu	Probably present, reports suggest large-diameter canes (probably this species) still common in remote parts of Dong Amphan NBCA but wholly harvested out from more accessible areas.

* NBCA = National Biodiversity Conservation Area, the highest level of protected area

Annex 5

Selected organizations involved in improving rattan management in Lao PDR

Darwin Initiative Lao Rattan Research Project
Dr Nick Brown, Project Leader
Oxford Forestry Institute, South Parks Road, Oxford OX1 3RB, UK
tel: +44-1865-275077
fax: +44-1865-275074
email: nick.brown@plants.ox.ac.uk

Mr Khamphone Sengdala, Lao Project Leader
PO Box 8916, Vientiane, Lao PDR (c/o FRC, below)
tel: +856-21-732298
fax: +856-21-413174
email: c/o NIC (below)

Forestry Research Centre (FRC)
Mr Khamphay Manivong, Director
PO Box 7174, Vientiane, Lao PDR
tel: +856-21-732298
fax: +856-21-413174
e-mail: frclao@laotel.com

FISS (Forestry Information Service Section) formerly the NTFP Information Centre (NIC)
Mr Bandith Ramangkoun, Director
PO Box 6957, Vientiane, Lao PDR
tel/fax: +856-21-732298
e-mail: frcfiss@laotel.com or frclao@laotel.com

IUCN-Non Timber Forest Products (NTFP) Project in Lao PDR
Mr. Sounthone Ketphanh, National Project Coordinator
P.O.Box 4340, Vientiane, Lao PDR
tel/fax: +856-21-732298
mobile: +856-20-511653
email: ntfplao@laotel.com

Mr. Joost Foppes, Project Advisor
P.O.Box 4340, Vientiane, Lao PDR
tel/fax: +856-21-732298 (Dong Dok)
tel: +856-21-216401 (IUCN Lao Country Office)
fax: +856-21-216127 (IUCN Lao Country Office)
mobile: +856-20-514661
home: +856-21-412488
email: jfoppes@loxinfo.co.th

IUCN Country Office
Mr Scott Perkin, Director
P.O.Box 4340, Vientiane, Lao PDR
tel: +856-21-216401
tel/fax: +856-21-216127
email: iucnlao@loxinfo.co.th or cro@iucnlao.laonet.net

Annex 6

Constraints to the sustainable management of wild rattans in the Lao PDR and some possible solutions

Constraint	Current trend	Possible solutions
Social Constraints		
Scarcity of cultural precedents for managing the forest	?	Increasing competition for resources and existence of successful pilot schemes may change public attitudes. Research to find and capitalize upon existing indigenous concepts which favour conservation has also been recommended (Enfield et al., 1998).
Imbalances in the political power of different actors	?	Structures in Lao PDR do not favour free and equal participation by all stakeholders. See Dove (1994) and Fisher et al. (1997) for further discussion.
Lack of trained staff or operating budgets in relevant government agencies.	Improving very gradually	Support for capacity building.
Policy constraints		
Open-access to resources	Gradually improving?	Ongoing land and forest allocation programme may establish new patterns of tenure. Input at policy level may build into the system incentives for sustainable harvesting.
Short, erratic quotas encouraging and permitting overharvesting	Not improving	Lengthen quotas, attach meaningful sustainability conditions. Shift emphasis to control of quotas by stakeholders with a long-term interest in the resource (e.g. villagers, concessionaires), starting with the less contested (small-diameter) species.
Uncertain legal status of diffuse trade in small canes, preventing investment or regulation	-	Establish a clearer legal framework that enables monitoring, regulation and taxation without wholly stopping the trade.
Complex and obscure regulatory framework in forestry sector	Not thought to be improving	More transparent and publicly available system of regulation. Better paid and trained regulatory staff.
Government imperative to maintain income in the short-term to allow national development	Worsening?	Macro-economic improvements outside the scope of this paper.
Economic and biological constraints		
Increased market access through road expansion and changes in economic policy	Worsening	No solution apparent. Tightened regulation is an option.

Constraint	Current trend	Possible solution
Intense rural poverty which drives villagers to overuse even those resources which they do fully control	Remaining a severe problem	Requires temporary alleviation until income stream from managed resource begins. Micro-credit schemes, rice banks ,etc., sometimes successful in the Lao People's Democratic Republic (Foppes and Ketphanh, 2000).
Physical difficulty of policing stocks dispersed across extensive, remote forests	Worsening as accessible stocks run out	Shifting harvest to easily policed forest near villages would be possible but would require a long wait and/or expensive silvicultural interventions for rattan stocks to re-establish.
Long travel times for harvesting, due to remoteness of remaining stocks	Worsening. Likely to become a worse problem if rural wages increase.	See above.
Low bargaining power of harvesters	-	Harvesters' associations, market information exchange networks. Modified quota system to remove monopolies.
Low unit value of wild cane which offers little profit margin to invest in management measures	Remaining a severe problem.	Dependent on changes in the international market. Within the Lao People's Democratic Republic, stricter enforcement of legal protection within NBCAs might increase the price of stocks in production forest. Certification is a possible future option, especially as part of the new 600 000 ha HIPA timber concession in the north, which is aiming for certification.
Annual productivity per hectare unknown but probably quite small in several key cases (Bogh, 1997; Bann, 1997; author's unpublished data).	-	Upper ceiling on productivity hard to remove without expensive techniques such as canopy opening or enrichment planting (which are probably better done in a plantation setting).
Substantial costs of designing a harvest regime, inventorying and monitoring stocks (Evans and Viengkham, in press).	Perhaps gradually improving.	Development of new, more efficient formal techniques, preference for low-information approaches reliant on local peoples' observations concerning the state of target populations, subsidized provision of technical forestry support.
Lack of technical knowledge about the ecology of commercial species	Gradually improving.	Continued research, focused on high-priority species. Technical information exchange networks to disseminate the results.

.

Annex 7

Rattan species of commercial importance in Viet Nam

Species	Diameter	Known distribution	Quality
Calamus poilanei	Large	Widespread, especially the south, 200-1100 m	Export
C. rudentum	Large	Almost every province, commoner in south.	Ordinary
C. palustris	Large (small in Lao PDR)	Widespread	Ordinary
C. platyacanthus	Medium-large	Every northern province, 100-1200 m (esp. 400-900 m)	Export
C. viminalis	Medium	South	Ordinary
C. tetradactylus	Small	Throughout, especially northern provinces, 100-800 m, especially 100-500 m	Export
C. tonkinensis (=*C. walkeri*?)	Small	Widespread, in coastal provinces	Ordinary
C. amarus (= *C. tenuis*?)	Small	North (provinces north of Hué)	Ordinary
C. dioicus	Small	Centre (Hué south to Dong Nai), 100-800 m	Ordinary

Source: Vu Van Dung and Le Huy Guang(1996), with *C. palustris* added by Ngo Thi Min Duyen and Nguyen Truong Thanh (1997). Some listed species probably now only found in small quantities.

DEGRADED TROPICAL FOREST AND ITS POTENTIAL ROLE FOR RATTAN DEVELOPMENT: AN INDONESIAN PERSPECTIVE

Toga Silitonga

Summary

Rattans are important non-wood forest products, naturally growing in a wide-range of habitats in the Indonesian tropical rainforest. Indonesia is the major rattan producing country in the world. A number of artefacts for daily human needs in both rural and urban areas are made from this raw material. Rattan is traded internationally and thus contributes significantly to foreign exchange earnings; it also generates ample labour opportunities for the country' population.

A number of studies that have been carried out in the last few decades indicate that rattan grows well in logged-over areas or secondary forests. A sizeable portion of the over 30 million ha of logged-over areas in Indonesia are considered degraded forests. These areas and others in similar conditions elsewhere in the world are suitable for growing rattan artificially.

The purpose of this paper is to discuss the present rattan situation in Indonesia and to explore the possibility of sustainable rattan cultivation for future development.

1. Introduction

Among a few hundreds non-wood forest products of the tropical forest of Indonesia, rattan mainly of the genera *Calamus*, *Korthalsia* and *Daemonorops* are commercially, socially and culturally important. To most low-income rural families, some 51 out of a total of 300 existing Indonesian rattan species are utilised in the making of furniture, tools, and artefacts, and the young shoots of some species being consumed as food.

Trade statistics indicate that rattan contributes significantly to foreign exchange earnings. In 1996, about 80 percent of the rattan on the international market originated from Indonesia. Over 90 percent of the production is obtained from the natural forests of the country's three major islands, i.e. Sumatra, Kalimantan and Sulawesi, the remaining 10 percent being mainly small-diameter canes harvested from plantations.

It is estimated that about 300 species are found in Indonesia. In terms of diameter, there are two distinct groups of rattan – small diameter and larger diameter. The commercially important small-diameter rattans are mostly found in the swampy areas and lowlands of the archipelago, and constitute the major part of available rattans in their natural habitat. Larger diameter rattans grow mostly in the dry lowlands and at higher elevation over 1000 m above sea level. The latter group can be found as solitary individuals or in clumps. However, 68 percent of the rattan demand from trade and market is directed to larger diameter rattans and 32 percent to small-diameter ones (Gintings *et al.*, 2000). In the past intensive research has been carried out in Indonesia embracing taxonomy, provenance and species trial, propagation, etc.

2. Methodology and Characterization of Rattan Development

2.1 Resources potential

Several attempts have been made by scientific groups, such as research institutes and universities, to introduce rattan inventory models. At best the only applicable method is the combination of cluster sampling techniques and direct observation. The reason for this is simply inventory cost. Aerial photo interpretation is impractical due to the growth characteristics of rattan since it grows both above and under canopies. At present, production fluctuation over a long period of time is the most reliable approximation of production potential.

2.2 Trade and industry

Rattan trade plays an important role in the rural economy, as well as in small and big enterprises. Most raw rattan traded on the market is obtained from natural growth in the primary and secondary forests.

2.3 Rattan research

Since the early 1970s, the rattan industry and related scientific institutions have been involved in research in the following areas:
  1. rattan distribution and *ex situ* conservation;
  2. silviculture and propagation;
  3. botanical study;
  4. physical and mechanical properties and processing techniques of marketable rattan; and
  5. trade and economics of rattan.

2.4 Poverty alleviation and social safety nets

In recent years, the economic crisis has had a negative impact on the poor and already economically marginalized people who live in or near the forest. The era of the growing forest industry in the last two decades has failed to bring prosperity to the people who live in these sensitive areas. The growing number of poor families in the area could have a direct impact on future rattan development.

2.5 Sustainable forest management

Experience has shown that sustainable forest management (SFM) in Indonesia is much easier said than done. A numbers of initiatives have been taken by international organizations and non-governmental organizations. The initiative to establish SFM in Indonesia was triggered by the International Tropical Timber Organization in 1992 with the establishment of criteria and indicators (C&I). Social and legal aspects are of paramount importance in the success of SFM. Rattan could play a significant role in that endeavour.

2.6 Rattan information sharing and research network

The distribution of rattans in the world is mostly concentrated in Southeast Asia and, to a lesser extent, in mid-west Africa. It is not surprising that rattans in the above areas are used in the making of household items such as sleeping mats, handicrafts, furniture, etc. In the early 1980s, Indonesia and the International Development Research Centre (IDRC) of Canada established a joint effort on rattan development. A rattan information centre in Kuala Lumpur (Malaysia) was the first of its kind ever established in the world. In 1993, IDRC created the International Network for Bamboo and Rattan (INBAR) aimed at harnessing a research network for both bamboo and rattan.

Figure 8: Rattan baskets in Kalimantan (Dransfield)

3. Discussion of Findings and their Importance for Future Development

3.1 The importance of rattan in Indonesia

Rattan plants occupy a decisive position in the natural forest of Indonesia. According to best available information (Jasni *et al,.* in Gintings *et al.*, 2000), 300 rattan species are found in Indonesia. Two genera (*Calamus* and *Daemonorops*) have important commercial value. Most species are still not well identified and documented. They have a wide range of distribution stretching from Sumatra Island in the west. They are well distributed in both primary and secondary forests. It is estimated that in the country there are over 3.0 million ha of forests containing rattans, and the annual output is 570,000 t. The highest output reached was in 1980 with 573,000 t. Several species of rattan are now planted. Rattan plantation is encouraged because of its economic value and broad utilization. To most forest dwellers, however, rattan planting is only a secondary crop in addition to the ordinary cash crops in their areas. Rattan collectors, normally work in teams of 5–7 persons. They spend days or even weeks in the forest before bringing back to the market their produce. As the distance to collect naturally grown rattan is increasing, growing rattan near their villages is preferable. It only takes 5-8 years before-small diameter rattans can be harvested, while it takes 15–25 years for large-diameter rattan, depending on the species.

The history of rattan trade and utilization in Indonesia goes back about 100 years. With the development of the rattan industry, many varieties of handicraft, furniture, and household appliance are made with this material for both the domestic and the international markets. Since 1986, the Indonesian Furniture Industry and Handicraft Association reported a continuous development trend of rattan furniture exports. In 1986, the total value of rattan furniture exports (combined with a small portion of wood) reached 112,359 t with a value of US$115.14 million. In 1999, the export volume increased over fivefold to 590,021 t and a value of US$1.147 billion (ASMINDO). It seems that the economic crisis experienced by the country had a positive impact on the rattan trade and industry.

3.2 Rattan and degraded forest

The growth of the Indonesian economy in the early 1970s started as a consequence of the acceptance of the foreign investment law in the forestry sector. Investments in the rattan industry were made in the period from late 1970s through to the 1990s. When forest concessionaires embarked on logging the natural forests, they also improved forest accessibility to remote areas and, as a result, rattan production increased sharply. In 1995, there were 548 rattan-based industries throughout the country, an enormous increase from 1988 when there were only 381. During that period, the number of people involved in rattan-related activities (rattan gatherers, farmers, household industries, etc.) more than doubled. It is estimated that over one million Indonesians were involved in such activities, either part-time or full-time.

The peak of rattan trade was reported in the period 1994–1997. This was partly due to the conducive investment and trade policies promulgated by the Government. Since then, rattan trade has been declining. Sizeable forest areas containing rattans were converted into other plantations, mostly of oil palm or rubber trees.

In the meantime, forest degradation has continued. This is partly due to factors such as malpractice in logging techniques, illegal cutting, forest fires, etc. It is not unwise to state that, if no immediate action is taken, the rattan culture in Indonesia will soon become history. It is also realized that the decline in economic activities of most forest dwellers would result in an acceleration of forest degradation. This phenomenon is already noticeable. Policy-makers should be made aware on this and encouraged to adopt corrective action immediately.

3.3 Rattan as an element in the Sustainable Forest Management Programme

Under the pressure of the economic crisis, forest degradation has become more intense and has in fact reached an alarming rate. A number of Government policies and an action plan in an effort to correct this situation were unpopular. An unfavourable economic and political situation has contributed to wider deviation from achieving sustainable forest management in Indonesia. Several conceivable options are available to improve the economic situation of forest dwellers, among others, introduction of medicinal plant cultivation under the forest canopy, cultivation of perennial crops among the forest stands, etc. Such practices could also provide additional value to the forests.

From an economic perspective, rattan has a place for development in the already degraded forest for a number of reasons. One important reason is that the rattan chain of activities extends from rural area households to medium trade and industries including international trade. Secondly, rattan with diversified growth characteristics could be propagated in degraded forest conditions. In addition, rattan has a wide range of habitats throughout the country and, therefore, connected activities could be distributed among a wide range of rural poor. If policy-makers fail to take solid action for improving the economy of the rural poor, efforts toward achieving SFM will lead to nowhere. Poverty is the root of forest degradation; therefore, poverty alleviation of those people who live in or near the forests is imperative in order to achieve SFM.

3.5 Rattan research, development and application

The great potential of rattan resources and their role in the livelihood of many Indonesians have attracted the attention of scientists in the last two decades. An integrated study recently carried out included: taxonomy, resources assessment, growth factor requirements, phenology, seed and propagation studies, nursery, harvesting techniques, post-harvest processing, rattan anatomy, workability and engineering, durability and preservation methods, policy studies, trade and cooperatives. Knowledge of the above aspects is available to some extent to support future rattan development.

Exchange of views and information-sharing between scientists and rattan-related institutions to ensure significant development in the future is of great importance. Networking on rattan-related subjects, established by INBAR, has demonstrated its great impact on the development of rattan-related activities.

4. Conclusions and general remarks

Rattan is an important non-wood forest product with inherent capacity for growth. It has long been used for domestic needs as well as an internationally tradable merchandise. As merchandise, it has found its way in international markets and contributes significantly to foreign exchange earnings. Rattan also has an important role to play in supporting the socio-economic conditions of many rural poor whose living solely depends on forests.

The poor socio-economic situation of forest dwellers has been identified as the underlying root cause of forest degradation in Indonesia and in many similar parts of the world. Rattan development in such areas is not only economically feasible but also socially acceptable, and environmentally desirable. Research findings are available that could support future rattan development. It is imperative that scientific groups encourage policy-makers to adopt strategic action development plans and to explore further the potential role of these important species.

REFERENCES

Alrasjid, H., 1999. *Teknik Penanaman Rotan*. Informasi Teknis Penelitian dan Pengembangan Hutan. Bogor.
ASMINDO. 2000. Indonesian Furniture Industry and Handicraft Association. Jakarta.
Dransfield, J. *et al.,* 2000. *A short guide to rattan*. Biotrop ITF/74/128. Bogor, Indonesia.
Gintings *et. al.,* 2000. *Himpunan Sari Hasil Penelitian Rotan dan Bambu*. Pusat Penelitian Hasil Hutan. Bogor.
INBAR. 1993. International Network for Bamboo and Rattan.
ITTO. 1992. *Criteria for the measurement of Sustainable Tropical Forest Management*. Policy Development Series.
Kalima, T., 1996. *Flora Rotan di P Jawa serta Penyebaran dan Populasi Rotan di Tiga Wilayah Kawasan TN Gunung halimun, Jawa Barat*. Thesis S2 University of Indonesia. Unpublished.
Pusat Penelitian Hasil Hutan. 2000. *Sari Hasil Penelitian Rotan dan Bambu*. Occasional paper. Bogor. ISBN: 97995743.5.8.

COUNTRY REPORT ON THE STATUS OF RATTAN RESOURCES AND USES IN MALAYSIA

Abdul Razak Mohd Ali and Raja Barizan R.S

1. Introduction

Rattan (family *Palmae/Arecaceae*) is considered to be the most important non-wood forest product in Peninsular Malaysia. It belongs to a large subfamily of the palms known as the *Calamoideae* (Uhl and Dransfield, 1987). About 600 different rattan species belonging to 13 genera are found in the world. In Peninsular Malaysia alone, 106 species (8 genera) occur naturally (Dransfield, 1979). However, of these, only about 30 species are utilized and exploited commercially. In Sarawak, the largest state in Malaysia covering an area of 12.5 million ha, rattan flora is very diverse. A total of 105 species with eight genera have been identified (Dransfield, 1992) and a great number of species are endemic. Such richness is partly due to the presence of diverse forest habitats.

The rattan industry in Sarawak is still at the developmental stage. The two government departments currently involved in research and trial plantations of rattan are the Forestry Department and the Agriculture Department. KERESA and Rajang Wood Sendirian Berhad are two private agencies currently involved in rattan plantations. However, some of the areas already planted with rattan are likely to be converted soon to oil palm due to the better prospects and higher economic value of oil palm.

About 84 species of rattan are found in Sabah (Dransfield 1984). The Forestry Department of Sabah began small-scale research on rattan cultivation in 1979. Research in this area was also prompted by a growing interest in creating a man-made rattan resource. Subsequently, several public and government organizations embarked on rattan plantations. As of mid-1996, the total area of rattan plantation in Sabah was about 23,157 ha. Innoprise Corporation Sendirian Berhad (ICSB) accounts for almost 43 percent of the total area planted, followed by the Sabah Forestry Development Authority (SAFODA) with 42 percent; Jeroco Plantation, 8 percent; and Sejati Plantation, 7 percent.

Malaysia is fortunate to be endowed with plenty of raw materials from its forests. However, the contribution of the rattan-based industry to the country's economy is relatively very small – US$21.7 million – compared to the wood-based industry which generated about US$666.1 million of export earnings in 1997 (Ministry of Primary Industries, 1998). This is meagre compared to countries like Hong Kong and Singapore that exported rattan products valued at US$53 million and US$23 million, respectively. It must be noted that the two latter countries do not have any natural resources of their own. This amount far exceeded Malaysia's exports, and something needs to be done in order to upgrade and establish rattan-based industries as a main industry in the development of forest product activities. This is in line with the Malaysia Industrial Master Plan (IMP) that calls for a rapid development in the downstream production of the forest-based industries.

2. Status of the Resource in Natural Forests

Since 1972, three national forest inventories (NFI-1, NFI-2, and NFI-3) have been carried out to assess and determine the status of the various natural forest resources in Peninsular Malaysia. The third inventory (NFI-3) was jointly carried out in 1991/92 by the Forestry Department Peninsular Malaysia (FDPM) and FAO under a UNDP project (Chin *et al.*, 1994). Besides estimating tree stocking, non-wood forest products such as rattan, bamboo and palm resources were also enumerated. Table 1 shows the estimated number of rattan clumps present according to forest strata in the Permanent Reserve Forests (PRFs) in Peninsular Malaysia. It was estimated that the total number of rattan plants (irrespective of age) found in PRFs in Peninsular Malaysia amounted to around 32.7 million, of which the most abundant (about 37%) were *Korthalsia* species. Of *Calamus* species, *C. manan* is the most abundant one with around 5.9 million (irrespective of age). In terms of distribution according to forest strata, it was found that the forest logged during the period 1971-1980 seem to have the highest number of rattan plants (40.2%). Table 2 shows that the state of Pahang, which has the largest PFRs, also has the largest rattan resource base (37.2%).

2.1 Development strategy

The growing commercial demand for rattan resources in Peninsular Malaysia warrants the need to formulate appropriate strategies to assess, manage, develop and conserve this valuable non-wood resource. Some of the strategies include:

- continue assessing the rattan resource in natural forests so that forest development and management plans can be drawn up to ensure its sustainability;
- replenish the rattan resource through large-scale cultivation with species of high commercial value in order to sustain the increasing needs of the rattan-based industry;
- determine the socio-economic contribution and impact of the rattan resource to the rural and urban communities and the rattan industry; and
- collect data on the rattan industry including, those pertaining to demand, supply and utilization of the rattan resource and employment opportunities.

Table 1. Estimated number of rattan clumps according to forest strata and commercial rattan groups in the Permanent Reserved Forests, Peninsular Malaysia based on NFI-3 (in thousand clumps)

| Forest Stratum | Rattan groups | | | | | | Total | % of Total |
	Calamus manan	*Calamus tumidus*	*Calamus caesius*	*Calamus scipionum*	*Calamus ornatus*	*Korthalsia spp.*		
Virgin:Superior	578.3	677.9	433.1	110.2	583.7	1 197.1	3 580.3	10.9
Virgin:Good	484.4	488.7	168.1	493.1	95.1	2 145.5	3 875.5	11.9
Virgin:Moderate	1 017.2	618.1	102.2	855.6	890.7	1 659.7	5 143.5	15.7
Virgin:Poor	86.4	107.6	154.4	331.5	22.6	596.5	1 299.0	4.0
Logged during 1971-1970	3 011.8	637.1	1 513.1	702.2	3 475.1	3 822.6	13 161.9	40.2
Logged during 1961-1970	296.5	507.3	233.9	108.7	378.8	1 676.9	3 202.1	9.8
Logged during and before 1960	436.4	190.5	261.2	414.8	165.9	971.1	2 439.9	7.5
Total	5 911.0	3 227.2	2 866.6	3 016.1	5 611.9	12 069.4	32 702.2	100.0
% of total	18.1	9.9	8.8	9.2	17.2	36.9	100.0	

Source: Yap and Hasnuddin, 1995

The state forest departments have been implementing rattan-planting programmes since the early 1980s in an effort to replenish the rattan resource in PRFs and also as an activity to rehabilitate the forests. Besides allocation of development funds by state governments, the Federal Government is also providing financial assistance from the timber exports levy to be used for rattan planting and management (Harnarinder & Chin, 1999). A sum of $M 17 824 550 has been granted to plant 9,430 ha over the five-year period 1998-2002. An additional $M 8 100 400 has been allocated to carry out maintenance and silvicultural treatments in 13,832 ha of planted area.

**Table 2. Estimated number of rattan clumps according to States and commercial rattan groups
in the Permanent Reserved Forests, Peninsular Malaysia based on NFI-3
(in thousand clumps)**

| State | Rattan groups | | | | | | Total | % of total |
	Calamus manan	Calamus tumidus	Calamus caesius	Calamus scipionum	Calamus ornatus	Korthalsia spp.		
Johor	443.2	262.8	229.1	278.1	432.0	1 027.2	2 672.4	8.2
Kedah	409.9	248.1	231.3	299.7	379.9	1 041.0	2 609.9	8.0
Kelantan	724.5	382.1	305.5	359.3	814.7	1 374.8	3 960.9	12.1
Meslaka	11.6	3.9	5.5	6.3	7.9	23.8	59.0	0.2
N. Sembilan	129.2	65.5	69.3	49.1	104.2	271.8	689.1	2.1
Pahang	2 288.7	1 247.6	1 087.7	1 046.9	2 022.4	4 460.7	12 154.1	37.2
Perak	1 020.1	585.0	469.7	602.1	985.1	2 192.2	5 854.2	17.9
Perlis	12.8	7.8	13.1	20.1	11.9	53.2	118.8	0.4
Pulau Pinang	10.0	3.4	4.9	9.0	4.5	25.1	56.9	0.2
Selangor	164.7	85.3	84.0	48.5	155.1	349.5	887.2	2.7
Terengganu	696.2	335.6	366.6	297.1	694.1	1 250.2	3 639.7	11.1
Total	5 991.0	3 227	2 866.6	3 016.1	5 611.9	12 069.4	32 702.2	100.0

Source: Yap and Hasnuddin, 1995.

2.2 Choice of rattan species

In Peninsular Malaysia, only 30 species are presently collected and utilized by the rattan industry for a variety of purposes (Appendix 1). A list of some important and popular species used in the rattan industry in Malaysia is shown in Appendix 2. Large-diameter canes are mainly used for making furniture frames while small canes are used for tying and other parts of furniture. The most important commercial canes come from the genus *Calamus*. The five most important species, in terms of utilization and cultivation are:

- *Calamus manan* (rotan manau) is the best large-diameter (>18mm) cane and is usually confined to the steep slopes of hill dipterocarp forests. It was once abundant at 600–1000 m altitude and grows well when planted on flat lowlands. It is a solitary and high climbing rattan reaching 100 m or more. For optimum growth, the species requires about 60 percent of light. It grows well under rubber trees, with growth rates around 0.3–3.0 m/a (Aminuddin & Nur Supardi, 1994).
- *Calamus tumidus* (rotan manau tikus) is classed under the large-diameter group but its canes are always smaller than those of *Calamus manan*. The cane is used locally in a way similar to that of *C. manan*. The habit is solitary, high-climbing and it is a rather rare rattan of freshwater swamp forest, peat swamp forest and on alluvial flats.
- *Calamus scipionum* (rotan semambu) is a widespread lowland species growing up to 200 m altitude. It is found on alluvial soils in flood plains of rivers and in secondary forests but not in primary dipterocarp forests. The cane is used for making walking sticks and umbrella handles because it has long internodes. The species is a clustering type with 5–10 stems per clump, climbing high up to 50 m or more. The growth rate of the cane is slower than that of *C. manan*, about 0.15–1.5 m/a.
- *Calamus caesius* (rotan sega) is the best smaller diameter (<18mm) cane. It is used for all types of binding and weaving in the furniture industry, and in the finest basket ware. The habit is clustering, with more than 100 stems per cluster, and high-climbing reaching about 100 m or more in length. The species is found in the lowlands such as alluvial flats, freshwater swamps, margins of peat swamp forests to hill slopes up to 800 m altitude. The clump tends to be rather close and dense. The advantage is its multiple-stem habit, which allows repeated harvests to be carried out without the need of replanting.
- *Calamus trachycoleus* (rotan irit) is a small-diameter cane (<18 mm). It is a clustering dioecious species with a more open type of clumping, producing additional stems via long stolons, which have the potential of increasing the number of aerial stems exponentially. It grows on seasonally flooded riverbanks on alluvial clays and margins of peat swamp forests. In general, the canes of this species have shorter internodes, smaller diameter and thinner layer of silica than *C. caesius*.

However, there is more demand for it for weaving purposes because its cane is softer, more pliable and easier to work with. Its habit is multiple-stem like sega, and needs no replanting. This species is not indigenous to Peninsular Malaysia and records are based on plants introduced for trials.

2.3 Resource management

The richest rattan habitats – the lowland dipterocarp forests – have mostly been converted to oil palm and rubber plantations. Furthermore, the remaining commercial forest areas are now highly accessible as a result of the construction of logging roads. This has resulted in heavy and unsustainable exploitation of rattans. In order to maintain the resource, large-scale rattan plantations offer a solution. Land identified as suitable for establishing rattan plantations is logged forests (newly logged or old logged forest) and existing plantations (tree forest, abandoned or commercial rubber plantations and oil palm plantations). Virgin forests with heavy canopy and low light levels on the floor are not recommended for rattan cultivation.

The choice of species for commercial cultivation will have to take into account numerous factors, of which profitability is probably the most important one. The quality of the cane must be acceptable to the industry either as raw cane or semi-processed or finished product. Other important factors are the gestation period, and whether the species is single or multiple-stemmed. This will determine whether single or multiple harvests can be obtained. Sufficient knowledge of the silviculture of the species chosen is crucial.

2.3.1. Cultivation in natural forests

In Peninsular Malaysia, large-scale cultivation of rattan, especially the most economically important species *C. manan*, has been undertaken since the mid-1980s. Many of the state forestry departments have been carrying out rattan planting in logged-over natural forests using yearly budget allocations from the Silvicultural Cess Fund or Forest Development Fund of each state. By the end of 1997, a total of 15,000 ha had been planted with rattan in a number of PRFs throughout the peninsula (Harnarinder & Chin, 1999). More than 80 percent of the area was planted with large-diameter cane, *C. manan* only, and the remaining area was planted with small-diameter canes, *C. caesius* and *C. trachycoleus*.

2.3.2. Rattan cultivation in forest plantation

A number of land development agencies, such as the Rubber Industry Smallholders Development Authority (RISDA), the Federal Land Development Authority (FELDA), and private companies such as Kurnia Setia, Guthries and Golden Hope Plantations, have also initiated rattan plantations. They interplanted rattan in rubber and oil palm plantations and smallholdings. In Peninsular Malaysia, about 91 ha of *Pinus caribaea* plantations were interplanted with rotan manau in the Kemasul Forest Plantation, Mentakab, Pahang in 1986. When the canes overtopped the pine trees, either whole trees or branches and shoots were bent or broken. The branches of the pine trees were not strong enough to support the weight of rotan manau. In the same area in 1986, about 10 ha of *Pinus caribaea* were planted with a clustering rattan species, *Calamus caesius*. *C. caesius* performed well and vigorously dominated the areas. This posed a difficulty for the main crop – how to manage the pines? It can be concluded that clustering small-diameter cane species are not suitable for interplanting in commercial plantation forests (Raja Barizan and Chong, 1999).

In Sabah and Sarawak, a number of public agencies and private companies have actively planted rattan to enlarge significantly the resource base in the country. Commercial planting of *C. caesius* in Sabah was started in the early 1980s. To date the species remains the most important small-diameter cane for planting in Sabah. In 1988, Innoprise Corporation Sendirian Berhad (ICSB), a commercial holding company of Sabah Foundation, started a rattan plantation in Luasong Forestry Centre, about 100 km northwest of Tawau, Sabah. The objective of this project was to enrich a logged-over forest by line-planting rattans and to provide cash income before the next timber harvesting cycle. Besides that, SAFODA and Sejati Plantation Sendirian Berhad Sabah have attempted to plant rattans under *Acacia mangium.* Two species of rattan, *Calamus manan* (a solitary species) and *C. merrillii* (a clustering species from the Philippines), have been planted under *A. mangium* on a trial basis at Ulu Tungud, Sabah.

2.3.3. *Rattan interplanting in abandoned or well-managed rubber plantations*

Rattan species have been intercropped with rubber (*Hevea brasiliensis*) trees in either well-managed commercial rubber plantations or smallholdings and abandoned or semi-abandoned plantations. During 1981/1982, SAFODA of Sabah planted several hundred hectares of rattan, mainly *C. caesius*, under 10-year old or older rubber plantations. The rubber trees were still being tapped rather irregularly. In Sarawak, *C. caesius* was interplanted with rubber trees in the semi-abandoned rubber holding along the Sungai Sebetan, Seratok. Here, rubber trees were not tapped regularly. The rattan plants were planted haphazardly and grew irregularly. Thus, rubber trees can undoubtedly be used as support or shade trees for growing rattans successfully, but only abandoned rubber holdings are suitable for multiple-stemmed small-diameter canes such as *C. caesius* and *C. trachycoleus* (Tan, 1992). In this case, rattan would be the main crop while rubber tapping the secondary one.

Rattan interplanting in well-managed commercial rubber plantations was established according to a concept similar to agroforestry. This was aimed at increasing the yield of land and supplement the income of smallholders/rural people. The income was estimated to be more than sufficient to cover the costs of replanting rubber (Salleh & Aminuddin, 1986). The survival and stem growth of rotan manau planted under rubber trees in plantations were reported to be better than when this species was planted under forests. The establishment of rattan in rubber plantation is more cost-effective than planting in forest areas. This is because the prevailing conditions in managed rubber plantation are almost ready-made for immediate establishment of rattan seedlings.

Rattan plants are usually planted in the middle of rubber tree rows. The planting distance of rattan under rubber varies according to the rubber spacing. Another approach has been to plant rattan in every other row of rubber trees. Another method was to group two or three rattan plants at each planting point. From various studies, it was observed that rotan manau seedlings required about 60 percent relative light intensity (RLI) that is considered to be a fairly open condition. This suitable light condition can be attained under rubber trees.

In Peninsular Malaysia, so far only three species have been found to be suitable for growing under rubber: *C. manan* (rotan manau), *C. scipionum* (rotan semambu), and *C. palustris* (rotan manau langkawi). The techniques of planting rotan manau under rubber trees have been well developed. The age of rubber trees at intercropping and planting densities per hectare are also important factors that need to be determined before embarking on planting. Four to seven–year old rubber trees were found to be best for intercropping with rotan manau.

Although intercropping rattan with rubber trees appears feasible, rattan should be viewed as a supplementary crop only. In the planning phase, rattan planting should be timed for harvesting when the rubber trees are reaching the stage when they need replanting, i.e. at around 25 years of age. This would minimize the difficulties encountered during harvesting of rattan, and prevent damage to the rubber tree that could occur if the cane is harvested earlier. A longer planting time would mean that the canes are allowed to reach maturity and are more suitable for commercial processing.

Different clones of rubber probably vary with regard to their suitability as support and shade trees because of differences in branching habits, maximum height attainable, strength of branches, adaptability to soil conditions and proneness to wind damage. RRIM 600 clone was found suitable for intercropping with rattan because it has low branches and is easy for rattans to cling on with their cirri or flagella (Aminuddin *et. al.*, 1991). Another clone, PB 260, has high and strong branches, is only suitable for supporting mature rattan, but can hold 2-4 rattan plants. Clones which are under groupings similar to RRIM 600 are RRIM 712, RRIM 701, PB 255 and PB 217, whilst rubber clones that bear criteria similar to PB 260 are RRIM 623, RRIM 901, GT 1, PB 235, PM 10. No studies have been carried out to investigate whether latex production of individual rubber trees is affected by the presence of rattan plants.

With intercropping of rattan and rubber, some management problems can occur: Rattan can hinder tapping operations (Raja Barizan and Chong, 1999). The dense crown of rattan can prolong the drying of the bole of the rubber trees after a rain. Rattan harvesting can damage the branches of the rubber trees as well.

2.3.4. Developing intercropping systems with other crops

Planting rattan under other crops such as oil palm is still under investigation. The growth of six-year old *Calamus manan* (rotan manau) planted under 13-year old oil palm at the Malaysian Palm Oil Board (MPOB) Paka plantation, in Terengganu, appeared good. Annual height increment was 1.5 m (Nur Supardi & Suboh, unpublished data). However, there are some management problems that have to be solved first. The rattan crown hindered the harvesting of oil palm fruit bunches and, consequently, caused a drop in the quantity of fruits collected. When the oil palm frond was pruned, the rattan crown fell to the ground, causing shoot damage.

Planting rattan (rotan manau and rotan sega) integrated with bamboo is being investigated in FRIM. *Gigantochloa laevis* (Buloh Beting) was chosen as an alternative support tree for rattan. In this trial, rattan was treated as the main crop. Using bamboo as support tree would ease rattan harvesting later. This integrated planting would also increase land yield with harvests of bamboo shoots (rebung) from the third year onwards. This study needs to be monitored further before the ability of Buloh Beting as a support plant can be established.

2.3.5. Raising of planting stock

Rattan can be propagated from seeds, wildings, suckers or tissue cultured material. Seeds are the most important planting material for large-scale rattan plantations. They are relatively easy to obtain in large quantities from established plantations, which are now good sources of seed. However, these plantations were not established using selected seed material. The Forest Department of Peninsular Malaysia has identified the potential seed production areas (SPAs) for the production of high-quality rattan seeds.

Small wildings collected from the forest floor also can be used as a source of planting material. However, wildings are not good for large-scale plantations due to their availability only in small quantities and scattered over a large area in the forest.

Tissue culture techniques using embryos or tissue from the shoot apex region have been used in propagating rattan plants. Although planting material developed from micropropagation techniques is too costly, a high proportion of the plantlets produced are identical to the original mother plant. Rattan plants can be micropropagated on a large scale. At present some work on micropropagation through somatic embryogenesis of large cane species – *Calamus manan*, *C. subinermis* and *C. merrillii* – has been carried out in the Plant Biotechnology Laboratory, Innoprise Corporation/CIRAD-Forêt, Tawau, Sabah (Goh, 1988). *C. merrillii* demonstrated most prospects for regeneration via somatic embryogenesis.

3. Rattan Collection

According to the National Forestry Act (NFA) 1984, a licence is needed to harvest rattan from the Permanent Forest Estate (PFE) and Stateland Forests. Under the provision in the NFA 1984 and its amended version of 1994, besides requiring a licence, there are also other regulations to be adhered to in rattan production, such as payment of premium, royalty, forest development cess, licence fee and registration fee to the state government for application as a contractor. Rattan is normally cut and gathered by rattan gatherers. These gatherers generally obtain permits for cutting rattan from the government through the state forest department. These permits are renewable annually and are given for 250 acres per permit.

The rate of collection of royalties and the implementation of the licensing procedure varies from state to state. On average, the royalty rate for *Calamus manan* (large-diameter cane) and *C. caesius* (small-diameter cane) is $M 0.20/m and $M 0.10 per 100/m, respectively (Poh *et al.*, 1995). The state of Pahang imposes its licence fee based on the size of the area applied for, whereas others impose a flat rate per licence. Depending on the different states, the registration fee that is charged varies from a highest $M 100.00/year to a lowest $M 20.00.

Although rattan is quite heavily exploited in Sarawak, there is no record of production levels. The main reason is that there is no royalty collected by the State. To collect rattan, a permit is required, for which a monthly fee of $M 1.00 is charged. However, no permit is required if rattan is collected for domestic use. In Sabah, the royalty collected from rattan is lumped under miscellaneous forest revenues and, as such, it is not possible to gauge the production level. The present royalty rate charged is $M 400/t. A permit or a licence is required for the extraction of rattan from the forest. The monthly fee is $M 5.00/person.

3.1 Developing harvesting techniques

At present, rattan harvesting from the wild is being done mainly by the Malay aborigines, the *orang asli*, who are skilled at climbing trees. A rattan harvesting trial using *orang asli* was carried out by FRIM in three 0.4 ha plots at Block D3, compartment 10, of the Sungai Buloh Forest Reserve. Ninety man-days were required for cutting a total of 232 mature rotan manau per hectare (Nur Supardi, unpublished data). On average, a person could harvest, clean and cut between two and seven rattan plants per day. Using this conventional harvesting method, on average one-third of rattan crowns were left hanging in the forest canopy. The prime concern here is loss of the uppermost part of the cane, which can reach up to 50 percent of total cane length. Improved harvesting methods are needed to reduce this loss.

Mechanical harvesting method is an alternative to the conventional method. For this purpose, four harvesting tools (prototypes A–D) were designed, fabricated and tested in FRIM. Since the harvesting sites are not accessible by vehicles, the tools were designed to be simple to operate and portable. The components were sourced off the shelf to keep the costs low. Further improvements to the cutting tools are needed before they can be operationalized (Chong *et al.*, 1999).

4. Rattan Processing Industry in Peninsular Malaysia

At present most of the raw canes are obtained from the jungle where they grow naturally. Once the rattans have been harvested, they are cut into poles – 3 m long for big diameter and 6 m long for smaller diameter canes. The methods used in processing rattans are mostly dictated by the type of rattan species, their initial physical condition and diameter.

The rattans that have been harvested are normally of mixed species, size and quality. Upon arrival at the processing depots, they are immediately sorted according to species and diameter. Large-diameter rattans (> 20 mm) are further sorted into five diameter classes, i.e. :

- 18 to 24 mm;
- 25 to 29 mm;
- 30 to 34 mm;
- 35 to 39 mm; and
- above 40 mm.

For quality determination, both large and smaller diameter rattans (< 20 mm) are sorted into two quality classes:
- good quality, with no or few defects (good); and
- heavily defective (inferior).

These rattans then undergo boiling in oil or curing process. The trend of rattan processing mills in Peninsular Malaysia is as reported by Razak *et al.* (1999). Rattan processing commonly refers to any activity involving cooking, drying and processing of rattan into semi-processed products such as peels, cores and skins either manually or by machines.

Boiling or curing: In Malaysia raw canes are boiled in diesel solutions with the aims to:
- remove moisture, waxy materials, resins and gums;
- to improve colour quality, texture and flexibility; and
- prevent, to some extent, fungal and insect attack.

The immersion period ranges from 10 to 30 minutes at a temperature 60°C–150°C. Immediately after boiling, the canes are either washed with pressurized water or scrubbed with sawdust or rag to remove any remaining dirt and excessive diesel present.

Drying of oil-cured rattans: After the oil curing and cleaning process, the rattans are air-dried in an open ground by leaning them on wooden frames (end-racking). Alternatively, the rattans are bundled and loosely tied at one end and placed in a wigwam-like structure with their upper ends in contact with the ground. This position helps to accelerate the drying process. Small-diameter rattans about 6 m long are hung over a wooden stand or spread over a wooden stacker placed on the ground. The drying period of oil-cured rattans ranges from 10 to 14 days depending on the species and weather conditions. They are considered dry and ready to be transferred to a shed when the stem surfaces turn yellowish, light in weight, and produce a high pitched sound when beaten on a hard surface or object.

Sulphur fumigation: Only good-quality large-diameter and peeled rattans undergo the process of sulphur fumigation whereas the smaller diameter ones are scraped with knives or processed in a splitting machine to remove nodes before further processing. Large-diameter rattans are washed and subsequently smoked overnight with sulphur dioxide (SO_2) fumes in an enclosed shed or chamber. Besides providing preservative treatment, this process also produces greater uniformity in colour.

Storage: Rattans that have been sorted and graded are straightened manually or by machine before being tied in bundles of 20–30 canes (large diameter) or in bundles of 30–60 kg (smaller diameter). They are now ready to be stored, marketed or prepared for further processing. Sufficient ventilation needs to be provided to ensure dryness and, at the same time, to reduce the probability of fungal attack.

5. Rattan Furniture Manufacturing Industry

The rattan furniture manufacturing industry in Malaysia still needs much research and development efforts in order to maintain its relevancy to other industries. The industry is dependent on skills and technologies. Rattan furniture manufacturing can be best run as a small and medium-scale industry (single ownership). It is also suitably operated as a community cooperative business since the villagers can become the workforce. To be successful, the industry needs a substantial amount of funding, constant supply of raw material and secured skilled workforce. Generally, rattan furniture manufacturing involves the processes described in the flow chart in Appendix 3.

Looking back at history, it can be said that the rattan furniture manufacturing industry in Malaysia has grown significantly only in the last decade. Before that, the rattan industry was dominated by exports of raw and semi-processed materials to other furniture manufacturing countries such as Hong Kong and Singapore. In 1986, these countries, even with no natural resources, exported US$19.6 million and

US$8.5 million worth of rattan furniture, respectively. Malaysia, in the same year, exported only a mere US$2.6 million. Realizing these discrepancies, the Malaysian Government took drastic measures by banning the export of raw canes as of December 1989. Among others, the ban was aimed at encouraging the development of rattan products manufacturing, especially furniture, by ensuring a constant and affordable supply of raw material. The value of rattan furniture exported from Malaysia has increased since then to about US$24.0 million per year.

6. Rattan Marketing and Trading

There are always markets for high-quality rattan furniture either locally or abroad. There are various ways for hunting potential buyers, and the first step is display of products in showrooms. Advanced information technology, virtual displays, and the web can be exploited toward this end. Many international rattan furniture companies have their own web sites and are selling their products through e-commerce. Another effective way of marketing is participation in furniture fairs held by various organizations in the world. Malaysia has been holding her own International Furniture Fair (MIFF) annually since early this decade and the effectiveness of the event is demonstrated by the sharp rise of the furniture exports in the last few years. To assist in marketing the products, the Malaysian Timber Industry Board (MTIB), the Trustee Council for Indigenous People (MARA), PKKM and MEXPO were commissioned to collect and disseminate information on markets for Malaysian products. The Ministry of National Rural Development (MNRD) organizes trade exhibitions to create market opportunities for the entrepreneurs.

7. Policy, Social Aspects, Facilities and Institutional Supports for the Promotion of Rattan Industry

As a means of social development in rural areas, full-time cultivation of small-diameter canes is a good option. It is labour-intensive and offers full employment for the family. Downstream processing of rattan canes employs a much larger workforce than rattan growing. The rattan furniture manufacturing industry could provide good sources of income to many people, regardless of age, gender, and educational background.

It would be ideal for the industry to operate within an industrial zone. Basic infrastructure facilities (as shown in Table 3) must be made available, without which production output would be hampered. In cases where facilities are not provided or not in operation, contingency plans should be proposed. For instance, water pumps with adequate filtration systems could provide the water needed by the factory from sources such as rivers and wells. In the case of electrical blackouts, generator sets should be made available as standby units.

In Table 4 the need of human resource development is illustrated. Where the industry is concerned, training of workers is mainly in-house through mentor and apprenticeship systems. Realizing this, the government agencies mentioned below have developed several in-house Research and development training programmes. The Government, through its agencies gives support to the industry in various ways. The Ministry of Trade and Industry (MITI) and the MNRD, together with various technical agencies, are directly involved in the development of small-scale enterprises in Peninsular Malaysia.

Financial assistance is extended in the form of loans. Under the programme, certain banks provide interest-free loans of between US$800–20,000 to qualified operators. In addition, government-supported institutions such as the Ministry of Youth and Sports (MYS), the Agricultural Bank, the Development Bank of Malaysia Limited, the MARA, and the Malaysian Industry Development Bank, also provide financing to small-scale entrepreneurs at below-market interest. Any owner with assets of less than US$100,00 and fewer than 50 full-time employees is eligible to apply.

Table 3. Basic facilities for a rattan furniture manufacturing factory

Facilities	Institution	Contingency plan
1. Water	Local Department of Water Supply	Water pumps
2. Electricity (3-phase)	Tenaga Nasional (National Power)	Generator set
3. Communication	Telekom Malaysia ((Telecommunication service company)	Hand phone
4. Road and easy access for 40 footer containers	Local Authority / Jabatan Kerja Raya (Public Works Dept)	-

Table 4. Training aspects supported by government agencies

Training aspects	Institution	Follow-up activities
Skill development	Kraftangan Malaysia	In-house training
Marketing	MTIB	Showroom
Research & Development	FRIM	In-house R&D unit

Notes: Kraftangan = Malaysian Handicraft Development Corporation
MTIB = Malaysian Timber Industrial Board
FRIM = Forest Research Institute Malaysia

Technical aid is extended by providing machinery, technical services and training. Under the village industry assistance programme, established in 1986, the MNRD has provided machinery to set up factories and sometimes even the building itself. Various technical agencies such as FRIM, MTIB, and Standards and Industrial Research Institute of Malaysia (SIRIM), provide technical assistance in production and basic design. They also organize seminars and training courses in management and production. FRIM has a core of experienced researchers able to train the entrepreneurs in the uses of rattan and bamboo.

8. Environmental, Occupational Safety and Health

Rattan furniture manufacturing does not impose as significant a threat on the environment as that of the wood-based manufacturing industries. Nevertheless, one major problem that needs to be looked into is the high wastage of rattan raw material, especially rattan poles, which often become a burden to the environment through open burning and illegal dumping. Rattan poles are wasted even before the furniture manufacturing process takes place because both ends of the poles supplied are normally unusable owing to fungal attacks and non-uniform diameter. It was reported that in some cases about 15 cm in length of each end must be cut and thrown away. At this stage, the wastage rate could be as high as 10 percent (2.7 m divided by 0.3 m). In terms of money, as much as US$0.20 is wasted from each pole (10 % of US$2.00 – average price of rattan pole). A productive way of managing the waste is by using them as fuel for the steaming chest.

Occupational safety and health problems (hazardous working conditions) would come from the usage of hand tools such as nail drivers, staplers and spray guns. Carelessness in operating the tools might cause serious injuries to the workers. Prolonged inhalation of finishing chemicals from spray guns could cause long-term health problems (especially to the respiratory system). Therefore, the workers must be made to wear proper personal protective equipment (PPE) while working.

9. Conclusions

Like any other resource-based industry, the constant supply of raw material is important to ensure the existence of the industry. Over the years, the industry has observed a decline in the quantity and quality of rattans. As a result, the industry is forced to accept inferior quality rattan. Hence, the industry encounters problems in meeting buyers/clients specifications. Cultivation of rattan through

Appendix 1

List of potentially available rattan species

Species	Vernacular name	Region
Large diameter (>18mm)		
Calamus manan	Rotan manau	P. Malaysia
C. tumidus	Rotan manau tikus	P. Malaysia
C. palustris	Rotan manau langkawi	P. Malaysia
C. ornatus	Rotan mantang/dok	P. Malaysia/Sabah
C. scipionum	Rotan semambu	P. Malaysia/Sabah/Sarawak
C. peregrinus	Rotan jelayan	P. Malaysia
C. optimus	Rotan sega	Sarawak/Sabah
C. subinermis	Rotan batu	Sabah
Daemonorops grandis	Rotan sendang	P. Malaysia
D. angustifolia	Rotan getah	P. Malaysia
Korthalsia rigida	Rotan dahan	P. Malaysia/Sabah
K. flagellaris	Rotan dahan	P. Malaysia/Sabah/Sarawak
K. laciniosa	Rotan dahan / merah	P. Malaysia
Small diameter (<18mm)		
C. caesius	Rotan sega	P. Malaysia/Sabah/Sarawak
C. trachycoleus	Rotan irit	Sarawak
C. axillaris	Rotan sega air	P. Malaysia
C. speciosissimus	Rotan sega badak	P. Malaysia
C. insignis	Rotan batu	P. Malaysia
C. laevigatus	Rotan tunggal	P. Malaysia/Sabah/Sarawak
C. densiflorus	Rotan kerai	P. Malaysia
C. diepenhorstii	Rotan kerai hitam	P. Malaysia/Sarawak
C. javensis	Rotan lilin	P. Malaysia/Sabah/Sarawak
D. propinqua	Rotan jernang	P. Malaysia
D. didymophylla	Rotan jernang	P. Malaysia/Sabah/Sarawak
D. micracantha	Rotan jernang miang	P. Malaysia/Sabah/Sarawak

Source: Aminuddin Mohamad (1991).

Appendix 2

List of the Malaysian major rattan
species and their use

Species	Local name	Uses
Korthalsia spp.	Rotan dahan	Rattan strip/split and furniture components
Calamus ornatus	R. mantang	Rattan peel, core, furniture components and walking sticks
Plectocomiopsis geminiflora	R. giling (R.rilang)	Handicraft items
Myrialepis paradoxa	R. kertong	Handicraft items
Calospatha scortechinii	R. demuk	The fruits are occasionally eaten (very rare species). Hardly of commercial interest
Daemoncrops calicarpa	R. lumpit	The leaves and stalk are used for making roof & handicraft items
D. leptopus	R. bacap	The leaves are used for making cigarette papers by the aborigines
D. kunstleri	R. bulu landak	The leaves are used for making thatch
D. angustifolia	R. getah	Rattan core and furniture components
D. melanochaetes	R. getah	Rattan core and furniture components
D. grandis	R. sendang	Furniture components
Calamus manan	R. manau	Rattan core, furniture components walking sticks
C. viridispinus	R. kerai gunung	Rattan core, skin and binding materials
C. longipathus	R. kunyung	The leaves are used for making cigarette papers by the aborigines
C. javensis	R. lilin (R.mendon)	Binding materials
C. tumidus	R. manau tikus	Rattan core, furniture components and walking sticks
C. exilis	R. paku	Binding materials
C. caesius	R. sega	Rattan core, skin, furniture components and handicraft
C. axillaris	R. sega air	Rattan core, skin, furniture components and handicraft
C. apeciosissimus	R. sega badak (R. semut)	Rattan core, skin, furniture components and handicraft
C. scipionum	R. semambu	Furniture components and walking sticks
C. paspalanthus	R. sirikis	The fruits are edible
C. didymophylla	R. jernang	Rattan strip/split rattan, fruits are a source of dragon's blood (dye)
C. propinqua	R. jernang	Rattan strip/split rattan and the fruits are a source of dragon's blood (dye)
C. micracantha	R. jernang	Rattan strip/split rattan and the fruits are a source of dragon's blood (dye)
C. castaneus	R. cucor	The leaves are use for making roof and the seeds can be used for medical purposes
C. lobbianus	R. cucor kelabu	The seeds is edible
C. erinaceus	R. bakau	Rattan core, skin and rattan strip/split
C. insignis	R. batu	Rattan core, skin and handicraft items
C. ornatus	R. dok	Furniture components
C. perakensis var. *perakensis*	R. duduk	Walking sticks
C. luridus	R. kerai	Rattan core and handicraft items
C. viridispinus	R. kerai gunung	Rattan core, skin, furniture components and handicraft
C. perakensis var. *crassus*	R. tekok gunung	Walking sticks
C. laevigatus	R. tunggal	Rattan core, skin, furniture components and handicraft items
C. balingensis	R. tanah	Rattan core, skin, furniture components and handicraft items

Source: Compiled by Razak bin Wahab, Mohd. Tamizi bin Mustafa and Arshad Omar (1998). Based on the primary source (Dransfield, 1979).

Appendix 3

Manufacturing Methods

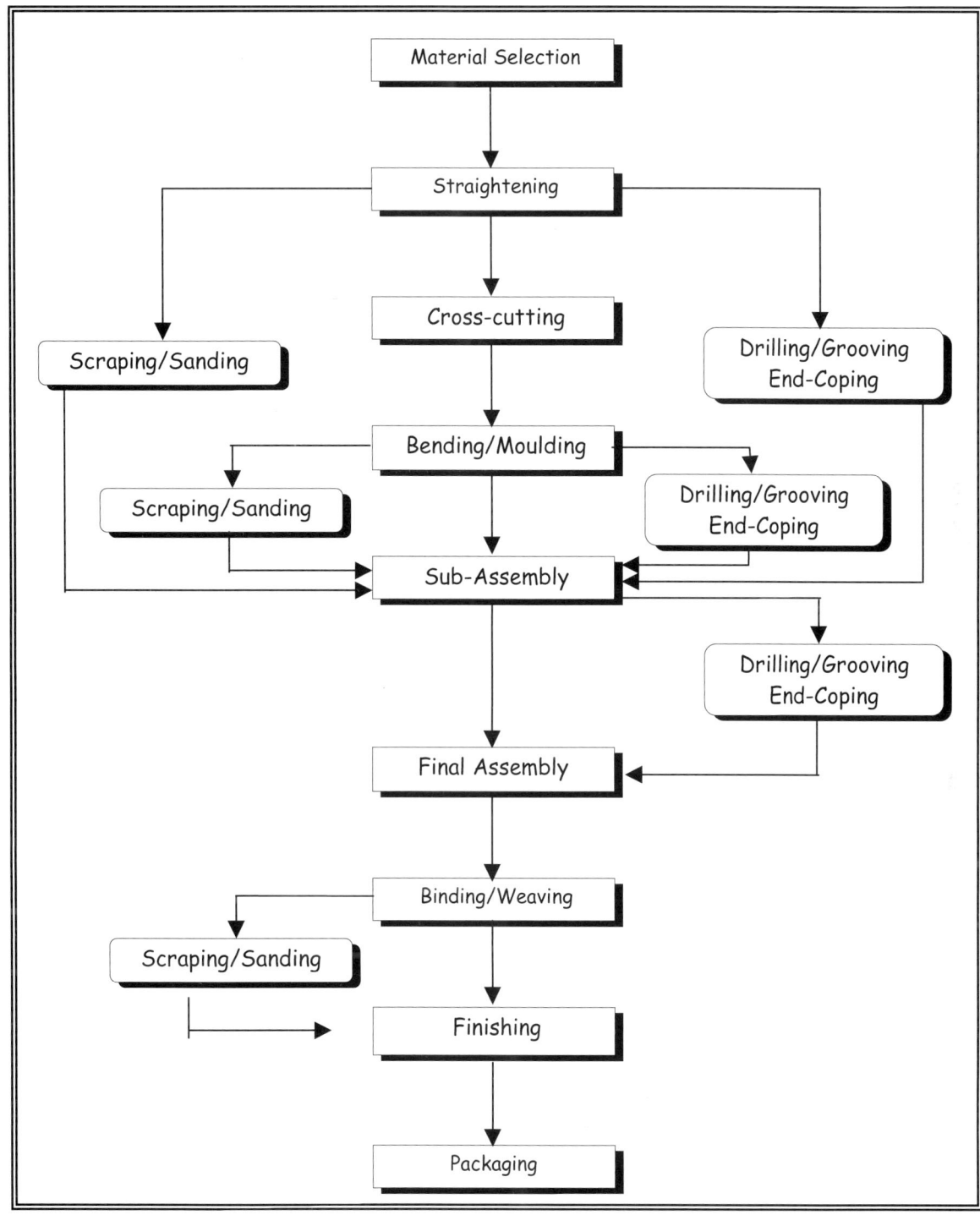

RATTAN RESOURCES OF THE PHILIPPINES
THEIR EXTENT, PRODUCTION, UTILIZATION AND
ISSUES ON RESOURCE DEVELOPMENT

Florentino O. Tesoro

1. Introduction

The total land area of the Philippines is approximately 30 million ha. As of 1997 the actual forest cover was about 5.4 million ha or 18 percent of the total land area of the country (Table 1). The dipterocarp forests still are the most important ones, accounting for 66 percent of the forest cover. Old growth or virgin forests cover about 0.81 million ha, while 2.7 million ha are residual forests. Of the other forest types, pine forests (mainly *Pinus kesiya*) extend over an area of 0.23 million ha and mangrove forests over about 0.11 million ha. Mossy forest covers about 1.04 million ha (DENR, 1998).

Table 1. Land use type in the Philippines

Land use type	Area (ha)		
	1997	**1988**	**1977**
Total	30 000 000	30 000 000	30 000 000
Forest	5 391 717	6 460 600	13 068 231
Dipterocarp	3 536 017	4 401 100	10 861 207
Old Growth	804 900	988 300	3 586 223
Residual	2 731 117	3 412 800	7 274 984
Pine	227 900	283 800	198 865
Closed canopy	123 900	129 600	
Open canopy	104 000	109 200	
Submarginal	475 100	544 200	
Mossy	1 040 300	1 137 400	1 759 021[1]
Mangrove	112 400	139 100	249 138
Brushland	2 232 300	2 525 100	2 155 766[2]
Other Land Uses	22 375 983	21 014 300	14 776 003

[1]   Including bamboo forest of 7 924 ha
[2]   Including submarginal areas

Source: 1998, 1988, 1977 Philippine Forestry Statistics, Forest Management Bureau, DENR

Based on the results of the latest inventory reported in 1987 and 1988 (DENR, 1988), the forest cover declined from 13.07 million ha in 1977 to 5.4 million ha in 1997, a decrease of almost 60 percent over a 20-year period. From 1988 to 1997, the decrease was 17 percent.

2. Rattan Resources, Issues and Constraints for their Management

2.1 Resource status

2.1.1. Rattans found in the Philippines

Rattans are naturally found in the Philippine dipterocarp forests and are distributed throughout the country. Four genera, namely *Calamus, Daemonorops, Korthalsia* and *Plectocomia*, with 64 species are found in the country (PCARRD, 1991) (see Annex 1).

Calamus, with 44 species and 23 varieties, is the largest of the four genera. It is widely distributed, but some species are narrow endemics and confined to specific islands or mountains. *Daemonorops* is the second largest group of Philippine rattans. There are 14 species and one variety under this genus. Like *Calamus*, they are widely distributed but many are confined to islands and mountains.

Korthalsia consists of only five species. Members of this genus have a more restricted distribution. Three species are only found in Palawan, two species in Mindanao and one species in Luzon, Polillo Island and in Mindanao. *Plectocomia* has the smallest number of members. It consists of two species. Members of this genus are confined to the primary rainforests of Palawan, Leyte and Mindanao.

2.2 Resource assessment

The only nationwide inventory of rattan resources in the country was conducted from 1983 to 1987 under a Philippine-German forest resources inventory project (DENR). The results of this inventory are shown in Table 2.

Table 2. Rattan resources in the Philippines, 1988

Species	D < 2 cm (million m)	D > 2 cm (million m)	Total (million m)
Apas (*Calamus symphysipus* Mart.)	462.2	57.8	518.0
Ditaan [*Daemonorops mollis* (Blanco) Merr.]	199.1	32.9	232.0
Limuran (*C. ornatus* Blue, var. *philippininensis* Becc)	550.2	591.4	1 141.6
Palasan (*C. merrillii* Bec.)	645.2	730.6	1 375.8
Sika (*C. caesius* Blume)	68.6	7.9	76.5
Sumulid (*D. ochrolepis* ecc.)	58.2	16.5	74.7
Tandulang-gubat (*C. microcarpus* Becc.)	340.7	69.7	410.4
Tumalim (*C. mindorensis* Becc.)	451.2	131.7	582.9
Others	92.3	68.3	160.6
Total	2 865.7	1 706.8	4 572.5

Source: Natural Forest Resources of the Philippines, Philippine-German Forest Resources Inventory Project Forest Management Bureau, Department of Environment and Natural Resources

Limuran, palasan and tumalim are the most demanded species by the furniture and handicraft industries. Not only are they commonly found all over the Philippines, but they also reach a diameter size of 2 cm or more. The combined length of poles of these species of diameter >2 cm was 1.45 billion linear metres or 85 percent of the total resources of this diameter class. Since the inventory was conducted the supply of these species have dwindled. Due to shortage of these prime species, other rattans are now being utilized by the industry.

No inventory of rattan resources was carried out after 1988. However, an estimate of the resources was made by the National Resources Accounts Programme in 1997 (NRAP, 2000). The estimate was based on net stand growth, reduction due to harvest, wastes and deforestation (Table 3).

Table 3. Rattan resources physical accounts (1997) in million linear metres

Account title	< 2 cm	> 2 cm	Total
Opening stock	4 243	1 126	5 369
Addition (net stand growth)	389	86	475
Reduction	144	92	236
Harvest	69	46	115
Wastes	50	33	83
Deforestation	25	13	38
Net change in stock	246	-6	240
Closing stock	4 488	1 119	5 607

Source: NRAP, DENR

The NRAP estimated that in 1997 the total stock was 5 607 million linear metres (mlm). This is a much higher figure than the one indicated in the 1988 inventory (4 573 mlm), considering the significant decrease in the dipterocarp forest cover which occurred from 1988 to 1997.

Another way of estimating the available rattan resources of the country is by using the annual allowable cut (AAC) of rattan licensees. DENR Administrative Order No. 04 Series of 1989 (dated 10 January 1989) prescribed the computation of sustainable AAC for rattan cutting permits.[6] With AAC, the rattan density in a given licence area can be estimated which, in turn, is used to determine available stock in the same area. Table 4 shows estimated rattan resources using this method.

Table 4. Rattan resource based on calculation of annual allowable cut (1997)

Region	AAC (lm)	Area (ha)	Rotation	% Recovery	Density[1]	Resource[2]
CAR	826 979	82 840	10	0.85	117	9 692 280
1	675 032	15 115	10	0.85	525	7 935 375
2	11 575 261	260 670	10	0.85	522	136 069 740
4A	6 480 364	100 330	10	0.85	760	76 250 800
4B	14 024 846	296 041	10	0.85	557	164 894 837
5	278 550	5 619	10	0.85	583	3 275 877
9	6 045 310	205 878	10	0.85	345	71 027 910
10	4 231 313	67 320	10	0.85	739	49 749 480
11	77 584 116	685 598	10	0.85	1331	912 530 938
12	5 020 872	63 851	10	0.85	925	59 062 175
13	28 102 135	284 561	10	0.85	1162	330 659 882
TOTAL	120 697 333	1 783 262				1 490 489 412

[1]metres per hectare
[2]linear metres

Using the above method, it is estimated that there are 1 490 mlm of rattan on the 1.78 million ha of licence areas. This is a low estimate considering that there are at least 3.54 million ha of dipterocarp forests. However, it provides the lower point in the estimate of rattan resources of the country while the NRAP calculation (5 608 mlm) provides the upper point of the estimate. With respect to diameter classes, it is estimated is that a maximum of 40 percent of the total resources belong to the 2 cm and above class, while a minimum of 60 percent belong to the less than 2 cm diameter class.

2.3 Rattan plantation development

The Philippines has some experiences in rattan plantation development, albeit on a small scale. In Ifugao Province *Calamus manillensis*, a solitary large-diameter rattan, is widely cultivated in home gardens for its fruits, which are edible and used as a cure against coughing (Fernando and Palaypayon, 1988).

Small rattan farms integrated in agro-forestry farms (2.27 ha in Aklan Province and 4 ha in Capiz Province) in Panay Island have also been documented. Species planted were palasan, limuran, sumulid (*Daemonorops ochrolepis*), ditaan (*D. mollis), C. dimorphacanthus var. halconensis* and *C. vidalianus.* Harvests, however, were not monitored since the rattans were mainly for household use. Outside sales were only occasional (Fernando and Palaypayon, 1988).

In 1983 the National Development Corporation (NDC) jointly developed with the Paper Industries Corporation of the Philippines (PICOP) a 4 000 ha rattan plantation in Mindanao. Palasan was planted under bagras (*Eucalyptus deglupta)* and falcata (*Paraserianthes falcataria*) plantations (Formoso, 1988). There is no record of the volume harvested.

---

[6]SYC = [A x D x f]/r where: SYC is sustained annual yield cut in lineal meters; 'A' is the forested area of rattan license; 'D' is the average density per hectare in lineal meters; 'r' is the rotation period which is 10 years; and 'f' is the recovery factor which is 85 percent.

The Government encourages and provides incentives for the development of rattan plantations (DENR, 1989). Incentives include reduced rental fees for the plantation area; reduced forest charges; provision of rattan seedlings to developers at production cost; free technical assistance; and the right to harvest, sell, convey or dispose of the rattan in any manner the owner sees fit.

Rattan plantations have also been developed under Government projects. Table 5 shows the reported rattan plantations developed under the Forestry Sector Project of DENR. Some of these plantations were established as early as 1989 using mainly palasan and limuran, the two important species for the furniture industry.

Table 5. Rattan plantations under the Forestry Sector Project

| Region | Area (ha) | |
	Under Loan I	Under Loan II
CAR	270	162
1	1 470	316
2	90	271
3	421	4
4	650	460
5	284	666
6	250	834
7	110	277
8	400	2 283
9	470	1 070
10	382	114
11	174	268
12	–	–
13	A	251
ARMM	2	3
Total	4 982	6 977

a = formerly under Regions 10 and 11
Source: National Forestation Development Office, DENR

3. Resource management

Management of rattan resources in the Philippines may be viewed under two aspects:
1. management of natural stands; and
2. management of plantations.

3.1 Management of natural rattan stands

DENR AO No. 04 governs the allocation of cutting areas through bids, harvesting, annual allowable cuts, and imposition of a special rattan deposit. Individual rattan gatherers, cooperatives, associations, corporations, partnerships, indigenous communities and owners or operators of rattan processing plants are qualified to bid. However, cultural communities have priority over other applicants on areas within their ancestral domain. Licences have a maximum duration of ten years.

To ensure sustainability, a sustained yield cut is prescribed (Annex 1) and rattan licensees are required to replant. To ensure such replanting, a special deposit is collected for every linear metre gathered, P 0.57 (US$0.011) for 2 cm diameter or above and P 0.46 (US$0.009) for below 2 cm diameter. The licensee may contract the services of private parties or Government entities to establish plantations using the rattan deposit.

3.2 Management of plantations

The few plantations of rattan established in the Philippines have provided some experiences in the management of plantations. It has been shown that palasan can grow under timber plantation of bagras or falcata, as well as under coconut trees and fruit trees (PCARRD, 1991). In one old plantation, palasan was grown in the open.

To establish the parameters for rattan plantation development, trial plantings were initiated in 1977 in Quezon Province using commercial species. Trial plantings were done under different canopy openings, various shade trees and residual forests. Planting materials from seeds and from wildlings were used. About 200 ha were planted with palasan and limuran (Lapis, 1995).

Since then several researches have been conducted (PCARRD, 1991). The ecology and phenology of commercial rattan species have been studied extensively (Fernando,1989). Seed collection, processing, germination and nursery cultural practices were also studied. Tissue culture of commercial rattan species was also studied as an alternative source of planting materials (Garcia and Villena-Sanches, 1988). Clones of palasan and limuran were successfully established in nurseries and in the field.

Plantation establishment practices were also developed. These included site selection, site preparation, out-planting techniques and plantation maintenance, soil requirements of various rattan species, canopy opening, spacing, weeding and fertilization regimes, as well as the application of mulch during dry spells. Growth and yield studies were likewise conducted for plantation grown rattan as well as for natural stands (Cadiz, 1989).

3.3 Biodiversity and gene pool conservation

There is very little conscious effort to preserve biodiversity of rattan in the Philippines despite its commercial importance. Biodiversity conservation of rattan would result from broader natural forest conservation. In 1992 the Integrated Protected Areas System (IPAS, RA 7586) was passed into law. It calls for the identification and establishment of IPAS whereby biological resources are protected and preserved. By operation, rattan resources found in the area are likewise protected and preserved. Table 6 shows four protected areas where rare and endemic rattan taxa are found (Lapis, 1995).

Table 6. Endemic/rare rattan species found in IPAS in the Philippines

Protected Areas	Species	Conservation Status	Information Source
Batanes Island	*Calamus batanensis*	Endemic	Baja-Lapis, 1987
	C. dimorphacanthus *var. batanensis*		
	C. mitis	Endemic	Fernando, 1990
Mangyan Heritage	*C. jenningsianus*	Endemic	IPAS Mgt. Plan, 1992
	C. dimorphacanthus *var. halconensis*	Endemic	Baja-Lapis, 1987
	C. mindorensis	Endemic	Fernando, 1990
	C. microcapus	Endemic	Fernando, 1990
	C. ornatus var.philippinensis	Endemic	Fernando, 1990
	C. reyesianus		Fernando, 1990
		Endemic	
Mt. Apo	*Plectocomia elmeri*	Endemic/rare	IPAS Mgt. Plan, 1992
Turtle Island	*Daemonorops mollis*	Endemic	IPAS Mgt. Plan, 1992
	C. filispadix	Endemic	Baja-Lapis, 1994
	C. merrillii	Endemic/threatened	Tan, Rojo & Fernando, 1994
	C. ornatus var. philippinensis	Endemic	Baja-Lapis, 1994
	D. mollis	Endemic	Baja-Lapis, 1994
	D. ochrolepis	Endemic	Baja-Lapis, 1994

Source: Lapis, 1995

One strategy for genetic conservation outside the natural habitat of rattan is through a gene bank. In 1983 a gene bank, where local and exotic rattan could be grown, was established in a 5 ha area in Mt. Makiling, in Los Baños, Philippines. Initially, 44 rattan taxa, collected all over the archipelago, were planted in the gene bank (Baja-Lapis and Santos, 1993). Growth accounting of each taxon is being assessed by measuring the length and the number of suckers produced.

4. Some Constraints in Managing Rattan Resources

4.1 Rattan disposition (lack of tenure)

The limited tenure of the rattan-cutting licence, which is ten years, does not provide an incentive for the licensee to manage the rattan resource sustainably. Neither is the imposition of an allowable cut a guarantee to a sustainably managed resource. The tendency is to gather as much as possible without regard to the succeeding harvest after the ten-year period.

To remove this constraint, it has become a policy to transfer all rattan-cutting licensees within community-based forest management (CBFM) areas to the jurisdiction and management of CBFM people's organizations or indigenous communities.[7] These communities have a 25-year tenure over the area, renewable for another 25 years. The long tenure and the benefits that communities derive are incentives for them to manage the resources sustainably.

4.2 Rattan deposit

The special rattan deposit does not guarantee that plantations will be developed within the cutting area or elsewhere unless it is strictly imposed that licences are terminated if planting is not undertaken in any given year. As it is, only limited plantations have been developed the deposit. In one region the rattan special deposit collected since 1991 now amounts to more than p 25 million (US$555,000), but no plantation has been developed.

Instead of collecting the rattan special deposit, the licensee should be required to submit a plantation development plan, submit evidence that an amount equivalent to one-year expenditure for plantation development has been deposited in a bank and that a contract has been entered into with a third party for the development of the plantation. Plantation development should be monitored and no rattan-cutting licensee should be allowed to operate in the succeeding year if the planned plantation development did not become operational.

4.3 Inadequate funds for rattan plantation development

Funds for rattan plantation development are inadequate. The Philippine Government has been able to obtain loans from the Asian Development Bank (ADB) and JBIC for tree plantation development. Rattans are sometimes included in the plantation development, but these are minimal. A loan similar to that for industrial tree plantations at favourable interest rates should be obtained to develop primarily rattan plantations. Government financial institutions should also open a window for private sector development of rattan plantations.

5. Rattan Processing and Product Development

5.1 Product harvesting

The rattan licensee usually engages a contractor who, in turn, hires rattan gatherers. The gatherers are the ones who actually go into the forest, cut the rattan and bring it down from the mountain to the stockyard (Rivera, 1988).

---

[7]DENR DAO No. 96-29. Rules and regulations for the implementation of Executive Order No. 263, otherwise known as the Community-based Forest Management Strategy.

The gatherers are paid according to the number of pieces of rattan brought to the stockyard. The price depends upon the species, diameter and length. Rattan poles are bundled and carried down the mountains by the gatherers themselves. Sometimes water buffaloes are used to carry the bundled rattan to the stockyard.

At the licensee's stockyard the rattan undergoes scraping, trimming and drying. Often, the rattan poles are treated with preservatives prior to drying. In some cases, the poles undergo splitting. After drying, the poles are sorted according to diameter, length and colour (stained or unstained). From there the licensee distributes the poles to furniture or handicraft manufacturers.

There are a number of issues and concerns at various stages in the harvesting, preliminary processing, transport and distribution of the rattan poles.

5.2 Harvesting

Rattan harvesting is still literally by hand. The stem is cut as close to the ground as possible and is pulled down. Since rattan clings to the canopy, pulling it down is a difficult task. The larger portion of the rattan stem is often left in the canopy as waste. No mechanical devices are used to pull down the rattan to obtain the maximum harvestable length.

Often immature stems are cut, which are later rejected by the contractor or the licensee. Since rattan gathering often takes weeks, the poles are left lying on the forest floor and often staining sets in. At the very least, the grade and value of the poles are reduced; at worst, the poles are discarded.

Harvesting rattan in the Philippines is generally done during the dry season. From late May to November or December, rattan gathering slows down or even stops except for gatherers who do not have other sources of livelihood. Areas where harvestable rattan is located are now much deeper into the forest, affecting the volume that each gatherer can bring out from the forest and, consequently, his income.

5.3 Hauling, transportation and distribution

Hauling the poles from the cutting area to the roadside is basically manual labour. Since gatherers do not own water buffaloes for carrying the poles down from the forest, they are constrained to minimize their load (Wakker, 1993). Thus poles are cut shorter than the desired length or are split, resulting in a lower value of the materials.

When transported, rattan often must pass through a number of checkpoints. In most instances these checkpoints constitute expense points. This increases its price, which is passed on to the buyers.

5.4 Sorting, scraping, drying and grading

Manual sorting is often inaccurate and leads to re-scaling by Government authorities and the buyers. Hand-scraping is a slow process and often non-uniform. There are no drying facilities at the contractors' or at the licensees' level. Sun-drying is the most common practice, but often leads to staining and damage by borers, thus degrading the materials.

A national standard (PNS 229-1999) has been approved in 1999 for rattan poles and rattan by-products. Product standards in the Philippines are not mandatory and the Bureau of Product Standards relies mainly on the acceptance of the standards by the stakeholders.

5.5 Preservative treatment

Although acceptable technologies for pole preservation exist, treatment of rattan poles at the cutting area or at the stockyard is seldom practised. This stems from lack of awareness of the importance of treating the materials. Licensees try to avoid the added cost of preservative treatment. However, lack of treatment often results in early deterioration of the poles (Pabuayon and Espanto, 1996).

6. Rattan Processing

6.1 Primary processing at the village level

The price and value of rattan poles can be increased by undertaking primary processing at the village level, or at the stockyard of the contractor or the licensee. Primary processing may take the form of trimming, scraping, treating, drying, straightening, grading/sorting and bundling (Wakker, 1993). These operations require minimum skills that could be obtained through training and experience.

Another primary processing is splitting. Rattan splits can find beneficial markets from handicraft and large rattan furniture factories. It is labour-intensive and requires little equipment outlay. Communities can also engage in rattan furniture manufacture, but this requires more extensive training. A strategy to pursue is for large furniture manufacturers to contract the production of furniture parts out to the community, which some manufacturers are already doing.

Presently, no certification as to the source of the raw materials exists; neither is chain of custody being undertaken in rattan furniture manufacture. But as foreign markets become more discriminating as to whether rattan raw materials come from sustainably managed forests, certification will become a practice. In the Philippines there is no certification even for wood products.

7. Rattan Marketing and Trade

The National Statistics Office (NSO) is responsible for collecting and collating trade data. These are collected from line agencies such as the Bureau of Customs, Bureau of Domestic Trade, Bureau of Export Trade Promotions, the Department of Trade and Industry and from other Government agencies concerned. DENR is responsible for data on natural resources harvesting and production, as well as for analysis of data on production and trade of forest products.

In 1999 (Table 7) the share of rattan furniture in the total furniture exports from the country was 32 percent with a value of US$112.89 million, while that of wood furniture was 38 percent with a value of US$132.67 million. Since 1990 the value of rattan furniture exports has been swinging up and down despite the 6.4 percent growth rate of the entire furniture industry.

The United States is still the most important foreign market for rattan furniture from the Philippines. The value of imports in the United States in 1998 was US$53.67 million or 50 percent of the total export value. Other major markets are Japan, Great Britain, Australia and France.

The greatest barrier to a better position of the rattan furniture industry in the export market is the decreasing supply of rattan of the right size and quality. The production of rattan poles during the last five years has been decreasing rapidly (Table 8). The requirement for the production of exported rattan furniture in 1998 was estimated at 36 to 108 mlm[8]. Production of rattan in 1998 was only 10.46 mlm. For the same year the country imported only 112,973 kg of rattan poles, equivalent to 9,400 poles (112,000 lm). Total production and imports of rattan poles fell far short of the amount used by the industry in 1998. Either these were under-reported or the actual values were not captured in the surveys. In any case, demand far exceeds supply.

---

[8]According to the Chamber of Furniture Industries of the Philippines, a lineal meter of rattan pole (2 cm or larger in diameter) will fetch an export value from US$1.0 to US$3.

Table 7. Exports of rattan furniture from the Philippines

Year	f.o.b. value (US$ million)	% Change
1999	112.89	+4.28
1998	108.26	-12.00
1997	123.02	+3.12
1996	119.30	-0.32
1995	119.69	-2.62
1994	122.91	+7.62
1993	114.20	-1.20
1992	115.58	-2.25
1991	118.24	-2.53
1990	121.31	

Source: Bureau of Export Trade Promotions

Table 8. Rattan production in five years

Year	Split rattan ton	Unsplit rattan (million lm)
1998	5	10.46
1997	2	19.52
1996	17	24.61
1995	24	17.46
1994	4	19.09

Source: 1998 Philippine Forestry Statistics, DENR

8. Policy and Institutional Related Aspects for the Promotion of the Rattan Sector

8.1 Social aspects

As of 1996 rattan licence areas within CBFM and those within ancestral lands are no longer put out to tender, but would be operated by the CBFM community or the ancestral community (DENR, 1989). This way the upland communities can benefit from the rattan resources in their areas.

In the present set-up, where rattan gatherers are engaged by contractors or licensees to harvest rattan, minimal benefits accrue to them. It was estimated that in 1996 the gatherers received only about 25 percent of the traders' price for large-diameter rattan, while they got about 36 percent from small poles (Pabuayon and Espanto, 1996). In 1998, the price at the source (gatherers) for 2 cm diameter rattan and above was only 17 percent of the traders' price, while it was about 20 percent for small-diameter poles (Rivera, 1998).

In terms of finished products, only 5–21 percent goes to the gatherers, while 63-87 percent of the value of the products goes to manufacturers (Pabuyon and Espanto, 1996). There is a need to add value to the rattan poles prior to delivery to the contractor or licensee in order to improve the benefits that accrue to the gatherers.

A recommended approach to providing more benefits to the gatherers and their communities is to undertake primary processing and eventually progress into production of furniture parts in joint ventures with furniture manufacturers.

8.2. Institutional aspects: Existing arrangements for resource management and conservation/ utilization

Responsibility for direct resource management and conservation/utilization rests with rattan licensees. DENR provides general supervision over the licensees. Rattan licensees are required to submit an annual cutting and replanting plan. The replanting plan is implemented through the payment of a Rattan Special Deposit. Actual planting may be done through organizations or groups sub-contracted by the licensee. For the CBFM communities, replanting is undertaken according to an annual resource management plan submitted and approved by DENR.

Rattan plantation development outside of the license areas is provided for by Government policies (DENR, 1989). However, there is no vigorous Government rattan development programme comparable to that for industrial tree plantation development where financing is provided. In Government reforestation programmes, rattan plantation development is but a small portion of the total area developed.

To ensure future rattan supply, the Government should mount a vigorous rattan plantation development programme. Sufficient incentives should be provided for both smallholder and large-scale plantation development.

In addition to access to sufficient financial resources at suitable interest rates, successful rattan plantation development in the Philippines requires some technological inputs. Among the technological requirements are:
1. Production of quality planting materials in sufficient quantity and at reasonable cost. Tissue culture would be an excellent alternative if only the cost of the seedlings could be greatly reduced.
2. Nursery management is also an important aspect of the technology that must be developed particularly in pest and disease management.
3. Growth and yield data of important commercial species such as palasan, limuran and tumalim should be established.

For the manufacturing sector, the Government should continue to provide assistance in product development, waste reduction (such as in pole drying and treatment) and waste utilization. The Government is now assisting furniture exporters through the organization of international furniture trade fairs. The Government in cooperation with the private sector should provide marketing assistance through the Internet.

8.3 Recommended revisions of existing policy and legal framework related to the rattan sector

The policy and legal framework for rattan resources management, development and utilization is still DENR A0 No. 04 Series of 1989, including its various amendments[9]. The administrative order needs to be revised in order to:
1. give priority of access to the resource to nearby communities for areas outside their CBFM areas or ancestral domains;
2. develop an alternative scheme that would abolish the special rattan deposit but would ensure that licensees undertake replanting or plantation development;
3. require periodic monitoring of replanting activities of licensees by the Community Environment and Natural Resources Officer (CENRO);
4. put out to tender only production areas that have been inventoried;

---

[9]DENR MC No. 15, Series of 1990. Updating of Cost Estimates for the Establishment of Rattan Plantation by Contract and Administration; DERN MC No. 18, Series of 1990. Prioritizing the Establishment of Rattan Plantation by Contract and Administration and Exploring the Possibility of Availing Management and Technical Services Agreements with Plantation Developers and other schemes; DENR MO No. 11, Series of 1990. Prescribing Guidelines for the Movement/Transport of Finished and Semi-Finished Rattan Forest Products.

5. provide more incentives to developers of rattan plantations to encourage rehabilitation of the resource;
6. for expiring licences outside of CBFM and indigenous community areas, base renewal on performance of plantation development and sustainable management; and
7. incorporate in DENR policies and guidelines considerations for certification of sustainably managed rattan resources and plantations.

Figure 9: Artisanal cleaning of rattan stems in Sarawak, Indonesia (Sastry)

In addition to revising the policies and guidelines on rattan resources management, development and utilization, strengthening of institutions in implementing the policies and regulations should be undertaken. Among these are:

1. yearly evaluation of the performance of licensees on replanting or plantation development;
2. establish within each CENRO a rattan management and development desk which would be responsible for implementing the policies and regulations on rattan management and development;
3. monitor strictly the declared harvest of each licensees to avoid under-reporting;
4. CENRO, through the rattan management and development desk, should have all rattan production areas inventoried prior to opening the area for bidding or awarding it to communities or indigenous peoples;
5. reduce documentation needed for the movement of rattan poles so as not to impede the transport and marketing of the product;
6. start preparations for eventual certification of sustainably managed rattan resources (natural forests) and plantations; and
7. Government should take concrete steps to provide financing for rattan plantation development in the same manner as it is doing for industrial tree plantations.

REFERENCES

Baja-Lapis, A. & Santos, G., 1993. *Establishment of rattan gene bank – a Philippine experience.* Post paper presented in the International Symposium on Genetic Conservation and Production of Tropical Forest Tree Seed held in Chiangmai, Thailand, 14-16 June. 18 pp. Unpublished.

Cadiz, R.T., 1989. *Growth assessment of rattan in existing natural stands and man-made plantations.* Los Baños, Laguna. PCARRD-IDRC Terminal Report.

DENR. 1988. *Natural forest resources of the Philippines.* Philippine-German Forest Resources Inventory Project, Forest Management Bureau, Department of Environment and Natural Resources.

DENR. 1989. *Revised regulations governing rattan resources.* AO No. 04 Series.

DENR. 1998. *Philippine Forestry Statistics.* Forest Management Bureau, Department of Environment and Natural Resources.

DENR. *Rules and regulations for the implementation of Executive Order No. 263, otherwise known as the Community-based Forest Management Strategy.* DAO No. 96-29.

DPNS. 1999. Rattan poles and rattan by-products – Specifications. Philippine National Standard

Fernando, E. S., 1989. *Phenology of the commercial species of Philippine rattans.* Terminal Report. College, Laguna.

Fernando. E.S., 1990. A preliminary analysis of the palm flora of the Philippine Islands. Principes 34: 28-45.

Fernando, E.S. & Palaypayon, W.R., 1988. Small-scale rattan farming: Notes on a case study of two sites in Panay. *Proceedings of the National Symposium/Workshop on Rattan. Cebu City, 1–3 June 1988.*

Formoso, G.R., 1988. Economics of rattan plantation development. Proceedings of the National Symposium/Workshop on Rattan. Cebu City, 1–3 June 1988.

Garcia, M.U. & Villena-Sanches, E., 1988. Tissue culture of rattan: Progress and prognosis. *Proceedings of the National Symposium/Workshop on Rattan. Cebu City, 1–3 June 1988.*

Lapis, A., 1995. *Rattan genetic resources in the Philippines.* UNDP/FAO Regional Project on Improved Productivity of Man-Made Forests through Application of Technological Advances in Tree Breeding and Propagation (FORTIP). Los Baños, Philippines.

NRAP. 2000. Rattan resources accounting update. In: *Forest resources accounts update, 1989–1997.* ENRAP 4 and FMB-PEENRA Counterparts.

Pabuayon, I.M. & Espanto, L.H., 1996. The Philippine rattan sector: A case study of an extensive production system. *Bamboo and Rattan Seminar/Workshop, 28 June 1996.* Sponsored by INBAR.

PCARRD. 1991. *The Philippines recommends for rattan production.* Philippine Council for Agriculture, Forestry and Natural Resources Research and Development. Department of Science and Technology.

Rivera, M.,1998. Personal communication.

Rivera, M.N., 1988. *Marketing system and price structure of rattan raw materials in the Philippines.* Terminal Report. Ecosystems Research and Development Bureau. DENR.

Wakker, E., 1993. *Towards sustainable production and marketing of non-timber forest products in Palawan, the Philippines.* Haarlem (The Netherlands), Tropical Social Forestry Consultancies (TSF). Statenbolwerk 16.

Annex 1

Checklist of Philippine Rattan

Genus	Species	Distribution	Endemic	Vernacular name
Calamus	C. aidae E. Fern	Luzon, Samar, Biliran, Dinagat Island, Mindanao	E	ulasi, ulisi, inihian
	C. arugda Becc.	Luzon	E	quwenlunhuy, arugda
	C. balerensis E. Fern	Luzon	E	rituuk
	C. batanensis (Becc) Baja-Lapis	Batames		valit
	C. bicolor Becc	Mindoro	E	sambunotan, lasi, rasi, obanan
	C. caesius Blume	Palawan	also found in Sumatra Malay P., Borneo and south Thailand	
	C. cumingianus	Luzon/Minidanao	E	ubut, dowung-dowung
	C. diepenhorstii Miq. var. exulans	Luzon/Palawan and Polilio Island	E variety is endemic	abuan
	C. dimorphacanthus Becc. var.	Luzon, Panay	E	
	var. montalbanicus Becc.	Luzon	E	
	var. zambalensis Becc.	Luzon	E	
	var. benguetensis Baja-Lapis	Luzon	E	lambutan, umbanan, oban-oban
	var. halconensis (Blanco) Baja-Lapis	Luzon, Mindoro Panay, Mindoro		
	C. discolor Mart. var. discolor	Luzon	E	kumaboi
	var. negrosensis Becc.	Negros, Surigao	E	
	C. elmericanus Becc.	Luzon, Danagat, Mindanao		tagiktik, panllis, sababai, samanid
	C. erinaceus (Becc) Dransf. var. erinaceus	Palawan (south Thailand, Malay Peninsula, Sumatra, Borneo)		
	C. filispadix Becc	Palawan	E	pangan-panganan, nokut, kangnobnob
	C. foxworthyi Becc	Palawan	E	
	C. grandifolius Becc	Luzon	E	saba-ong
	C. javensis Blume	Palawan	S. Thailand, Malay P., Sumatra, Borneo, Java	
	C. jeningsianus Becc	Mindoro	E	
	C. malawaliensis J. Dransf	Palawan	Also in Malawali	

Genus	Species	Distribution	Endemic	Vernacular name
	C. manillensis (Mart) H.A. Wendl	Luzon, Dinagat Is., Mindanao	E	lituku, giiwi, lintukan
	C. marginatus (Blume) Mart.	Palawan	Also in Sumatra and Borneo	labsikan
	C. megaphyllus Becc	Leyte, Mindanao	E	magbagaki, banakbo
	C. melanorhynchus Becc.	Mindanao	E	dalimban
	C. merrillii Becc var. merrillii	Luzon, Masbate, Palawan, Mindanao, Masbate	E	palasan, quwen, babuyan, pasan, nanga, acab-acab
	var. merrittianus Becc.	Luzon, Mindoro	E	
	var. nanga Becc.	Mindana	E.	nanga
	C. meyenianus Becc.	Pangasinan, N. Viscaya		
	C. microcarpus Becc	Luzon, Polillo	E	korayot
	var. microcarpus	Mindoro, Leyte, inxno		tandulang-gubat, potian, obanon
	var. diminutus Becc.	Luzon	E	kamlis
	var. longiocrea Baja-Lapis	Luzon		cham-may, damayon
	C. microsphaerion Becc.			
	var. microsphaerion	Luzon, Culion, Palawan	Also in Borneo	kulakling-labit, pinpin, siksik, sika-sika
	var. spinosior Becc.	Palawan	E	
	C. mindorensis Becc.	Luzon, Mindoro	E	tumalim
	C. mitis Becc.	Batanes, Luzon, Babuyan	E	tevdas, matkong
	C. moseleyanus Becc	Mindanao, Basilan	E	sarani
	C. multinervis Becc.	Mindanao	E	bugtungan, balala, ubli
	C. ornatus Blume			
	var. philippinensis Becc.	Luzon, Polillo, Mindoro, negros Minidanao	E	limuran, Quwen, Gamngan, kalapi
	var. pulveruletos E. Fern	Palawan, Mindoro	E	mananga, borongan
	C. ramulosus Becc.	Luzon	E	panlis
	C. samian Becc.	Luzon, Mindanao	E	apa, lukuan
	C. scipionum Lour.	Palawan	Also in S. Thailand, Malay P., Sumatra and Borneo	bastonan
	C. siphonospathus Mar.			
	var. siphonosphathus	Luzon	E	biri, tallawan
	var. dransfieldii Baja-Lapis	Mindanao		pasan-pasan
	var. farinosus Becc.	Luzon	E	
	var. oligolepis Becc.	Luzon	E	sukol
	var. sublevis Becc.	Luzon, Mindanao	E	sipay, papakin
	C. spinifolius Becc.	Luzon, Panay, Mindnao	E	kurakling
	C. trispermus Becc.	Luzon	E	
	C. subinermis H.A. Wendl. Ex Becc.	Palawan	Also in Borneo	bugtung

Genus	Species	Distribution	Endemic	Vernacular name
	C. symphysipus Mart	Palawan	Also in Celebes	apas, bolanog
	C. usitatus Blanco	Luzon, Visayas, Mindanao	Also in Borneo	tandulang-parang, talora, termarura, butarak, taguiti, Quwen lantos
	C. vidalianus Becc.	Luzon	E	
	C. vinosus Becc.	Mindanao	E	akal
	C. viridissimus Becc.	Mindanao	E.	bag-bag
Daemonorops	D. affinis Becc.	Mindanao	E.	
	D. cleminsiana Becc.	Mindanao	E	
	D. curanii Becc.	Palawan	E	pitpit, saranoi
	D. gracilis Becc.	Palawan	E	labsikan
	D. longipes (Griff.) Mart.	Palawan (Malay P., Sumatra, Borneo)		
	D. loheriana Becc.	Luzon	E	
	D. margaritae (Hance) Becc. var. palawanica Becc.	Palawan	Variety endemic. Species also found in S. China	pinpin
	D. mollis (Blanco) Merr.	Luzon, Visaya, Mindanao	E	eitaan, quwen, mangnaw, nanga, gatasan, sumulid
	D. ochrolepis Becc.	Luzon, Polillo, Leyte, Mindanao	E	sumulid, palaklakanin, ditaan, nokot, taletoi
	D. oligolepis Becc.	Mindanao	E	ragman
	D. pannosa Becc.	Mindanao	E	sabilog
	D. pedicellaris Becc.	Leyte, Mindanao	E.	hanmham, delot, logman, hiyod, oban-oban, rogman
	D. polita E. Fern	Mindanao, Zamboanga		lapa-utong
	D. urdanetanas Becc.	Mindanao	E	sahaan
Korthalsia	K. lacinosa (Griff.) Mart.	Luzon, Polillo, Mindanao	Also in SE Asia and Indonesia	danan, tambuanga, planug, miling-piling
	K. merrillii Becc.	Palawan	E	buragat
	K. rigida Bl.	Palawan	Also in S. Thailand, Malay P., Sumatra, Borneo	
	K. robusta Bl.	Palawan	Also in Borneo and Sumatra	kalalias
	K. scaphigeroides Becc.	Mindanao	E	kaprigid
Plectocomia	P. elmeri Becc.	Mindanao	E	ungang
	P. elongata Mart. ex Bl. var. philippinensis Madulid	Palawan, Mindanao, Leytu	The variety endemic. Species found in Thailand, Palay P., Borneo, Sumatra	panson, panog, kalaanan, laaana, binting dalaga, paang dalaga

Source: Fernando, 1990

THAI RATTAN IN THE EARLY 2000

Isara Vongkaluang

1. Introduction

Rattan is a very important non-wood forest product of Thailand. The products from rattan are mainly used for making household goods, decoration and furniture. For years rattan has played a major role in providing job opportunities and enhancing the economic situation of some groups of the population, both in rural and urban areas. Nowadays, rattan products attract the interest of consumers inside and outside the country and it is believed that this interest will grow in the future.

Large quantities of raw material are needed by industry to produce rattan goods, and at the moment, there is some difficulty concerning the supply of rattan. Many people (especially those involved in the rattan industry, researchers and Government officials) began to realize that there was a possibility of the shortage of rattan canes in the future and are looking for possible ways to compensate for the rattan canes that are now being exploited from the forest. Unfortunately the basic knowledge about rattan in Thailand is quite limited; therefore, all possible information about this material should be collected so that there would be a good indication for better management in order to increase production in the future.

2. Thai Rattan in General

In Thailand rattan can be found mostly in the southern part of the country. Eighty-three species of rattans have been recorded for the whole country (Dransfield, *pers. comm.*). These rattans grow mostly in the moist evergreen forests from the lowlands to the top of the hills. Although seven genera can be found in the country (*Calamus, Ceratolobus. Daemonorops, Korthalsia, Plectocomia, Plectocomiopsis* and *Myrialepis*) (Vongkaluang, 1986), most belong to genus *Calamus* while the others occur in very limited quantity. The size of the canes varies from 3 mm to 15 cm in diameter, depending on the species.

Calamus longisetus, C. rundentum, C. peregrinus, C.s caesius, C. manan, and *C. wailong.* (Wai Nampueng) are the species of greatest economic importance. All these species are decreasing in quantity and it is becoming increasingly difficult to find them. One of the reasons for this is that in the past the required supply of rattans was collected exclusively from the natural forest. The forested area in Thailand has been constantly decreasing due to the increasing need of land for agriculture, as well as illegal encroachment on forested areas. In addition, rattans are considered to be only minor forest products and not given much importance compared to wood. There is almost no record of any action being taken in the past to prevent rattan from being exploited from the forest, or of any conservation measures being adopted with regard to species of economic importance. It can be expected that in the future the shortage of rattan canes will become a major problem and appropriate measures should be taken now or at the earliest possible opportunity.

3. Past Management

In the last 30-40 years, concessions to exploit rattans were given without limit. The canes were extracted from the forest easily and in large quantity, without a precise replanting scheme. A very limited number of rattans was planted by the Royal Forest Department (RFD) in the following areas:

1. The rattan plantation at I-sa-tia national forest Amphoe Rangae, Narathiwat. The total plantation area covers 1 230 rai (1 rai =0.16 ha or 196.8 ha). *Calamus caesius* has been planted in this area since 1968.
2. The rattan plantation under the King Project at Sukirin, Narathiwat. *Calamus caesius* was also planted in this area starting in 1980 on a total area of more than 2 000 rai (320 ha).
3. The rattan plantation at Ngoa waterfall, Amphoe Muang, Ranong, on a total area of 100 rai (16 ha) since 1979.
4. The rattan plantation at Khoa Tha Petch, Amphoe Muang, Surathani on a total area of 100 rai (16 ha) since 1979.
5. The rattan plantation at Kapoh waterfall, Amphoe Thasae, Chumporn, on a total area of 100 rai (16 ha) since 1979.

These plantations were mainly established on a preliminary trial basis at 4 x 4 m spacing.

At that time there were very few people interested in carrying out research on rattans. No basic knowledge was generated in spite of the huge amount of rattan being harvested from the forest. Lack of rattan research was one of the key factors since most researchers were more interested in carrying out research on forest trees rather than on minor forest products.

Later, when population pressure increased to turn forestland into agricultural use, rattan became evidently overexploited, but it was more difficult to replant it in the secondary forest because of unsuitable conditions of the land in comparison to the virgin forest in the past. However, there was still a demand for the canes on the part of the manufacturing industry and, therefore, certain quantities of rattan canes had to be bought from neighbouring countries so that the production line could continue.

The private sector and the Government sector then became aware of the shortage in supply of rattan canes. The Royal Forest Department and the Department of Agricultural Extension started again a planting programme in 1989. There were two projects at that time:

1. Seed orchards were set up in the following areas:
   (a) Kao Soi Dao, Chantaburi – 300 rai (48 ha);
   (b) Nai Chong, Krabi – 1 150 rai (184 ha).

2. Rattan plantations were set up in the following areas:
   (a) Kao Sok, Surattani – 2 300 rai (368 ha);
   (b) Kao Bunthat 1, Trang – 750 rai (120 ha);
   (c) Kao Bunthat 2, Trang – 750 rai (120 ha);
   (d) Sakaerat, Nakornrajsima – 200 rai (32 ha).

The RFD expects to plant more rattan in the future, while the Department of Agricultural Extension is distributing rattan seedlings to the people to grow them in their homestead. Additionally, the private sector is looking into the possibility of setting up rattan plantations in the forests or in rubber plantations and hope that this can materialize very soon. Much research has now been carried out in the area of:
- rattan ecology;
- species collection and identification;
- seed germination;
- growth study;
- tissue culture;
- natural durability;
- *ex-situ* conservation;
- seed orchards, etc.

However, research is being carried out on a short-term basis due to lack of funding. Therefore, basic knowledge is available but still very limited. It could be expanded in the future once proper management requirements and gaps in research are identified.

4. Present Management

In Thailand there is no special management practice for rattan. The Royal Forest Department is the only organization that has the direct responsibility to look after this resource, especially in protected areas. In the past the people received concessions to extract rattan from the forest, but nowadays the forest is officially closed and such extraction is forbidden. Therefore, in terms of *in situ* conservation, the genetic resource and its biodiversity should presumably be plentiful. However, this is not the case because large quantities of rattan are periodically harvested illegally from the forest. The Royal Forest Department is well aware of the situation and since 1991 has tried to establish enrichment planting of economically valuable rattan species, e.g. *Calamus rudentum, C. longisetus, C. latifolius, C. caesius,* etc., on at least 1 000 rai (160 ha) per year (Appendix 7). This is, however, much less than the resources depleted. Tests were also carried out by the Forest Industry Organization and Office of the Rubber and Replanting Aid Fund on intercropping rattan with rubber trees. Early results looked very promising but the system did not find acceptance among the farmers. There is the plan in the future to educate the farmers through extension programmes on the benefits gained from intercropping so that it can be accepted by them.

In the northeastern part of Thailand, many people cultivate rattan (*C. viminalis*) for shoot as food, along with their cash crop. Rattan is planted and harvested when it is about 18 months old. The shoot, which is about 1 m tall, is cut to the ground level, the sheath is peeled off and the uppermost part of the shoot is used for consumption. However, this kind of rattan is cultivated on small plots and its shoots are consumed mainly by the local people. The Royal Forest Department is now launching an extension programme on growing edible rattan in the north and aims at extending it to the central plain and also to the south because this species of rattan can be grown using the same practices as growing vegetables.

5. The Role of Rattan in the Country's Economy

Rattan is a plant whose parts are almost all useful. For example, the skin is good for weaving and also as a binding material and for furniture making; the canes are used to make household materials, furniture and sport equipment; the cores of the canes are good for weaving and the leaves are used as curtains, roofing material, for weaving, etc. In general all the parts serve as basic raw material for making furniture and utensils that generate good income for the population and also for the country.

The important role of rattan in the country, according to its utilization and economic value can be classified as follows:
(a) Domestic utilization for
   * furniture making;
   * weaving and household materials;
   * decorative purposes;
   * construction, especially in rural areas;
   * food, etc.
(b) Utilization outside the country.

Rattan is exported only in the form of furniture and production of rattan furniture for export is supported by the Board of Investment. In 1999, the total export value of rattan furniture amounted to B 121 million.

Rattan utilization, both locally and for exports, has a positive economic impact for the following reasons:

- Generates earnings;
- Increases job opportunities for the population, extending from cottage industry level to the manufacturing plant level;
- Decreases imports of furniture made of any material;
- The trend to establish rattan plantations for raw material supply in the near future will help increase the forested area and also provide better job opportunities for the local people.

6. Resource Base

Even though rattan products generate a good income for the country, there is always the problem concerning the supply of raw materials, especially from the forest. As mentioned before, collection of rattan from the natural forest is prohibited by a number of Forest Acts. Therefore all rattan used by the industry is mainly imported from abroad. Domestication would seem necessary but, since the Forest Acts impose many constraints to obtaining permission to establish plantations and the right to harvest the rattan from the plantations, the private sector is not willing to invest in such an uncertain venture and, therefore, prefers to order the raw materials from abroad for their regular need and processing in the factories. The Government has become aware of this problem but its efforts to plant more rattan in the natural forest every year, mainly for conservation purposes, has not alleviated the raw material problem. Intercropping with rubber trees has also been tried, but it does not seem to be popular among the farmers. Actually the area of rubber plantations in the country [12 million rai (1.92 million ha)] has great potential and would be a good opportunity for the farmers to get value added from their plantations. In this case, proper extension programmes and management from the Government sector would be very important and should be a good chance to increase the rattan plantation area of the country. The only domestication which seems successful is small-scale production of edible rattan shoots (*C. viminalis*) in the northeastern part of Thailand, but this kind of production is of local importance since it meets only the specific taste of the people living there.

7. Harvesting and Processing

In the past rattans were harvested from the natural forests, mainly in the protected areas, by the villagers living along the border of the forest. Harvesting was done in groups of 10–20 people who would go into the forest and find the rattan, harvest by cutting it at the base and pulling it down from the tree, cut it into pieces and pull the cut canes, sometime by elephant, out of the forest. The harvesting period could be any time in the year except during the rainy season.

After rattan was harvested from the forest, it was piled together. The small stems were put to dry in the sun while the large-diameter canes would be transported directly to the manufacturing plant. When the material was used in local households or at the village level, the large canes would be boiled in diesel or used lubricating oil while the smaller canes would be dried in the sun and fumigated with sulphur. At the industrial level, processing was done by boiling the cane in a mixture of diesel and used lubricating oil. Then they were dried in the sun for one to two weeks and kept in storage until use. But nowadays, since the raw material used in rattan manufacturing industries is mostly imported as semi-processed material ready to be used in the production process, the canes are no longer treated in any factory. Treatment to prevent borer and stain fungi, if needed, consists in adding insecticide, mostly of the synthetic pyrethroid group, and fungicide, such as IF-1000 and TCMTB, to the boiling oil.

8. Export and Import Situation

Thailand imports rattan in the form of cane and furniture. In both cases the quantity fluctuates from year to year, especially for rattan canes. Export of local rattan canes has been prohibited since 1978. The quantity and value of manufactured products are also decreasing (Appendix 3).

Since rattan furniture from Thai manufactures is of high quality, the demand for it is high but production is decreasing due to lack of supply of the raw material. Major consumers are worldwide (Appendix 6). Major rattan manufacturing companies, working for the export market, were granted investment promotion certificates. According to the Government policy, the company that is granted an investment promotion certificate will obtain certain benefits and rights in order to promote its products. There is, however, an important restriction so that the company is not permitted to sell its products within the country.

In Thailand there are skilled labour and manufacturers with proper technical knowledge to operate and assemble the various product lines of rattan. Rattan products, mainly in the form of furniture, are exported to Germany, United States, France (Appendix 4) and various other countries worldwide, totalling 115 countries in 1999 (Appendix 5). Statistics from the Custom Department indicate that in 1999 the top five nations that imported rattan from Thailand were Germany, United States, France, Maldives, and Belgium. Altogether they spent B 67 million out of a total export value of B 121 million of rattan products. Therefore, if Thailand could have a sufficient and steady supply of raw material, the manufacture of products from rattan could bring benefits to the country's economy and also create more job opportunities for the population. Unfortunately, since the raw materials have to be imported and the supply is sometimes limited and uncertain, the income generated from this product fluctuates accordingly.

The key stakeholder in rattan trade is the private sector. A few companies, such as Hawaii Thai Co. and O.S.C. Industry Co., are among the major exporters and have been present on the international market for decades. Trade in rattan and access to the international market is currently very complicated due to competition among countries and inadequate raw material supply. As a consequence, many manufacturers are turning to the utilization of other raw materials that are more easily accessible, such as straw, water hyacinth or even plastic.

To remedy the above situation, promotion of rattan plantations for sustainable raw material supply is recommended. However, since rattan plants need at least 8-10 years before harvesting, consideration should be given to looking into the utilization of other raw materials as rattan substitutes and action should be taken to maintain a proper level of production.

9. Policy and Institutional Aspects

9.1 Social aspects

In the past rattan played an important role for the villagers from a social point of view. Normally groups of people in the village would gather and produce various household products from rattan. These products were beautiful and also showed the art and culture of the nation; they were either sold locally or sent to the markets somewhere else within the country. Later on, because of their good quality, they were shipped abroad and became very popular. Orders came from countries both in Asia and Europe. Trade in rattan products was prosperous for some time and it made a good return both at the local and national level.

However, in the past few years, the trend of production and marketing of rattan at the village level has been declining. Most of the people in the younger generation have a different attitude; they are not interested in the art and craft of the rattan as was the older generation. They feel that rattan production does not provide adequate income in terms of number of hours spent to make one piece of product. Therefore the only people in the village who are still working with rattan are the old and, therefore, production is decreasing and, in some villages, this activity is being replaced by others.

9.2 Institutional aspects

In the past rattan was harvested from the wild by various groups of people who had been given the right from concessions by the regional forest offices. Presently the forest is officially closed and no more concessions are given. The rattan being used by the industry comes from abroad. In order to promote rattan manufacturing, some manufacturers are supported by the Government through the Board of Investment which grants them some privileges for their investment in this business.

However, since the resources are diminishing and skilled labour is also decreasing, the Government may need to consider ways and means to overcome these problems. To date the Department of Industrial Promotion, Ministry of Industry, has tried to educate the people through extension programmes while the Royal Forest Department is planning to plant more rattan every year. However, participation from the people is lacking and the result generally is not satisfactory.

Presently the Government sector which mainly deals with rattan is the Department of Industrial Promotion, while the Royal Forest Department is the only organization that looks after the national forest where most of the rattan grows . Since most of the Thai forests are Government-owned, they are under the supervision of the forestry officials under the Forest Act, the National Park Act and the Wildlife Sanctuary Act. According to these laws, it is almost impossible to extract rattan from the forest. In the past the private sector seemed eager to plant rattan but there were limitations, such as long-term investment needed, inappropriate area for planting, strict laws and regulations regarding rattan extraction from the forest, etc. For these reasons, the industry prefers to import cane from neighbouring countries (Appendix 6). In the near future the Forest Plantation Act and the Community Forestry Act are expected to be put into effect and this may be a good opportunity to enhance rattan growing since these acts would allow landowners to plant rattan on their land; however, the area where rattan can be planted and the allowable quantity are not known.

Figure 10: Processing of rotan manau stems in oil, Jambi Sumatra (Dransfield)

10. What about the Future?

The question is, how can we manage to get enough supply of canes to meet demand? A possible answer could be the sustainable management of natural stands. But how? Research and management could provide the answer.

In view of the increasing depletion of the natural forest, the socio-economic importance of rattan for the villagers and the high export value of rattan products, it is necessary to identify the areas where new research needs to be carried out, as well as to continue research along the following lines:

- basic research on species, ecology and silviculture;
- plantation techniques, especially intercropping;
- provenance trials;
- establishment of seed orchards;
- biotechnology; gene fusion;
- preservation;
- product development, etc.

When appropriate knowledge and techniques are available through research, then management plans could be set up. Plans must be drawn up both for short-term and long-term objectives along, with their operational procedures. The plans must indicate how to achieve the goal that whenever a quantity of the rattan is harvested, at least the same quantity of rattan is grown in the shortest possible time.

11. Conclusions

In the future the demand for rattans will increase, but the difficulty will be to ensure that supply will meet demand. In order to cope with this problem, various types of research should be carried out. However, organizations and researchers interested in rattan are very few and, therefore, very little research on rattan has been carried out in Thailand so far.

There is a great need for basic research, especially with regard to growth, seeds, regeneration methods, appropriate silvicultural methods, including properties of potentially valuable species. Results from such research could lead to the successful establishment of rattan plantations that could satisfy the increasing need for the raw material. In addition, appropriate management schemes must be set up in order to ensure that there are enough funds, manpower and proper operational procedures for sustainable production of rattan from the natural forest.

REFERENCES

Custom Department. 2000. *The import and export statistics of Thailand.*

Uhl, N.W. & Dransfield, J., 1987. *Genera palmarum: a classification of palms based on the work of H.E.Moore Jr.* pp610. The International Palm Society & the Bailey Hortorium, Kansas.

Vongkaluang, I., 1986. General morphology of rattan In: *Proceedings of the National Rattan Seminar, Bangkok, Thailand, 13-14 Nov. 1986.* pp. 55-72.

Appendix 1
Checklist of Thai rattans based on an unpublished list by J. Dransfield (*pers. comm.*)

Genus	Species
1. *Calamus*	1. *C. acanthophyllus* Becc.
	2. *C. acanthospathus* Griff.
	3. *C. arborescens* Griff.
	4. *C. axillaris* Becc.
	5. *C. balingensis* Furtado
	6. *C. blumei* Becc.
	7. *C. bousigonii* Becc. var. *bousigonii*
	7a. *C. bousigonii* Becc. var. *smitinandii* J. Dransf.
	8. *C. burkillianus* Becc. ex Ridley
	9. *C. caesius* Blume
	10. *C. castaneus* Griff.
	11. *C. concinnus* Mart.
	12. *C. densiflorus* Becc.
	13. *C. diepenhorstii* var. *diepenhorstii* Miq.
	14. *C. erectus* Roxb.
	15. *C. erinaceus* var. erinaceus (Becc.) J. Dransf.
	16. *C. exilis* Griff.
	17. *C. flagellum* Griff.
	18. *C. godefroyi* Becc.
	19. *C. griseus* J. Dransf.
	20. *C. guruba* Buch. Ham. ex Mart.
	21. *C. henryanus* Becc.
	22. *C. insignis* var. *insignis* Griff.
	22a. *C. insignis* var. *longispinosus* J. Dransf.
	22b. *C. insignis* var. *robustus* (Becc.) J. Dransf.
	23. *C. javensis* Blume
	24. *C.* latifolius Roxb.
	25. *C. laevigatus* var. *laevigatus* Mart.
	26. *C. leucotes* Becc.
	27. *C. longisetus* Griff.
	28. *C. luridus* Becc.
	29. *C. manan* Miq.
	30. *C. nambariensis* Becc.
	31. *C. oligostachys* Evans et al.
	32. *C.* ornatus Blume var. ornatus
	33. *C. oxleyanus* var. *montanus* Furtado
	33a. *C. oxleyanus* var. *oxleyanus* Teijsm. & Binn. ex Miq.
	34. *C. palustris* Griff.
	35. *C. pandanosmus* Furtado
	36. *C. peregrinus* Furtado
	37. *C. poilanei* Conrard
	38. *C. rudentum* Lour.
	39. *C. scipionum* Lour.
	40. *C. sedens* J. Dransf.
	41. *C setulosus* J. Dransf.
	42. *C. siamensis* Becc.
	44. *C. solitarius* Evans et al.
	45. *C. speciosissimus* Furtado
	46. *C. spectatissimus* Furtado
	47. *C. tenuis* Roxb.
	48. *C. tetradactylus* Hance
	49. *C. viminalis* Willd.
	50. *C. viridispinus* Becc. var. *viridispinus*
	51. *C. wailong* Pei & Chen

191

Genus	Species
2. *Ceratolobus*	1. *C. subangulatus* (Miq.) Becc.
3. *Daemonorops*	1. *D. angustifolia* (Griff.) Mart.
	2. *D. didymophylla* Becc.
	3. *D. geniculata* (Griff.) Mart.
	4. *D. grandis* (Griff.) Mart.
	5. *D. jenkinsiana* (Griff.) Mart.
	6. *D. kunstleri* Becc.
	7. *D. leptopus* (Griff.) Mart.
	8. *D. lewisiana* (Griff.) Mart.
	9. *D. macrophylla* Becc.
	10. *D. melanochaetes* Blume
	11. *D. monticola* (Griff.) Mart.
	12. *D. propinqua* Becc.
	13. *D. sabut* Becc.
	14. *D. sepal* Becc.
	15. *D. verticillaris* (Griff.) Mart.
3. *Korthalsia*	1. *K. flagellaris* Miq.
	2. *K. laciniosa* (Griff.) Mart.
	3. *K. rigida* Blume
	4. *K. rostrata* Blume
	5. *K. scortechinii* Becc
4. *Plectocomia*	1. *P. elongata* Mart. ex Blume
	2. *P. kerrana* Becc.
	3. *P. pierreana* Becc.
5. *Plectocomiopsis*	1. *P. geminiflora* (Griff.) Becc.
	2. *P. wrayi* Becc.
6. *Myrialepis*	1. *M. paradoxa* (Kurz) J. Dransf

Appendix 2
Economic rattan species of Thailand

A. Large size
1. *Calamus longesitus*
2. *Calamus wailong (Wai Nam-pueng)*
3. *Calamus erectus*
4. *Calamus manan*
5. *Calamus peregrinus*

B. Small size
1. *Calamus caesius*
2. *Calamus blumei*
3. *Calamus javensis*
4. *Calamus pandanosmus*
5. *Calamus densiflorus*

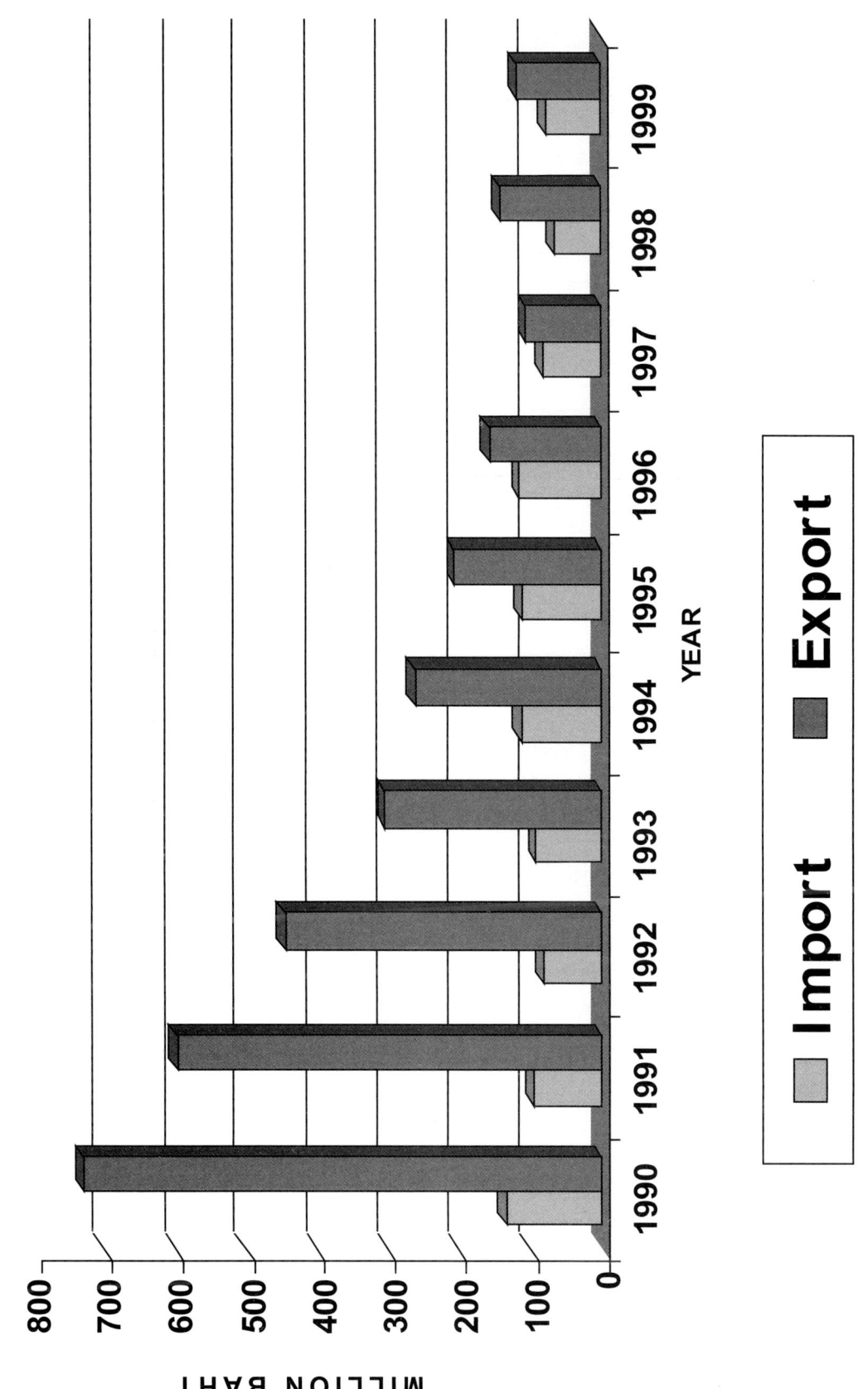

Appendix 3 : The import & export value of rattans (1990-1999)

Appendix 4

Top ten countries importing rattan from Thailand (1999)

Number	Country	Value (B)
1	Germany	23 441 797
2	United States	16 164 652
3	France	16 060 507
4	Maldives	6 023 945
5	Belgium	5 631 323
6	Japan	5 549 321
7	Spain	4 744 176
8	Netherlands	4 436 793
9	Singapore	3 763 146
10	Australia	3 306 804

Top ten countries exporting rattan to Thailand (1999)

Number	Country	Value (B)
1	Indonesia	14 036 960
2	Singapore	12 003 945
3	Myanmar	11 262 512
4	China	11 084 927
5	Malaysia	6 898 585
6	Lao, People's Dem. Rep.	3 694 282
7	Kampuchea	3 623 348
8	Hong Kong	3 519 066
9	Viet Nam	885 907
10	China's Province of Taiwan	577 691

Appendix 5

Countries importing Thai rattan

1	Afghanistan	30	Finland	59	Maldives	88	Seychelles
2	Angola	31	France	60	Malta	89	Singapore
3	Argentina	32	French Polynesia	61	Mauritius	90	Slovakia
4	Australia	33	French Southern and Antarctic Territories	62	Mexico	91	Slovenia
5	Austria	34	Gambia	63	Morocco	92	Solomon Islands
6	Bahrain	35	Georgia	64	Mozambique	93	South Africa
7	Bangladesh	36	Germany	65	Myanmar	94	Spain
8	Belgium	37	Greece	66	Nepal	95	Sri Lanka
9	Bolivia	38	Guadeloupe	67	Netherlands	96	Sudan
10	Brazil	39	Guam	68	New Caledonia	97	Swaziland
11	Brunei	40	Guatemala	69	New Zealand	98	Sweden
12	Bulgaria	41	Hong Kong	70	Niger	99	Switzerland
13	Cambodia	42	Hungary	71	Norway	100	Tanzania
14	Canada	43	Iceland	72	Oman	101	Togo
15	Chile	44	India	73	Pakistan	102	Trinidad and Tobago
16	China	45	Indonesia	74	Panama	103	Tunisia
17	Colombia	46	Ireland	75	Papua New Guinea	104	Turkey
18	Cook Islands	47	Israel	76	Peru	105	Uganda
19	Costa Rica	48	Italy	77	Philippines	106	United Arab Emirates
20	Cuba	59	Japan	78	Poland	107	United Kingdom
21	Cyprus	50	Jordan	79	Portugal	108	Uruguay
22	Czech Republic	51	Kenya	80	Puerto Rico	109	United States territories in the Pacific
23	Democratic People's Republic of Korea	52	Kuwait	81	Qatar	110	United States of America
24	Denmark	53	Lao, People's Democratic Republic	82	Republic of Korea	111	Venezuela
25	Djibouti	54	Lebanon	83	Réunion	112	Viet Nam
26	Dominica	55	Libyan Arab Jamahiriya	84	Russian Federation	113	Yemen
27	Egypt	56	Lithuania	85	Samoa	114	Yugoslavia
28	Estonia	57	Luxembourg	86	Saudi Arabia	115	Other countries
29	Fiji	58	Malaysia	87	Senegal		

Appendix 6

Countries exporting rattan to Thailand

1	Australia		20	Kyrgyzstan
2	Austria		21	Lao, People's Democratic Republic
3	Bangladesh		22	Malaysia
4	Barbados		23	Mongolia
5	Belgium		24	Myanmar
6	Cambodia		25	Netherlands
7	Canada		26	Norway
8	China		27	Pakistan
9	Colombia		28	Philippines
10	Comoros		29	Republic of Korea
11	Denmark		30	Singapore
12	France		31	Spain
13	Germany		32	Sri lanka
14	Hong Kong		33	Sweden
15	Iceland		34	Switzerland
16	India		35	United Kingdom
17	Indonesia		36	United States of America.
18	Italy		37	Viet Nam
19	Japan		38	Other countries

Appendix 7

Rattan plantation under the supervision of Silvicultural Division, Forest Technical Office, Royal Forest Department.

Year	Rattan plantations (rai)[1]
1991	1 000
1992	1 000
1993	1 000
1994	1 000
1995	1 000
1996	1 000
1997	1 000
1998	1 000
1999	1 000

1000 rai = 160 ha

Rattan seed production area:	Krabi	1 300 rai
	Chuntaburi	350 rai
Rattan plantations:	Suratthani	2 800 rai
	Trang	1 800 rai
	Nakhonrachasima	400 Rai

(1 ha   =   6.25 rai)
(1 rai   =   1 600 m^2)

RATTAN IN EAST KALIMANTAN, INDONESIA: SPECIES COMPOSITION, ABUNDANCE, DISTRIBUTION AND GROWTH IN SOME SELECTED SITES

Johan L.C.H. van Valkenburg

1. Introduction

Apart from timber, rattan is the most important forest-derived commodity in East Kalimantan. The multitude of species, that all have their specific use in both the traditional way of life and commercial trade, makes rattan stand out among non-timber forest products.

Despite the longstanding exploitation of rattan and early records of rattan gardens in East Kalimantan (Endert, 1927; Van Tuil, 1929), the resource remains poorly known. Whereas rattan floras for Sabah and Sarawak exist (Dransfield, 1984, 1992a) none is available for East Kalimantan. This is a serious handicap since rattan species tend to display a rather high degree of endemism, e.g. of the 107(--109) species and varieties reported for Sarawak, 63 species and 4 varieties are endemic in Borneo (Dransfield, 1992a, 1992b), and 23 species and 1 variety are endemic in Sarawak (Pearce, 1989; Dransfield, 1992a). For the island of Borneo as a whole, so far 146 rattan species have been recorded (Dransfield, 1992c). For Sabah, 82 species and varieties are reported, with 8 endemic species and 2 endemic varieties (Dransfield & Johnson, 1989). Species richness in West Kalimantan is assumed to be similar to that of Sarawak, and in East Kalimantan similar to that in Sabah (Dransfield, 1992c).

If rattan is to be used as a renewable resource, knowledge of the growth of rattan plants is essential to devise sustainable exploitation practices. The growth of non-exploited rattan in forest of various successional stages (see 3.2) is compared with the growth of rattan plants subjected to harvesting (see 3.3). Knowledge with respect to impact of harvesting on the rattan resource is essential for sustainable exploitation.

The important role of light for establishment and growth of rattan plants was observed in planting trials of four species (*Calamus caesius, C. manan, C. scipionum* and *C. trachycoleus*) in Malaysia and Indonesia (Wan Rhazali Wan Mohd. *et al.*, 1992; Wong & Manokaran, 1985). The effect of logging resulting in an increase of light, however, appeared to have negative effects on the rattan resource (Abdillah Roslan & Phillips, 1989; Kiew & Hood, 1991), whereas harvesting intensity affects the vitality of plants (Nandika, 1938; Kiew & Hood, unpublished)

In this paper attempts are made to answer the following questions:

- What is the species composition and abundance of rattan in the various sites?
- Does logging affect the species composition and abundance of rattan?
- Does logging have an effect on the subsequent increment of rattan both at the population and individual plant level?
- What is the effect of harvesting on the performance of a rattan clump?

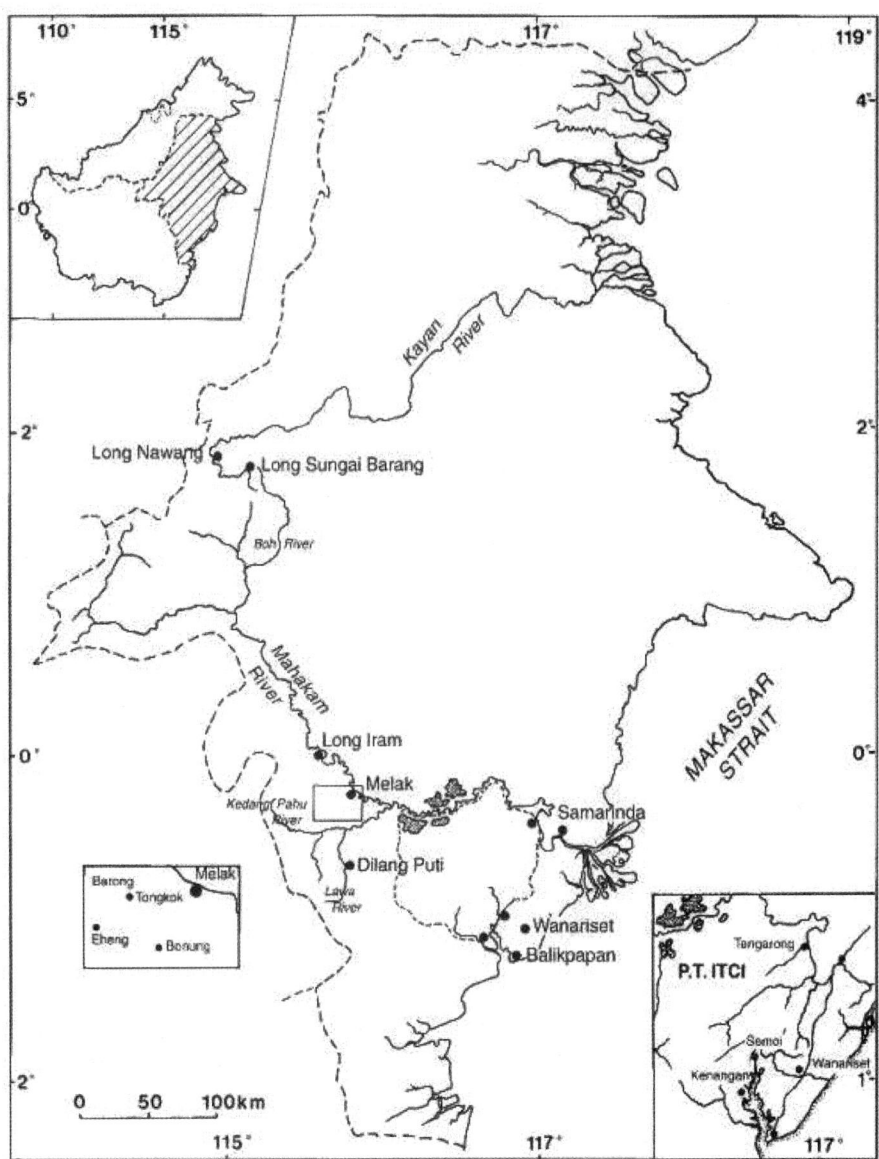

Figure 1.   Map of the study area

2. Methods

2.1 Permanent plots

Permanent plots were set up in 1992 to assess species composition, abundance and growth of rattan. Plots were located in the Wanariset area, the ITCI concession, and the Apo Kayan (Figure 1; for details see Van Bremen *et al.*, 1990; Van Valkenburg, 1997). Within each site the plots were situated in a topo-sequence from ridge top to valley floor. In the ITCI concession both primary and logged-over plots were studied.

In the Wanariset forest part of plot Matthijs (5100 m²) was inventoried. In the ITCI concession permanent plots established by the concessionaire were used, and either the entire plot or part of the plot was inventoried as follows: 72-1 (5000 m²), 72-2 (5000 m²), 76-3a (5000 m²), 76-3b (7500 m²), 76-4 (3200 m²) and 77-2 (1600 m²). In the Apo Kayan four plots of 1600 m² were established.

At all sites rattan plants in the plots were labelled with numbered aluminium tags using either nylon fish-line or iron-wire. Within a population, only plants with a minimum cane length of 50 cm were included. The growth stage of all shoots was recorded.

The abundance of rattan in the permanent plots was studied from three different aspects:
- number of plants/hectare
- number of plants with mature canes/hectare
- total number of mature canes/hectare

Although from an ecological point of view the total number of plants is most important, from the economic point of view the number of mature canes is more relevant to this study.

2.2 Additional plots

As the total area of the permanent plots in the Apo Kayan was considered not sufficient to obtain a good impression of rattan abundance over larger areas (i.e. the Apo Kayan in general), a line plot of 4x1200 m was set up in October 1992. A greater variation in topography and growth stages of the forest was included and effects of the often very local occurrence of rattan plants were compensated. Rattan plants with mature canes of all species and the total number of mature canes were recorded. Plants belonging to (potentially) commercial species were labelled and the number of shoots belonging to the different growth stages counted.

Also in the ITCI concession some additional inventory work was carried out to compensate for effects of the often local occurrence of rattan plants and species. In 1994 two plots of 1600 m^2 each were inventoried inside the permanent plot 72-8 situated in primary forest. All rattan plants with a minimum cane length of 50 cm were labelled and growth stage of all shoots was recorded.

2.3 General collecting

In addition to recording rattan species in the permanent plots, supplementary collections and observations were made in the three main research sites, Wanariset forest, ITCI concession and Apo Kayan in the vicinity of Long Sungai Barang. Other qualitative inventories of varying intensity were also conducted in the villages of Dilang Puti (June/July 1993) and Eheng (May 1994) upstream from the lakes on the Mahakam river and in the Bahau region. In the Bahau region, the inventory was carried out during a visit at the basecamp of World Wildlife Fund Indonesia, in the Kayan Mentarang Nature Reserve near Long Alango (November/December 1991). The listing of rattan species in the Meratus Mountains is based on herbarium specimens available at the Wanariset Herbarium.

Solving taxonomical problems was not the aim of this research. However, three new species were described (Van Valkenburg, 1995). Plants belonging to the species complex of *Calamus pogonocanthus, C. erioacanthus*, and *C. semoi* have for convenience been divided into three artificial taxa and are referred to as *Calamus pogonocanthus* 1, 2 and 3. The use of all three taxa is the same: namely as strips for good quality binding

2.4 Growth measurements

As a rattan seedling grows, there is a gradual increase in stem diameter, before the stem starts significantly to grow upwards. This process can take many years depending on the species and the light condition in which it grows. This initial phase is known as establishment growth, and a vigorous plant in this initial phase is referred to as rosette stage/plant.

Since rattan stems often become entangled in the canopy, the length of the canes could not be accurately recorded without pulling them down. This was not done because it would result in an altered light environment for the crown-leaves and would thereby influence subsequent growth of the plant. An alternative method for recording growth over a 2-year period at time intervals of 12 months was therefore developed. All shoots were classified as belonging to a specific growth stage as defined in Table 1. Growth was defined as the passage from one stage to the next.

At the beginning of the study only plants with a minimum cane length of 50 cm were recorded. If in the following years a plant no longer had a shoot with a minimum cane length of 50 cm, but suckers were present, they were still monitored. Not all mature canes are harvestable, since a minimum length of 3 meters of sufficiently mature cane is required. The distinction into harvestable and non-harvestable canes was only made at the end of the study, when canes were harvested in the Apo Kayan.

However, shoots of some species could not be classified in this way; e.g., *Calamus conirostris* and *Daemonorops hystrix* often flowered when cane length was still less than 200 cm and the lower leaf sheaths were still green. For these plants the flowering shoots were classified as mature. The lower leaf sheaths of *Calamus laevigatus* are normally still green when the cane has reached the canopy and is more than 30 m long. Since the cane can be harvested in this state, it was classified as mature.

The mature plants present in the Apo Kayan line plot were remeasured after 17 months.

Table 1. Growth stages of rattan

Stage	Length of cane	Appearance of lower leaf-sheath
Sucker	0 - 49 cm	green
Juvenile	50 - 199 cm	green
Immature	> 200 cm	green
Mature	> 200 cm	dead or bare cane exposed

2.5 Effect of disturbance on recruitment

In order to judge the significance of disturbance on the recruitment of rattan, the twelve plots were compared using a procedure fit-curve (Genstat 5, 1990). Each plot was classified as either disturbed or undisturbed depending on the presence of degrading canopy trees, a chablis, or serious effects of drought. Furthermore plot 72-2 was studied in detail, whereby all 50 subplots were tallied according to the degree of disturbance (three classes distinguished: 0, 1 and 2, with disturbance increasing from 0 to 2).

2.6 Harvesting experiment

In a non-exploited state a rattan clump might invest in elongation of already mature canes that have reached the canopy. Or in response to changes in the environment, an acceleration of the growth of dominant shoots might prevail. If the mature canes are harvested, growth of the remaining shoots may be released. Acceleration of the growth stage of the minor/young shoots might be induced, or new suckers could be formed.

In addition to the harvesting of canes in the Apo Kayan permanent plots and line plot at the end of the study in 1994, rattan plants outside these plots were subjected to maximum harvesting. The experiment was not only conducted to obtain a better estimate of harvestable length per plant but in particular to supply information on the effect of harvesting on the growth of remaining shoots.

In the Apo Kayan rattan plants of three locally important and (potentially) commercial species: *Calamus javensis* (*n*=31), *Calamus ornatus* (*n*=33), and *Daemonorops sabut* (*n*=33) were harvested in May 1993. Each plant was labelled with an aluminium tag. The total number of shoots belonging to the different growth stages was recorded and habitat of the plant was classified in terms of silvigenetic stage (Oldeman, 1980, 1990) of the surrounding forest, geomorphology/topography (valley floor/middle slope/upper slope/ ridge crest) and steepness of slope. Harvestable mature canes, with a minimum length of 3 m, were cut and their total length measured. The remaining shoots were tagged according to their growth stage. After 12 months the growth response of the plants and the individual shoots was recorded, and canes that had attained harvestable size were cut.

3. Results

3.1 Species composition, abundance and distribution

In general neither species composition nor abundance of rattan are evenly distributed.

Although in the Middle Mahakam region distribution and abundance of rattan species is strongly influenced by man, in other areas of the province it is presumed to be the result of natural phenomena. Influence of man on abundance and distribution is increasing due to the extensive logging activities and accompanying agricultural development.

Species can be geographically and/or ecologically (e.g. *Korthalsia flagellaris* confined to peat swamp forest) limited in their distribution. Variation in climate, natural boundaries or recent speciation may be reasons for their geographical limitation. As yet no information on distribution of rattan in East Kalimantan is available, because the province has been poorly collected. Of the 59 species I collected, only 21 were already represented by herbarium specimens from East Kalimantan in the collection at Leiden or Wanariset Herbarium. The information on species distribution presented in Table 2 is, therefore, based on the surveys in a limited number of sites in East Kalimantan.

Personal observations indicate that the following species are of common occurrence in East Kalimantan: *Calamus javensis, C. ornatus, C. pilosellus, C. pogonocanthus* 3, *Ceratolobus concolor, Daemonorops didymophylla, D. fissa, D. hystrix, D. korthalsii, D. sabut, Korthalsia echinometra, K. ferox* and *Plectocomiopsis geminiflora*. However, there is a difference in preferences for moisture regime and degree of disturbance.

Some species are restricted to one of the areas where they were locally common whereas other species were simply very rare. *Daemonorops atra, D. crinita, D. pumila, Calamus nigricans* and *Ceratolobus subangulatus* were common species where they occurred. While *Calamus gonospermus, C. hispidulus, C. paspalanthus, C. sarawakensis, Korthalsia hispida* and *Plectocomiopsis mira* occurred at very low densities.

Calamus mattanensis and *C. tomentosus* appear to be confined to the western part of East Kalimantan whereas *C. blumei, C. pandanosmus, C. rhytidomus, C. fimbriatus* and *Korthalsia furtadoana* were only encountered in the eastern part.

Some species, e.g. *Calamus laevigatus*, and *C. ornatus*, are more or less evenly distributed. While others are always encountered in groups e.g. *Calamus convallium, C. pandanosmus, C. caesius* and *Korthalsia cheb*. On dry ridges *Calamus conirostris* can become the dominant species; while in the Wanariset area *C. marginatus* takes this dominant role.

Table 2. List of rattan species recorded in different areas of East Kalimantan in the period 1991-1994 (for details see text)

	Wanariset	ITCI	Apo Kayan	Meratus	Bahau	Mahakam
Calamus blumei	*	*		□		
C. caesius	□	*			□	□
C. conirostris		*	*			
C. convallium			□			
C. fimbriatus	*	*				□
C. flabellatus	*	*				□
C. gonospermus			□			
C. hispidulus			*			
C. javensis	*	*	*		□	□
C. laevigatus	*	*	*		□	
C. manan						□
C. marginatus	*	*				□
C. mattanensis			*		□	
C. muricatus			*	□	□	
C. nigricans	*					
C. optimus						□
C. ornatus	*	*	*		□	□
C. pandanosmus	□	*				□
C. paspalanthus				□		
C. pilosellus	*		*		□	□
C. pogonocanthus 1			*			
C. pogonocanthus 2	*					
C. pogonocanthus 3	□	*	□		□	
C. praetermissus	□					
C. pseudo-ulur					□	
C. rhytidomus		*				□
C. sarawakensis	□					
C. scipionum		*				□
C. tomentosus			*		□	
C. trachycoleus						□
Ceratolobus concolor	*	*	*		□	□
C. subangulatus						□
Daemonorops atra			*			
D. collarifera				□		
D. crinita						□
D. cristata		*				
D. didymophylla	*	*	*		□	
D. elongata	□					
D. fissa	□		*		□	□
D. hystrix	*	*	□		□	
D. korthalsii	*	*	□		□	
D. periacantha	□					
D. sabut	*	*	*		□	□
D. pumila			*			
Korthalsia cheb			□			
K. debilis						□
K. echinometra	*	*	□			□

	Wanariset	ITCI	Apo Kayan	Meratus	Bahau	Mahakam
K. ferox	□	*	*		□	□
K. flagellaris	□					
K. furtadoana	*	*				□
K. hispida					□	
K. rigida	*	*				
K. rostrata					□	□
Korthalsia sp.						□
Plectocomia mulleri			□			
Plectocomiopsis geminiflora	*	□	*		□	□
P. mira	*					
Total no. species	30	24	26			

□ :	Only present outside permanent plots
Wanariset:	Plants present in the Wanariset research forest (incl. plot Matthijs) and vicinity including Sungai Wain area and Inhutani I concession
ITCI :	Plants present in the permanent research plots and vicinity of these plots
Apo Kayan:	Plants present in the Apo Kayan primarily vicinity of Long Sungai Barang, and the primary forest eastward of the Kayan River
Meratus :	Some sporadic collections of the Meratus mountains in the PT ITCI concession area
Bahau :	Plants collected in the Kayan Mentarang Nature reserve, during a visit at the WWF basecamp near Long Alango
Mahakam:	Plants collected in the vicinity of Dilang Puti and Eheng

Wanariset forest

In plot Matthijs 20 rattan species were found on 5100 m^2. Of these 18 species had mature canes (Table 3). Three species were not encountered in the Apo Kayan or ITCI permanent plots: *Calamus nigricans, C. pogonocanthus* 2, and *Plectocomiopsis mira*.

Calamus marginatus, usually a single-stemmed species of dry upper and middle slopes, was the dominant species accounting for 36 % of all plants. More than half of these plants had mature canes representing 25 % of total mature canes present in the plot (Table 3). *Korthalsia rigida* is second in frequency (15 % of all plants) and almost half of its plants had mature canes, representing 15 % of total mature canes.

Five species each represented 5-10 % of all plants: *Daemonorops sabut, Korthalsia furtadoana, Calamus flabellatus, C. nigricans, and C. pogonocanthus* 2. Four clustering species each represented 5-10 % of the mature canes in the plot: *C. javensis, C. flabellatus, K. furtadoana* and *D. sabut*.

ITCI concession area

A total of 24 rattan species (see Table 1) were encountered in the ITCI permanent plots (30,500 m^2). Six of these species (*Calamus caesius, C. pandanosmus, C. pogonocanthus* 3, *C. rhytidomus, C. scipionum* and *Daemonorops cristata*) were present neither in the Apo Kayan or Wanariset permanent plots. The primary and logged-over plots had 14 species in common.

Table 3. Presence of rattan species in primary forest, Wanariset, plot Matthijs (in 1994)
Number of plants ha^{-1} (a), plants with mature canes ha^{-1} (b) and total number of mature canes ha^{-1}(c)

	a	b	c
Species :			
Calamus blumei	4	2	2
C. flabellatus	31	24	33
C. javensis	18	16	41
C. laevigatus	2	0	0
C. marginatus	192	98	104
C. ornatus	2	0	0
C. pilosellus	16	10	18
C. pogonocanthus 2	41	14	18
Calamus nigricans	31	10	18
Calamus fimbriatus	12	6	6
Ceratolobus concolor	12	6	8
Daemonorops didymophylla	8	4	4
D. hystrix	8	2	2
D. korthalsii	2	2	2
D. sabut	25	14	22
Korthalsia echinometra	10	6	16
K. furtadoana	27	14	39
K. rigida	80	37	61
Plectocomiopsis geminiflora	4	4	6
P. mira	2	2	2
Total no. of plants/canes	528	268	400
Total no. of species	20		

Primary plots

A total of 18 species were encountered on 18,900 m^2 (Table 4). *Daemonorops cristata* was only found in the ITCI primary plots. Two other species were restricted to the ITCI permanent plots: *Calamus pandanosmus* and *C. rhytidomus*.

On ridge crest and upper slope, the abundance of rattan plants appears to be lower than on middle slopes. The total number of plants in the plots was low, and differences between plots in similar habitats were considerable.

In plot 76-3b a dense under-storey of *Borassodendron* palms was present in 21 % of the 100 m^2 subplots. Eighty-seven percent of the subplots with *Borassodendron* was void of rattan as compared with 47 % of the other subplots. The dense leaf litter that prevents seedling establishment and competition for light are plausible explanations for the absence of rattan.

Table 4. Presence of rattan species in permanent plots in primary forest in the ITCI concession area (in 1994) Number of plants ha⁻¹ (a), number of plants with mature canes ha⁻¹ (b) and total number of mature canes ha⁻¹(c) Plots arranged in a topo-sequence from ridge crest to valley floor

Species :	76-3a			76-3b			72-8a			72-8b			76-4a			76-4b		
	a	b	c	a	b	c	a	b	c	a	b	c	a	b	c	a	b	c
Daemonorops hystrix	4	4	4	3	1	1												
Calamus marginatus	22	4	4	22	15	15												
Calamus flabellatus	12	6	6	31	16	22	6	0	0	6	0	0	6	6	6	6	0	0
Calamus conirostris	2	0	0	5	3	3	38	19	25	19	6	6	6	6	6	13	6	6
Calamus rhytidomus	2	2	4	3	3	3	19	13	13	6	6	13	6	6	6	25	0	0
Calamus blumei				1	1	1							6	0	0			
Daemonorops cristata				3	1	1												
Korthalsia echinometra				1	1	4												
Calamus pogonocanthus 3				3	3	3				6	0	0	6	0	0	13	0	0
Ceratolobus concolor				7	1	1							13	0	0	6	0	0
Daemonorops sabut							13	6	13	6	6	6	13	0	0	6	0	0
Korthalsia rigida							50	13	13	6	0	0	69	0	0	31	0	0
Korthalsia furtadoana							19	0	0	19	0	0	19	0	0			
Korthalsia ferox							6	6	6	6	0	0						
Daemonorops didymophyl-la							6	6	6	6	0	0						
Calamus pandanosmus							6	6	6									
Calamus ornatus							6	0	0									
Calamus fimbriatus																6	0	0
Total no. plants/canes	42	16	18	77	44	55	163	69	81	81	19	25	144	19	19	106	6	6
Total no. species		5			10			10			9			9			8	

Logged-over plots

A total of 20 species were encountered on 11,600 m^2 (Table 5). Two species, *Calamus caesius, C. scipionum* were found only in these plots, although they are reported to be common in Southeast Asia/Borneo (Dransfield, 1992b). The low number of species in plot 77-2 as compared with the other plots can be ascribed to the smaller size (1600 m^2 versus 5000 m^2).

The abundance of rattan plants in the logged-over plots was highest in the subplots with a canopy of pioneer trees or in the transitional zone with primary forest. On the other hand, some subplots with a canopy of pioneer trees were void of rattan (in plot 77-2).

The logged-over and primary plots at ITCI have many species in common. Only four species present in the primary plots were not recorded in the logged-over plots, two of which were only seldom encountered (*Calamus fimbriatus, Daemonorops cristata*); one species was present in all of the primary plots and is typical of dry upper and middle slopes (*Calamus conirostris*). The fourth species (*Calamus ornatus*) was very rare and only represented by a single plant.

The similarity between logged-over and primary plots is caused by/has to be ascribed to the fact that a logged-over forest consists of patches of logged and primary forest. The balance between logged and primary patches depends on logging intensity and previous distribution of timber trees. In the logged-over plots, rattan species associated with primary forest can still survive. The difference is in an invasion of rattan species preferring more open growth conditions.

Table 5. Presence of rattan species in permanent plots in logged-over forest in the ITCI concession area (in 1994). Number of plants ha^{-1} (a), number of plants with mature canes ha^{-1} (b) and total number of mature canes ha^{-1}(c)

	72-1			72-2			77-2		
	a	b	c	a	b	c	a	b	c
Species :									
Daemonorops sp.	2	2	4						
Daemonorops hystrix	4	2	2						
Calamus flabellatus	4	2	6						
Calamus marginatus	2	2	2						
Korthalsia ferox	6	2	2						
Daemonorops didymophylla	14	8	8						
Calamus laevigatus	4	0	0	4	0	0			
Calamus rhytidomus	12	8	10	8	0	0			
Calamus pandanosmus	2	0	0	180	78	158			
Calamus caesius	10	2	2	72	34	46			
Calamus scipionum	2	0	0	8	8	8			
Calamus blumei	2	2	18	10	4	6	13	6	19
Daemonorops sabut	18	4	18	32	8	8	25	19	44
Ceratolobus concolor	8	2	2	6	6	10	6	6	13
Korthalsia rigida	22	4	4	24	8	8	44	13	13
Korthalsia furtadoana	2	2	26	32	8	28	38	13	75
Calamus pogonocanthus 3	20	18	30	18	12	32	13	13	13
Korthalsia echinometra	2	0	0				6	6	169
Calamus javensis				2	2	4	13	0	0
Daemonorops korthalsii				2	0	0			
Total no. of plants/canes	136	60	134	398	168	308	156	75	343
Total no. of species		18			13			8	

The abundance of plants in primary forest is on average lower than in logged-over forest (102 versus 230 plants ha^{-1}, see Table 7). The difference is higher when plants with mature canes are compared (29 versus 109 plants ha^{-1}), and highest when the total number of mature canes is compared (34 versus 262 canes ha^{-1}). The high contrast in mature cane numbers is the result of the presence of profusely clustering species (*Calamus pandanosmus*, *Korthalsia echinometra* and *K. furtadoana*) in the logged-over plots.

Apo Kayan

In total 26 rattan species were found in the Apo Kayan area. Local variation in abundance and species composition of rattan in apparently similar habitats was encountered in primary as well as secondary forest. Areas with reported fertile soils in the vicinity of the village of Long Sungai Barang are all used in the agricultural cycle. Primary forest is therefore no longer found on fertile land. Young secondary vegetation is poor both in species composition and abundance of rattan, as a result of burning that follows clearing of the land for a swidden. No seed-bank is present as rattan seeds quickly loose their viability (Yap, 1992) and the weeding during the rice cycle prevents regeneration by means of resprouting or establishment of new plants (*pers. observ.*).

Typical species of older re-growth are *Plectocomia mulleri, Daemonorops fissa, Ceratolobus concolor* and *Calamus pogonocanthus* 3. On riverbanks bordering the Kayan river *Daemonorops hystrix* was found, but the species was absent in the primary forest plots situated on upper slope and ridges. In swampy sites locally dense stands of *Korthalsia cheb* are found. This species is sought for its durable cane that is particularly suited as binding material for fish-traps, and the shoot is edible.

The nearest stands of the highly prized *Calamus caesius* are two days' (by boat/on foot) Southeast of the village, along the Boh river. In Long Sungai Barang, *Daemonorops sabut* is used as a substitute for *C. caesius* in the weaving of backpacks and rice-mats. Rattan is neither planted nor protected by the people of Long Sungai Barang.

The permanent plots

The four permanent plots harbour 18 species on 6400 m^2. Of these 16 species have mature canes (Table 6). Eight species present were not encountered in the ITCI or Wanariset plots: *Calamus hispidulus, C. mattanensis, C. muricatus, C. tomentosus, C. pogonocanthus* 1, *Daemonorops atra, D. fissa,* and *D. pumila.*

Species composition and abundance of individual species gradually change going down from ridge crest to valley floor. The abundance of rattan plants is lowest at the valley floor even when plants of *Calamus conirostris* (the dominant species of upper and middle slope) are excluded. The abundance of plants with mature canes at the valley floor is only 16 % or 36% (if *C. conirostris* is excluded) of the average abundance on upper and middle slope. The difference is smaller when the total number of mature canes is compared, 23 % and 47 % respectively.

When all permanent plots are combined in four groups (Apo Kayan, ITCI primary, ITCI logged-over and Wanariset) and compared, only three species are seen to occur in all four groups: *Ceratolobus concolor, Daemonorops didymophylla* and *D. sabut.* Both *C. concolor* and *D. sabut* respond to disturbance by more vigorous growth. If the two ITCI groups are combined, the number of species that all sites have in common is six with the addition of *Calamus javensis, C. laevigatus* and *C. ornatus* (Table 2). Both *Calamus laevigatus* and *C. ornatus* are species that are often encountered in primary forest as robust plants in the rosette stage. *Calamus javensis* is actually a complex polymorphic species that occurs from almost swampy valley floors to moist depressions on slopes.

The fact that 8 out of 18 Apo Kayan, and 6 out of 24 ITCI species were not found in the other research sites further emphasises the great regional variation in species composition.

Table 6. Presence of rattan species in four permanent plots in primary forest in the Apo Kayan (in 1994)

Number of plants ha^{-1} (a), number of plants with mature canes ha^{-1} (b) and total number of mature canes ha^{-1}(c)
Plots arranged in a topo-sequence from ridge crest to valley floor

Species:	A			B			C			D		
	a	b	c	a	b	c	a	b	c	a	b	c
Korthalsia ferox	6	6	13									
Calamus muricatus	163	163	169									
Calamus pilosellus	13	6	25									
Daemonorops fissa				13	6	6						
Calamus hispidulus				13	0	0						
Calamus laevigatus				6	0	0						
Daemonorops pumila				25	13	13						
Calamus tomentosus				156	150	156						
Plectocomiopsis geminiflora	69	69	175	88	44	150	50	50	206			
Daemonorops atra	100	100	106	156	150	163	163	138	144			
Calamus conirostris	469	300	369	1056	675	888	675	363	419	13	6	6
Ceratolobus concolor	6	0	0	130	13	13	25	13	13	63	13	19
Daemonorops sabut				25	25	50	6	6	6	6	6	6
Calamus ornatus				13	0	0	44	25	25	50	19	19
Calamus pogonocanthus 1				13	0	0	31	19	19	13	13	13
Calamus javensis							19	13	19	69	44	150
Daemonorops didymophylla							6	6	6	88	31	31
Calamus mattanensis							6	6	6			
Total no. of plants/ canes	825	644	856	1569	1081	1442	1025	638	862	300	131	244
Total no. of species	7			12			10			7		

Comparison of species composition and abundance

Species composition

The *Calamus pogonocanthus* complex represented by three taxa also illustrates regional variation. *C. pogonocanthus* 1 is restricted to primary forest in the Apo Kayan. *C. pogonocanthus* 2 is restricted to plot Matthijs and other parts of the Wanariset research forest. *C. pogonocanthus* 3 (the taxon that most resembles *C. pogonocanthus* Becc. ex H. Winkl.) occurs in primary as well as disturbed forest in all sites (at Apo Kayan and Wanariset outside permanent plots), with a higher abundance in disturbed forest.

As regards the (locally) more common species, some general comments on ecology can be given. The species that are almost exclusively encountered in logged-over/old secondary forest are *Calamus caesius*, *C. pandanosmus*, *C. scipionum* and *Daemonorops fissa*. A number of other species which are also present in primary forest show a clear positive response (in numbers of mature canes) to disturbance namely *Calamus pogonocanthus* 3, *Ceratolobus concolor*, *Daemonorops sabut*, *Korthalsia echinometra*, *K. furtadoana*, *K. rigida* and *Plectocomiopsis geminiflora*. Certain species are restricted to primary forest, some preferring dry sites like *Calamus conirostris*, *C. marginatus*, *C. muricatus*, *Daemonorops atra*, *D. hystrix* and *D. pumila*, while others, like *Calamus javensis*, *C. ornatus* and *C. tomentosus*, prefer moist conditions.

For species encountered on only a few occasions, it was difficult to define their preferences i.e.: *Calamus hispidulus*, *C. mattanensis*, *Daemonorops korthalsii*, *Daemonorops cristata* and *Plectocomiopsis mira*.

Calamus pilosellus although only found in two of the Apo Kayan plots and in plot Matthijs, locally formed dense entanglements in old canopy gaps on well-drained slopes in the Apo Kayan. Mature plants of *Calamus blumei*, *C. laevigatus*, *C. tomentosus* and *Korthalsia ferox* always occurred at very low densities.

The species richness in primary forest in the Apo Kayan plots and plot Matthijs is similar. Species richness, of a primary forest plot of comparable size, in the ITCI concession is lower. However, logging appears to have resulted in an increase in species richness in the ITCI plots (18 species on 18,900 m^2 in primary forest versus 20 species on 11,600 m^2 in logged-over forest) due to an influx of species adapted to disturbed sites.

Abundance

The abundance of rattan differs considerably between plots of the same site as well as between sites. This variation between plots may be attributed to the limited size of the plots. The size of the plots was not sufficient to comprise the variation in site conditions and successional stages, present in the area. Differences between research sites however point to a common trend.

The Apo Kayan plots have the highest abundance of rattan. This applies to the total number of plants, the total number of plants with mature canes, and the total number of mature canes (Table 7). Plot Matthijs is a good second with roughly half the number of plants and half the number of mature canes. The number of plants in the ITCI primary plots is only a tenth of the Apo Kayan value, and the total number of mature canes is a mere 4 % of the Apo Kayan value.

Calamus conirostris dominates the 'dry' Apo Kayan plots, where it accounts for over 50 % of all plants. Even when *C. conirostris* is excluded, the Apo Kayan plots still have the highest abundance of plants with mature canes and total number of mature canes.

Although the abundance in primary forest of the Apo Kayan and plot Matthijs is higher than the ITCI plots, the logging has (probably) resulted in an increase in abundance. However, it still needs to be emphasised that local differences in abundance are substantial, even in similar forest types.

Table 7. Rattan abundance in East Kalimantan (in numbers ha^{-1})

	Rattan all species			Rattan excluding *Calamus conirostris*		
	No. of plants	No. of plants with mature canes	Total no. of mature canes	No. of plants	No. of plants with mature canes	Total no. of mature canes
Wanariset						
plot Matthijs	528	268	400	528	268	400
ITCI primary						
plot 76-3a	42	16	18	40	16	18
plot 76-3b	77	44	55	72	41	52
plot 72-8a	163	69	81	125	50	56
plot 72-8b	81	19	25	63	13	19
plot 76-4a	144	19	19	138	13	13
plot 76-4b	106	6	6	93	0	0
average primary	102	29	34	88.5	22	26
ITCI logged-over						
plot 72-1	136	60	134	136	60	134
plot 72-2	398	168	308	398	168	308
plot 77-2	156	75	343	156	75	343
average logged-over	230	101	262	230	101	262
Apo Kayan						
plot A	825	644	856	356	344	488
plot B	1569	1081	1442	512	406	556
plot C	1025	638	862	350	275	446
plot D	300	131	244	288	125	238
average	930	633	851	377	288	432

3.2 Growth of rattan

The following account is based on observations on the number of plants and the number and growth stage of shoots over a 24-month period between the first recording (1992) and the last (1994). A rattan population is dynamic both from the standpoint in changes within the population (Table 8) and also within an individual clump.

Multi-stemmed rattans have canes in various growth-stages, the same plant can therefore be represented in more than one growth-stage class, e.g. one plant can be scored as both a plant with suckers and as a plant with immature canes.

Wanariset forest

Figure 2 represents the growth dynamics in plot Matthijs. The rattan population of plot Matthijs is very dynamic and showed a clear increase in number of plants. The percentage of plants showing an increase in the number of shoots was highest compared with plants showing no change or a decrease, for all growth stages, except for the mature class where the percentage of stable plants was higher. Possible causes of death of the shoots can be old age, herbivory or damage by fallen branches. The number of shoots increased for all growth stages except for the mature stage (Figure 5). The net increase in plants amounted to 20 %.

**Table 8. Changes in the number of rattan plants and number of mature canes
between 1992 and 1994.
Percentages are based on 1992 values for number of plants and total number of mature canes per plot.**

	Plants			
	Recruitment (%)	Mortality (%)	Net increase (%)	Mature canes (%)
Wanariset				
plot Matthijs	26	6	20	-1
ITCI primary				
plot 76-3a	57	0	57	+28
plot 76-3b	9	1	8	-10
plot 76-4a	78	14	64	-40
plot 76-4b	5	5	0	0
ITCI logged-over				
plot 72-1	23	1	22	-1
plot 72-2	35	5	30	+18
plot 77-2	8	0	8	-8
Apo Kayan				
plot A	11	5	6	-2
plot B	21	6	15	+2
plot C	16	7	9	-2
plot D	13	2	11	-32

The great number of plants that showed an increase in the number of shoots in the juvenile class (64 plants) was primarily the result of recruitment of 60 new plants since 1992. The increment in the number of both juvenile and immature shoots is most likely caused by/can be ascribed to plants that have appeared since the 1983 forest fire, and may also be influenced by the dry spell of 1991. The increase in light in the remaining forest, following the fire of 1983 may have caused a rejuvenation and stimulated growth of the rattan that had survived.

PT ITCI concession area

Primary plots

Since the number of plants in the plots was rather small (1994: n= 17, 21, 23, 59), changes in a single plant have a relatively high impact on the results obtained for the total population. Two plots (76-3a, 76-4a) show clear changes in the rattan population, whereas the others do not (Table 8).

The rattan population in plot 76-3a is very dynamic and showed a clear increase in the number of plants (Table 8). The absolute number of shoots for all classes increased. The number of plants with an increase in number of shoots in the sucker and juvenile class was more than 60 % (Figure 3). For all the classes it is more than twice the number of plants showing a decrease. Increase in the juvenile class was primarily caused by recruitment.

Ridge crest forest in the area is characterised by a very open under-storey. Periodic severe drought tends to kill saplings of trees during a dry spell and the same might apply to rattan plants. This could explain the low density of rattan plants and the relatively high increase in all growth stages in response to the better growth conditions since the severe dry spell of 1983 and the dry spell of 1991 (Anon, 1994a). So despite the fact that the plot is situated in primary forest, without major changes in the canopy, the rattan plants in fact experience a highly dynamic environment.

In plot 76-4a, there is a relatively high increase in the number of juvenile shoots and 60 % of the plants of the juvenile class show an increase in shoot number. This is primarily caused by recruitment. The number of mature canes, however, showed a considerable decline. Pigs that frequent the site, ploughing the soil and eating rattan shoots, probably cause these changes.

In plot 76-3b there was a net increase in sucker and juvenile shoots but a decrease in immature and mature shoots. The percentage of plants showing no change was always high. In plot 76-4b, rattan plants showed even less vigour. The increase in the number of sucker shoots was balanced by a decrease in juvenile shoots. Changes within the immature and mature class were negligible (Figure 3)

Logged-over plots

In two plots (72-1, 72-2) the rattan population increased considerably (22% and 30 %, see Table 5.8), in the third plot (77-2) the population was rather stable.

The number of plants and shoots increased in both plots 72-1 and 72-2. The population in plot 72-2 is the more dynamic of the two plots with the percentage of plants showing no change always smaller than in plot 72-1, except for the sucker class (Figure 2). The high increase in both the number of plants and number of shoots in plot 72-2 occurred in an area where the canopy of *Macaranga* trees was dying.

In plot 77-2, the population of rattan plants increased by 8 %, but the number of shoots in all growth-stages except the sucker stage showed a net decrease (Figure 5). For all classes, except for the sucker class, the percentage of plants showing no change was relatively high (Figure 2). The decrease in the number of mature canes was the result of one large *Korthalsia echinometra* plant, which was regenerating profusely, with a net increase in shoot number from 35 in 1992 to 40 in 1994.

Apo Kayan

In all plots an increase in the number of plants and a net increase in the number of sucker and juvenile shoots was found (Table 8, Figure 5), although the increment in the latter was smaller. Apart from the ridge and upper slope plot, the number of immature canes decreased and three of the four plots showed a net reduction in the number of mature canes.

Plots C and A showed a stable rattan population with little dynamism. Plot C showed the highest percentage of plants with no change for each growth stage class, except for the sucker stage class (Figure 4).

The valley floor plot (plot D) showed little change in the total number of plants but there was a strong reduction in the number of mature canes (Table 8), while at the same time the majority of plants in the sucker and juvenile class showed an increase in the number of shoots. This was primarily the result of large clumps of *Calamus javensis* and *Ceratolobus concolor* in which the number of mature shoots decreased and simultaneously the number of either sucker or juvenile shoots increased.

Plot B, situated on upper and middle slope was the most dynamic plot. The number of plants which showed no change in a given growth-stage class was always less than 30 % except for plants with mature canes where it was 70 % (Figure 4). A net increase in the number of suckers and juvenile shoots was evident (Figure 5) and can probably be ascribed to increased light reaching the forest floor as a result of a chablis (Oldeman, 1980) that was formed in 1991. The dynamic nature of this population is further supported by the recruitment of 48 new plants since 1992 resulting in a net increase of 15 % in the rattan population by 1994 (Table 8).

Comparison of the regions

With respect to growth of rattan, both in terms of an increase in the population and changes in total number of shoots a pattern can be discerned.

In all three sites increase in the population is highest in those plots experiencing disturbance (Table 8). The forest of plot Matthijs is still recovering from the 1983 forest fire. In the ITCI logged-over plots, both plot 72-1 and 72-2 harbour patches of degrading canopy trees. The primary plots 76-3a and 76-4a were subject to disturbance. And finally in the Apo Kayan in plot B a chablis was recently formed.

For the analysis of the influence of disturbance on population growth, the primary ITCI plot 76-4a was considered undisturbed since wild pigs caused the disturbance.

Besides initial population size, disturbance proved to be a significant factor ($p < 0.05$) in the recruitment of rattan. A procedure fit-curve (Genstat 5, 1990) for the twelve plots as well as for the 50 subplots of plot 72-2 in detail (all 50 subplots tallied according to degree of disturbance) confirmed that disturbance results in a significant increase in recruitment (see Figures 6 and 7).

Disturbance also results in a relatively high increase in the number of juvenile and immature shoots.

Logging, being a form of disturbance, appears to result in an accelerated increase in rattan population and in promoting the growth of individual plants.

3.3 Harvesting of selected rattan clumps in the Apo Kayan

In general, commercial collection of all harvestable canes has a negative effect on the vitality of a plant after one year. If harvesting leaves no suckers in a clump, no new suckers were formed with the exception of one *Calamus javensis* plant.

But if only shoots in sucker stage remained, the overall effect is not obviously negative but a difference between the species can be discerned (Figure 8). *Calamus ornatus* sucker shoots are more tolerant as compared with *C. javensis* and to a lesser extent *Daemonorops sabut*.

On a per clump basis, the dynamics of the various classes for the three species is clearly different (Figure 9). For *Calamus ornatus* the percentage of clumps that are stable is always highest for all classes. In *C. javensis* clumps are very dynamic and the percentage of stable clumps is always smaller than those showing either an increase or a decrease, except for clumps belonging to the mature class. For *Daemonorops sabut* an overall negative effect can be discerned, with the percentage of plants showing a decrease in shoot number always being greater than those showing an increase.

Looking at the number of shoots for the various growth stages, a similar pattern can be discerned (Table 9). In *Calamus ornatus* hardly any change was observed in the number of shoots compared with an increase in juvenile and mature shoots in *C. javensis*, which is only achieved at the expense of sucker and immature shoots, and with *Daemonorops sabut* where there is a reduction in all growth stages, except for the mature stage.

Table 9. Net changes in the total number of shoots for each growth-stage class, 1993-1994.
Harvested and non-harvested clumps of three selected species.

	Calamus ornatus				*Calamus javensis*				*Daemonorops sabut*			
	S	J	I	M	S	J	I	M	S	J	I	M
Harvested												
Apo Kayan	-4	+1	-1	-1	-18	+3	-18	+8	-37	-4	-1	0
Non-harvested												
Apo Kayan	+4	0	+1	-1	0	+3	+1	-11	-2	0	-1	+1
Matthijs	0	-1	+1	0	0	+2	0	0	0	0	-1	-1
ITCI logged-over					+4	-1	+2	0	+7	-8	-1	-1
ITCI primary									+3	0	0	0

S = Sucker stage; J = Juvenile stage; I = Immature stage; M = Mature stage

It is concluded that for *Calamus ornatus* the harvesting does not lead to an increase in the number of shoots and an acceleration in the development of the remaining shoots, as compared with the non-harvested clumps (Figure 9), could not be discerned after one year. For *C. javensis* harvesting resulted in a net decrease in the total number of shoots (Table 9) but an accelerated maturation of immature and mature canes was obvious compared with non-harvested clumps (Figure 9). For *Daemonorops sabut*, harvesting resulted in a considerable decrease in the number of shoots (Table 5.9). However, an acceleration in the development of the remaining shoots, as compared with non-harvested clumps (Figure 9), could not be detected.

The difference in dynamics of *C. ornatus* and *C. javensis* might be an indication for a different survival strategy. Whereas *C. ornatus* plants invest in a small number of relatively tolerant shoots, *C. javensis* plants invest in a large number of shoots with a shorter life expectancy.

4. Discussion

The inventory of rattan species in primary forest at the various sites provided an index of species richness and abundance of rattan. Differences in geographical distribution of species are apparent (see 3.1), as well as the influence of disturbance (see 3.1) and moisture regime (see 3.3).

Water seems to be the prime factor influencing both species composition and abundance of rattan in the Apo Kayan and ITCI concession area. With either an excess or a shortage of water playing a role.

The preferences of the various species were mentioned in the general section 3.1.

In the Apo Kayan where there is an annual rainfall of over 4000 mm (Voss, 1982), rattan abundance is lowest at the valley floor where rattan is apparently adversely affected by the standing water. Contrary to the situation in the Apo Kayan, rattan abundance in the ITCI concession, with an annual rainfall of 2000-2500 mm (Voss, 1982), is lowest on ridge crests and upper slopes. This is probably due to water deficit that occurs during periodic dry spells.

The three new species (*Calamus fimbriatus, C. nigricans, Daemonorops pumila*) that were discovered (Van Valkenburg, 1995) have a limited distribution thereby confirming the trend of a rather high degree of endemism in rattans, with e.g. more than 20 % of the Sarawak species being endemic in Sarawak (Dransfield, 1992a, 1992b). The economically important species belong to widespread *(Calamus javensis, C. ornatus, Ceratolobus concolor, Daemonorops fissa)* as well as geographically confined species (*Ceratolobus subangulatus, Daemonorops crinita*). The survival of the latter species is theoretically more threatened by forest conversion or over-exploitation.

The common occurrence in East Kalimantan of *Calamus javensis* and *C. ornatus* is in accordance with Dransfield (1992b). The presence of *Calamus hispidulus* in the Apo Kayan and *Daemonorops cristata* in the ITCI concession, is an extension eastward of their reported area of endemism (Pearce, 1989; Dransfield, 1992a). The presence of *Calamus trachycoleus* in the Middle Mahakam area is probably an old introduction from South or Central Kalimantan of this commonly cultivated species, while *Calamus manan*, also cultivated, might occur naturally in the area. Another widely cultivated species, *Daemonorops crinita*, was reported only to occur in Southern Sumatra (Dransfield & Manokaran, 1993).

Difference in tolerance of disturbance is a major factor limiting the potential of a commercial species for rattan plantations or enrichment planting in logged-over forest. In the logged-over ITCI plots, logging resulted in a shift in species composition and an increased species diversity with the invasion of secondary species. The species adapted to disturbance that invaded the plots are of considerable economic value. (For a more detailed comparison of the economic value of primary versus logged-over forest, see Van Valkenburg, 1997.)

Local variation in rattan abundance was observed throughout East Kalimantan and is further illustrated by the results of the line survey in the Apo Kayan. The average abundance was 362 mature plants and 902 mature canes per hectare, but variation ranged from 0 to 18 plants with 79 mature canes per 200 m^2.

Sixteen to twenty three years after logging the average abundance of rattan in the logged-over ITCI plots is higher as compared with the primary forest plots at ITCI, and was highest in old canopy gaps resulting from the felling. Elsewhere, shortly after logging a drastic reduction in abundance was observed in ridge dipterocarp forest in Sabah (Abdillah & Phillips, 1989). Felling and extraction were major causes of a reduction by 73% of an initial population of 148 plants in one hectare. Apparently the initial adverse effects of logging are later compensated by recruitment.

Although the disturbance by logging promotes the growth of rattan, five years after logging rattan canes are not sufficiently mature to be harvested (Kiew & Hood, unpublished). Furthermore, in the case of an absence of large support trees rattan canes will coil on the forest floor resulting in poor quality canes. Logging can also have long lasting negative effects on rattan growth. In Malaysia Orang Asli informants (Kiew & Hood, unpublished) claimed that on bulldozed areas there is no rattan regeneration. This was also found in part of ITCI plot 71-1 that was logged in 1987. On the old skid roads only some pioneer trees and large numbers of Cyperaceae and *Rubus* plants were encountered but no rattan seedlings.

Natural as well as man-induced disturbance significantly ($p < 0.05$) increased the recruitment of rattan plants in the research plots. Disturbance also resulted in a relatively high increase in the number of juvenile and immature shoots when compared with non-disturbed plots. The prime factor for promoting the growth of remaining plants and the recruitment of new plants is presumed to be light. Whereas in forest with a closed canopy a relative light intensity (RLI) of 0.1 to 5 % is measured, the establishment of *Calamus manan* required a 50 % RLI in a controlled light environment (Mori, 1980). By opening up the canopy logging results in higher RLI values. A higher survival and growth of rattan plants in forest with a partially opened canopy was also observed in various planting trials (Wan Rhazali Wan Mohd. *et al.*, 1992; Wong & Manokaran, 1985).

Logged-over forest does not necessarily show an accelerated growth at any point in time. The development stage of a patch of forest or eco-unit (Oldeman, 1980, 1990) is essential. During succession an eco-unit will go through several cycles of innovation, aggradation, biostatic and degradation phases. During innovation and aggradation, the total biomass rapidly increases including rattan plants. As the pioneer trees become fully developed and are dominating the eco-unit a biostatic phase is reached and growth of rattan is not different from that in primary forest. This can be observed sixteen years after logging activities in plot 77-2. When pioneer trees become senescent, the eco-unit reaches the degradation phase. More light reaching the rattan plants combined with possibly additional nutrients from the dying trees results in new opportunities for the rattan. The acceleration in growth observed in plot 72-1 and 72-2 was most pronounced under a canopy of senescent *Macaranga* trees. Exchangeable nutrients of forest floor litter in logged forest, ten years after logging, were found to be higher than in primary forest in Sabah, although leaf-litter fall was similar (Burghouts *et al.*, 1992).

Whereas logging results in disturbance at the population level, harvesting causes disturbance at plant level. A process similar to the dynamics observed following logging may occur at individual plant level following harvesting.

The maximum harvesting experiment had an overall negative effect on the rattan clumps after one year. The traditional management system in the Bahau area may overcome this effect by, after large-scale commercial harvesting of rattan, closing a tributary for 10 years so as to allow the rattan population to recover.

Whereas maximum harvesting apparently has a negative effect, especially on clustering species that form large clumps with numerous mature canes, a limited harvesting may promote the growth of remaining shoots or stimulate the formation of new shoots. Removal by harvesting of a limited number of mature canes is similar to the natural process of the mature canes dying. In a large *Korthalsia echinometra* clump, the death of five mature canes coincided with an increase in sucker, juvenile and immature shoots that was considerably higher than in clumps without dying mature canes. A method of limited harvesting may well be ecologically sustainable in the long run. A management system of limited but continuous harvesting is practised by Orang Hulu people in Johore, Malaysia (Kiew, 1989; Kiew & Hood, unpublished). There *Calamus caesius* in primary forest is harvested on a 4-5 month rotation. Each time only a small number of canes are cut and the clump retains its vitality. This management system is difficult to implement when density of rattan plants is high as in the case of rattan plantations. High frequency of harvesting of the often intertwined canes will cause too much damage to the remaining canes. Intervals between harvests therefore have to be several years.

Nandika (1938) already mentioned the importance of maintaining vitality of a clump. With respect to management of rattan gardens he recommended harvesting the canes in stages and not to cut the canes closer then 1-1.5 m from the clump. The remaining leaf surface area after harvesting might be an important factor influencing the vitality of a cluster (where vitality is capacity to recover from the harvesting). This vitality and resilience after harvesting differs between species, but a minimum remaining surface area or number of shoots is essential for a clump to be able to recover.

Regional variation in species composition and abundance of rattan in East Kalimantan is considerable. Response to disturbance and reactions to harvesting differs between species. Therefore detailed information on the rattan resource is a prerequisite for ecologically sustainable exploitation. Both primary forest as well as logged-over forest harbour (potentially) commercial rattan species that can be sustainably exploited.

Acknowledgements

The study on which this paper is based was undertaken from 1991-1995 as part of the International MOF-Tropenbos Kalimantan Project based at the Wanariset Research station in Samboja, East Kalimantan, Indonesia. This programme aims to develop appropriate techniques and guidelines for sustainable forest management. It is being implemented by the Indonesian Agency for Forestry Research and Development of the Ministry of Forestry and Estate Crops, the Institute for Forestry and Nature Research IBN-DLO, and the National Herbarium, in the Netherlands, together with the Indonesian state forestry enterprises P.T. Inhutani I and P.T. Inhutani II.

REFERENCES

Abdillah Roslan & Phillips, C., 1989. Preliminary report on rattan damage due to selective logging of ridge Dipterocarp forest in Sabah. Paper presented to: Persidangan Perhutanan Malaysia, Ke 10, Kuantan, Pahang, 24-29 Julai, 1989 (10th conference on Malaysian Forestry, Kuantan, Pahang, 24-29 July, 1989.

Burghouts, L.P.A., Ernsting, G., Korthals, G.W. & De Vries, T.H., 1992. Litterfall, leaf-litter decomposition and litter invertebrates in primary and selectively logged dipterocarp forest in Sabah, East Malaysia. Philosophical Transactions of the Royal Society of London series B 335: 407-416.

Dransfield, J., 1984. The rattans of Sabah. Sabah Forest Records 13. Forest Department, Sandakan, Sabah.

Dransfield, J., 1992a. The rattans of Sarawak. Royal Botanic Gardens, Kew & Forest Department, Kuching, Sarawak. 233 pp..

Dransfield, J., 1992b. The ecology and natural history of rattans. Pp. 27-31 in: Wan Rhazali Wan Mohd., Dransfield, J. & Manokaran N. editors. A guide to the cultivation of rattan. Malayan Forest Record no. 35. FRIM, Kepong, Malaysia.

Dransfield, J., 1992c. Rattans in Borneo: botany and utilisation. Pp. 22-31 in: Ghazally, I., Murtedza, M. & Siraj, O. editors. Proceedings of the International Conference on Forest Biology and Conservation in Borneo, July 30 - August 3, 1990, Kota Kinabalu, Sabah, Malaysia. Yayasan Sabah, Kota Kinabalu, Sabah.

Dransfield, J. & Johnson, D., 1989. The conservation status of palms in Sabah. Malayan Naturalist 43(1&2): 16-19.

Dransfield, J. & Manokaran, N., editors, 1993. Plant Resources of South-East Asia No 6: Rattans. Pudoc, Wageningen, the Netherlands. 137 pp.

Endert, F.H., 1927. Pp. 117-312 in: Midden - Oost - Borneo expeditie 1925. Weltevreden :[s.n.], 1927. 423 pp. (Publication Indisch comite voor wetenschappelijke onderzoekingen: no. 2).

Genstat 5, Release 2.2. 1990. Lawes Agricultural Trust (Rothamsted Experimental Station).

Kiew, R., 1989. Utilization of palms in Peninsular Malaysia. Malayan Naturalist 43 (1&2): 43-67.

Kiew, R. & Hood Salleh, 1991. The future of rattan collecting as a source of income for Orang Asli communities. Pp. 121-129 in: Appanah, S., Ng, F.S.P. & Roslan Ismail editors. Malaysian Forestry and forest products research. Proceedings of the conference October 3-4, 1990. FRIM, Kepong, Malaysia..

Kiew, R. & Hood Salleh. Unpublished report.The role of rattan in the economy of Orang Asli communities. Unpublished report for WWF-International.

Nandika, O., 1938. De rotanhandel, -cultuur en -winning en het voorkomen van rotanboeboek in het landschap Koetai. Het Bosch 6: 199-235.

Oldeman, R.A.A., 1980. Grondslagen van de bosteelt. Agricultural University Wageningen, Department of Silviculture, Wageningen, the Netherlands.

Oldeman, R.A.A., 1990. Forests: Elements of Silvology. Springer Verlag, Berlin, Heidelberg, New York. XXI + 624 pp.

Pearce, K.G., 1989. Conservation status of palms in Sarawak. Malayan Naturalist 43 (1&2): 20-36.

RePPProT. 1987. Regional Physical Planning Programme for Transmigration. Review of Phase I: Results East and South Kalimantan. Direktorat Bina Program. Direktorat Jenderal Penyiapan Pemukiman, Departemen Transmigrasi, Jakarta, Indonesia.

Van Bremen, H., Iriansyah, M. & Andriesse, W., 1990. Detailed soil survey and physical land evaluation in a tropical rain forest, Indonesia: A study of soil and site characteristics in 12 permanent plots in East Kalimantan. (Tropenbos Technical series no.6). The Tropenbos Foundation, Ede, the Netherlands. 188 pp.(94 + 94 appendices).

Van Tuil, J.H., 1929. Handel en cultuur van rotan in de zuider- en oostafdeling van Borneo. Tectona XXII: 695-717.

Van Valkenburg, J.L.C.H., 1995. New species of rattan (Palmae: Lepidocaryoideae) from East Kalimantan. Blumea 40(2): 461-467.

Van Valkenburg, J.L.C.H., 1997. Non-Timber Forest Products of East Kalimantan. Potentials for sustainable forest use. Tropenbos Series 16, the Tropenbos Foundation, Wageningen, the Netherlands. 202 pp.

Wan Rhazali Wan Mohd., Dransfield, J. & Manokaran N., editors 1992. A guide to the cultivation of rattan. Malayan Forest Record no. 35. FRIM, Kepong, Malaysia.

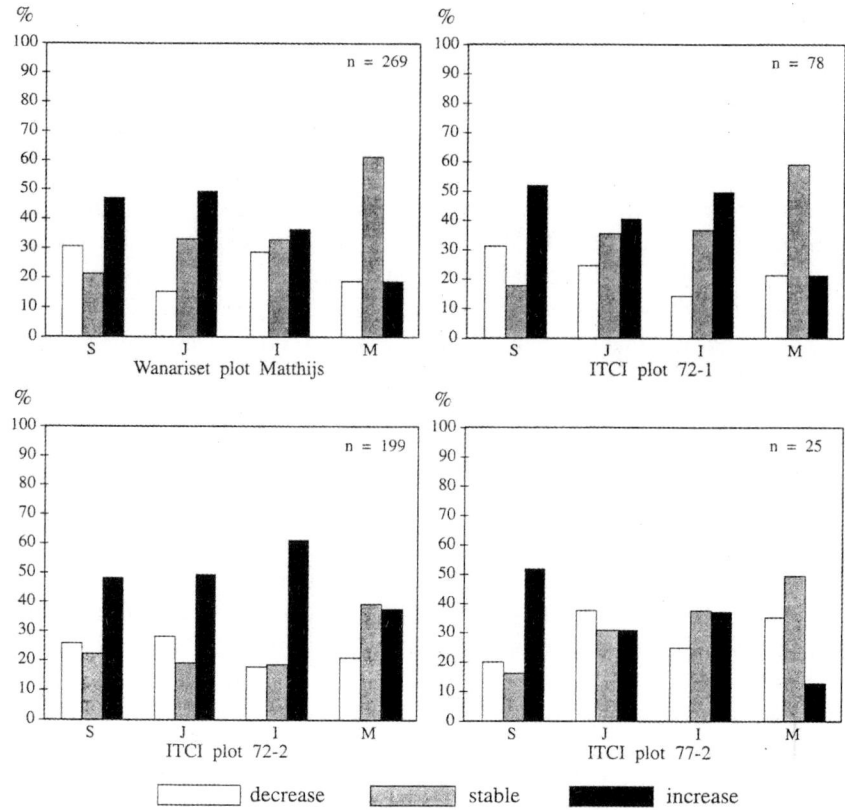

Figure 2. Dynamics of rattan growth in plot Matthijs and logged-over forest in the ITCI concession, illustrated by changes in the number of shoots per clump for each growth stage. Values in % of total clumps in each class. Changes between 1992 and 1994. – S = sucker class; J = juvenile class; I = immature class; M = mature class.

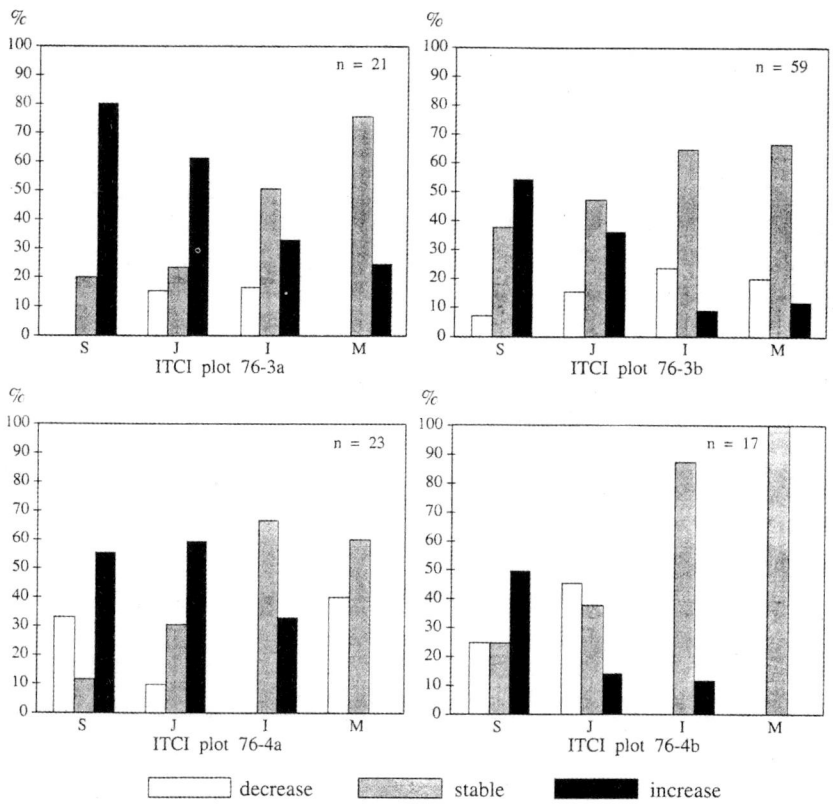

Figure 3. Dynamics of rattan growth in primary forest in the ITCI concession, illustrated by changes in the number of shoots per clump for each growth stage. Values in % of total clumps in each class. Changes between 1992 and 1994. – S = sucker class; J = juvenile class; I = immature class; M = mature class.

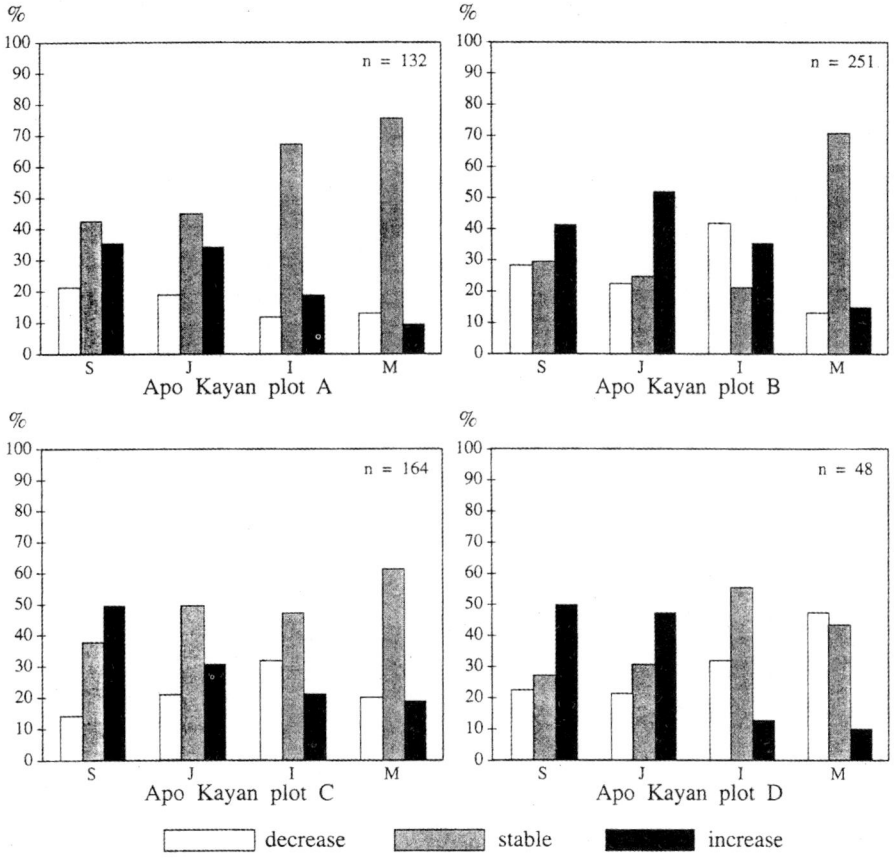

Figure 4. Dynamics of rattan growth in primary forest in the the Apo Kayan region, illustrated by changes in the number of shoots per clump for each growth stage. Values in % of total clumps in each class. Changes between 1992 and 1994. – S = sucker class; J = juvenile class; I = immature class; M = mature class.

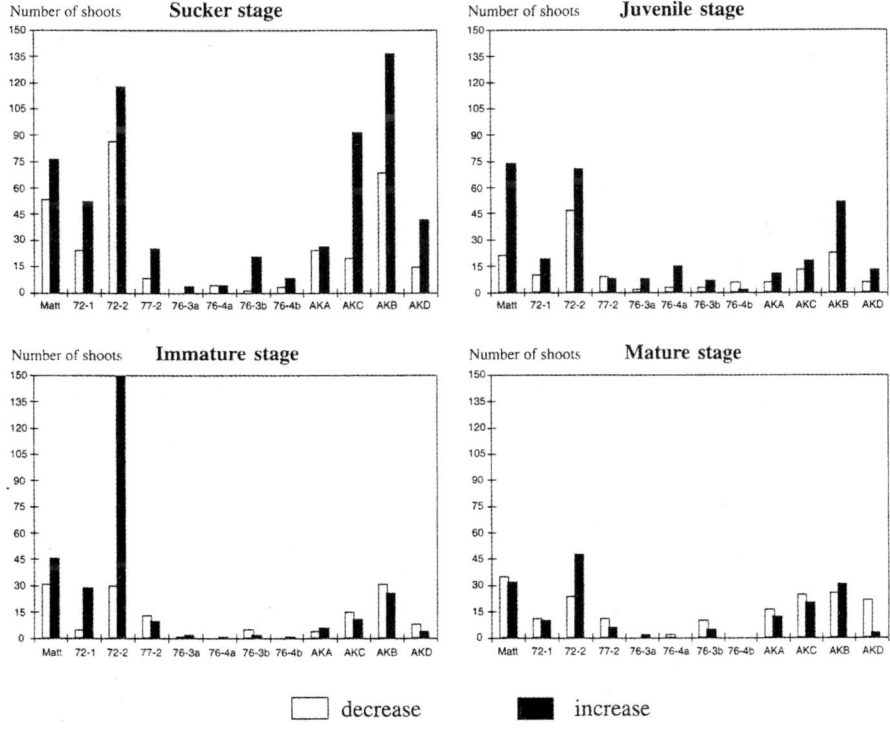

Figure 5. Increase and decrease in total number of shoots for four growth-stage classes of rattan populations (all species) in different plots in East Kalimantan. Changes between 1992 and 1994. – Matt = plot Matthijs; ITCI logged-over: 72-1, 72-2, 77-2; ITCI primary: 76-3a, 76-3b, 76-4a, 76-4b; Apo Kayan: AKA = plot A, AKB = plot B, AKC = plot C, AKD = plot D.

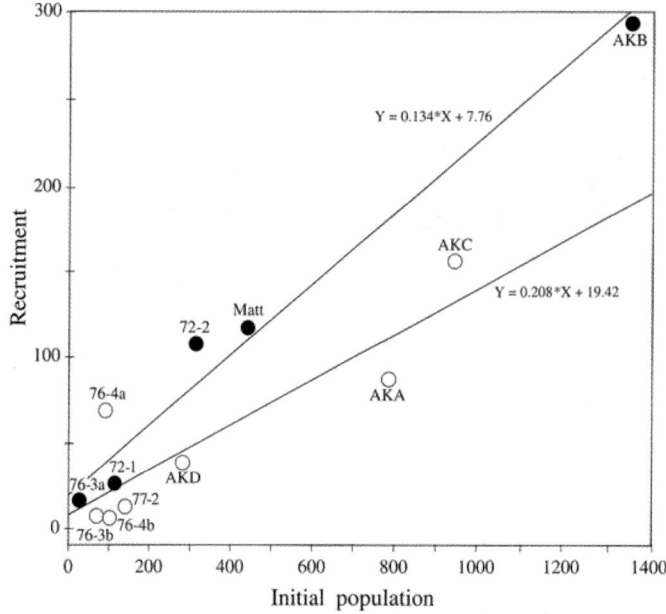

● = disturbed (Y = 0.134*X + 7.76)   ○ = undisturbed (Y = 0.208*X + 19.42)

Figure 6. Correlation between recruitment, initial population size of rattan (all species) and disturbance. Comparison of twelve plotsin East Kalimantan. – Matt = plot Matthijs; ITCI logged-over: 72-1, 72-2, 77-2; ITCI primary: 76-3a, 76-3b, 76-4a, 76-4b; Apo Kayan: AKA = plot A, AKB = plot B, AKC = plot C, AKD = plot D.

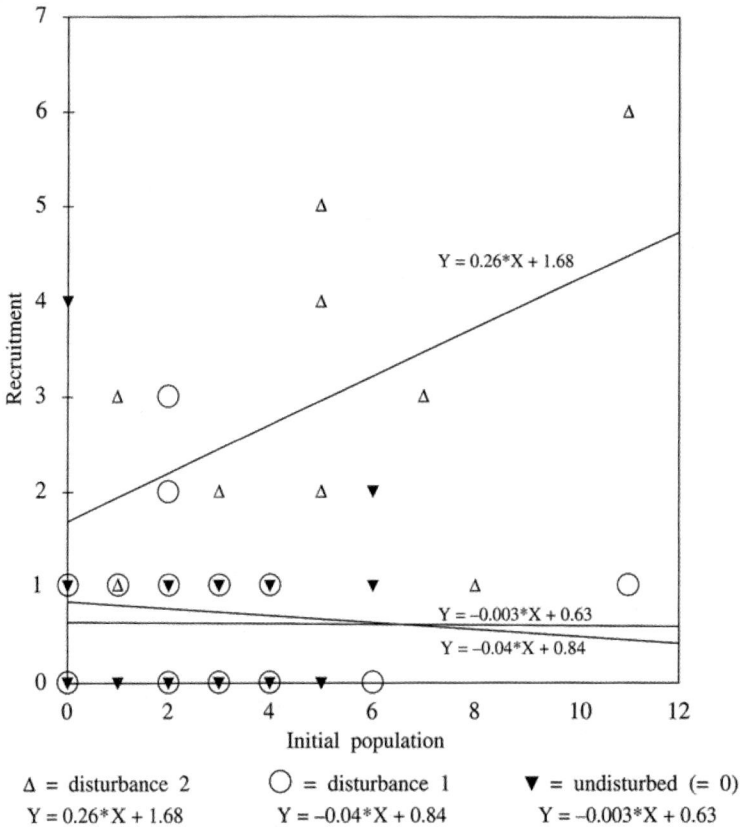

Δ = disturbance 2   ○ = disturbance 1   ▼ = undisturbed (= 0)
Y = 0.26*X + 1.68   Y = –0.04*X + 0.84   Y = –0.003*X + 0.63

Figure 7. Correlation between recruitment, initial population size of rattan (all species) and disturbance. Comparison of fifty subplots (100 m² each) in plot 72-2, situated in logged-over forest in the ITCI concession area, East Kalimantan. Level of disturbance increasing from undisturbed = 0 to disturbance = 2.

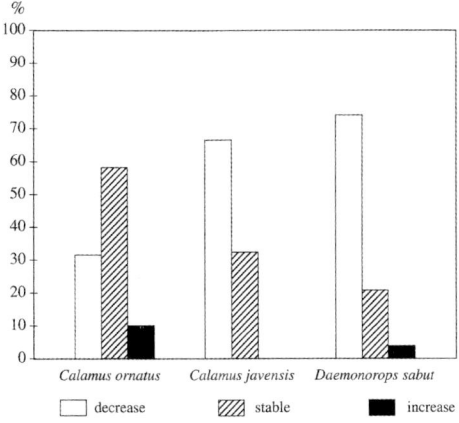

Figure 8. Change in growth of sucker shoots after 12 months for experimentally harvested clumps with only one sucker-shoot remaining. Comparison of three rattan species.

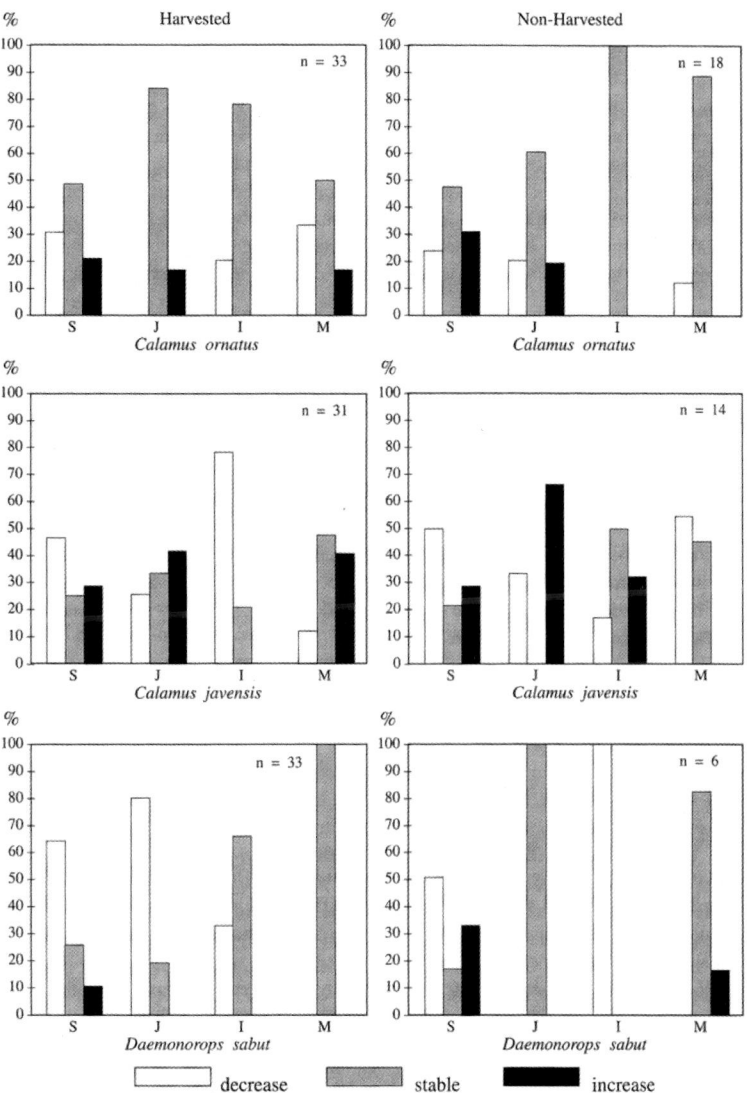

Figure 9. Growth performance of harvested and non-harvested rattan clumps of Calamus ornatus, C. javenasis and Daemonorops sabut, in the Apo Kayan region, illustrated by changes in the number of shoots per clump for each class. Values given as % of total clumps in each class. Changes between 1993 and 1994. – S = sucker class; J = juvenile class; I = immature class; M = mature class.

REVIEW OF POLICY, INSTITUTIONAL AND SOCIO-ECONOMIC ASPECTS GOVERNING THE RATTAN SECTION

HARVESTING WILD RATTAN: OPPORTUNITIES, CONSTRAINTS AND MONITORING METHODS

Stephen F. Siebert

1. Introduction

Rattan, one of the world's most important non-wood forest products (NWFPs), vividly illustrates the difficulties and uncertainties inherent in ascertaining sustainable extraction levels and impacts associated with harvesting. Rattan is collected almost exclusively from wild populations and market demand for cane is strong. Indonesia supplies over 90% of the world's commercial rattan cane (Dransfield and Manokaran, 1993) and the majority is gathered from forests in which management has been largely absent or ineffective (Barr, 2000). There has been little or no monitoring or management of wild rattan harvesting and virtually nothing is known about ecological effects associated with extraction.

Sustainable harvesting of NWFPs has been advocated as a means to simultaneously conserve forests and encourage economic development (Anderson, 1990; Freese, 1997) and is now an integral component of most tropical forest conservation and management efforts (CIFOR, 1999). However, many ecologists contend that NWFP harvesting is neither ecologically sustainable nor economically viable (Kramer *et al.*, 1997; Rice, *et al.*, 1997) and that the notion of sustainable extraction is an ecological oxymoron (Struhsaker, 1998). Is sustainable wild rattan harvesting an oxymoron? If not, under what biophysical, socio-economic and institutional conditions might harvesting be viable? What methods could be employed to assess and monitor effects associated with cane extraction? Finally, if harvesting wild rattan results in adverse effects to other flora or fauna, how do those impacts compare with alternative economic activities (e.g., converting forests to annual or perennial cash crops)?

In this paper, I document opportunities and constraints to rattan cane harvesting, and methods to assess and monitor ecological effects associated with extraction. I focus on *Calamus zollingeri*, a commercially important large-diameter rattan used in furniture manufacturing, and base my analysis on five years of on-going research in Central Sulawesi, Indonesia. Given the wide range of possible direct and indirect ecological effects associated with harvesting in primary forests, I focus on only two of the most important: (i) direct impacts to the plant itself (i.e., to genets, ramets and ramet growth); and (ii) the use of floater logs to transport cane to market.

2. *Calamus zollingeri* Cane Harvesting and Monitoring Methods

Calamus zollingeri is a robust, clustering rattan found throughout Sulawesi, the Moluccas and other eastern Indonesia islands and is one of the most important large-diameter canes in the furniture industry (Dransfield and Manokaran, 1993). Virtually all *C. zollingeri* cane is collected from unmanaged, wild populations in primary forests. Despite prohibitions, huge quantities of *C. zollingeri* cane are collected from Lore Lindu National Park (LLNP) and cane harvesting shows no signs of declining (*pers. obs.*). Indeed, given the collapse of central Indonesian governmental authority to prohibit collecting and expanding export demand, rattan harvesting is likely to accelerate.

One of the challenges in assessing ecological sustainability is determining and establishing monitoring protocols for the vast array of possible direct and indirect biological effects. Principal effects associated with rattan cane harvesting at the species level include: effects to genets, ramets, and ramet production and growth; at the ecosystem level: effects on ecosystem nutrients, forest structure, forest succession, and vertebrate food resource availability; and indirect effects from transporting cane to market and incidental hunting of birds and mammals. While assertions by biologists that it is impossible to reliably and comprehensively ascertain all possible ecological effects is undoubtedly true, continued rattan harvesting necessitates that efforts be made to monitor and manage cane harvesting. Table 1 summarizes the primary direct and indirect effects associated with *C. zollingeri* cane harvesting in Central Sulawesi and the methods used to assess and monitor them in this study.

Table 1: Potential Effects of Rattan Cane Harvesting and Monitoring Methods

Potential Effect	Monitoring Methods
Species level	
Genet Survival	Resample permanently marked plants
Genet population Structure	Resample permanently marked plants and replicated sampling of
Cane (i.e. ramet) production	random transects
Cane growth	" " " "
Ecosystem level	
Nutrient stocks	Determine nutrients in foliage and canes and volume cane
	extracted/unit area
Forest structure	Long-term monitoring of sample plots
Forest succession and composition	" " "
Understory trampling	" " "
Vertebrate food resources	Biweekly sampling of marked plants
Invertebrate use	Not investigated in the study
Other	
Cane transport (floater logs)	Determine weight of rattan extracted, tree species and volume used as
	floater logs, and location trees extracted
Hunting	Not investigated in this study

A wide variety of methods can be used to assess and monitor NWFP harvesting (Boot and Gullison, 1995; Godoy and Bawa, 1993; Hall and Bawa, 1993; Peters, 1994; Wong, 2000). I experimented with sample plots and transects of various sizes and lengths and found that the use of replicated belt transects measuring 10 x 500 m or 10 x 1000 m, randomly established perpendicular to the contour provided an effective, quick and easy way to assess and monitor *C. zollingeri* populations. Belt transects can also reliably sample the abundance and distribution of rattan and other lianas that tend to have patchy or clumped populations (Hegarty and Caballe, 1991).

I actively involved local rattan collectors in this project as key informants, research assistants, and in helping to identify important research questions and appropriate field sampling methods. Working with the same three experienced rattan collectors, we sampled an average of 500 m of transects per day. We sampled upslope in 10 m segments with two assistants recording the identity of all rattan species, the number of ramets/genet, cane lengths and evidence of harvesting within 5 m of either side of the transect line. We simultaneously gathered slope, elevation, light regime, soil, canopy height, and dominant tree species data along each transect (voucher specimens at the Herbarium, Tadulaku Univ., Palu, Sulawesi).

When first sampling the abundance and distribution of *C. zollingeri* in 1996, neither I nor my assistants knew how populations varied across the landscape or how they responded to site-specific environmental conditions (e.g., light and soil moisture). In the initial survey we found that *C. zollingeri* was restricted to lower montane forests below approximately 1150 m elevation. Consequently, we report *C. zollingeri* populations on a per hectare basis based on those portions of each transect below 1150 m. When recensusing the population in 2000, we sampled only forests below 1100 m using random transects of 10 m x 500 m length and again report the data on a per hectare basis.

To monitor potential harvesting effects on individual plants, we permanently marked (with flagging and metal tags) 106 mature *C. zollingeri* plants along the transects in 1996 and recorded the number of ramets, ramet lengths, evidence of cane harvesting and associated environmental conditions (as noted above). We also established three permanent 1 x 1 m sample plots around each plant to monitor the effects of cane harvesting on understory vegetation (i.e., extent and persistence of trampling). We resampled the marked plants and associated plots annually for four years. The field assistants monitored rattan flowering and fruiting phenology, and evidence of their use by birds and mammals on the marked plants on a biweekly basis for two years. We noted any evidence of damage to forest vegetation along the transects (i.e., branches broken or trees cut to harvest cane) each year and monitored their persistence for four years (e.g., natural and cut tree falls, trampling while harvesting cane).

Leaf and cane nutrient contents and losses associated with harvesting were assessed in paired leaf and cane samples (Siebert, 2001). Impacts of transporting cane to market were investigated by identifying the species, volume and location of trees harvested to float cane down river over a two-year period. Throughout the duration of this study, rattan (including marked plants) were harvested as desired by local collectors. In general, long canes (i.e., greater than 20 m) that could be readily pulled from supporting vegetation were preferred and no canes shorter than 10 m were cut. In the following sections I review only harvesting effects on *C. zollingeri* genets and ramets, and effects associated with transporting cane to market.

3. Ecological Effects Associated with Cane Harvesting

Calamus zollingeri is abundant and widely distributed throughout lower montane forests and is one of the most common rattans in and around LLNP. At approximately 1150 m it is replaced by *Daemonorops sarasinorum* and possibly *C. sp "ahliduri"*, but all three species are marketed under the trade name "batang". We recorded an average of 149 *C. zollingeri* genets with 1431 ramets and 66 canes of harvestable length (i.e., > 10 m length) per hectare in forests below 1150 m when first sampling the case study watershed in 1996 (Table 2). In 2000 we found an average of 143 genets with 1595 ramets and 46 canes per hectare in the same watershed. Populations of *C. zollingeri* exhibited extremely patchy distribution with massive plants (i.e., greater than 20 ramets) dominating the understory and canopy along some portions of the transects, while no *C. zollingeri* were observed in other areas.

Table 2: Differences in *Calamus zollingeri* Genet and Ramet Populations over 4 Years

	Mean number ha^{-1} (+/- std.dev.) based on 100 m^2 sample plots along 3 random transects	
	1996 (n = 205)	2000 (n = 150)
Genets	149 (+/- 103)	143 (+/- 118)
Ramets	1431 (+/- 1402)	1595 (+/- 1437)
Harvestable Canes (> 10 m)	66 (+/- 120)	46 (+/- 84)

Calamus zollingeri occurred in both high-light (e.g., canopy gaps) and densely-shaded environments, but was absent in poorly drained or seasonally flooded sites in both this site and a previous study in North Sulawesi (Siebert, 1993). Finally, we observed no evidence of mortality or dieback to *C. zollingeri* genets, irrespective of the frequency or number of canes harvested. For example, on marked plants we recorded no cane harvesting-induced mortality even though approximately 33% were harvested each year and in 2000 canes were frequently cut upon reaching 10 m in length.

On marked *C. zollingeri* plants we found that the mean number of ramets/genet was significantly greater in 2000 than in 1996 (Table 3). However, the mean number of harvestable ramets/genet (i.e., canes > 10 m length) was significantly lower in 2000 than in 1996, as was the mean length of harvestable cane and the total amount (i.e., length) of harvestable cane. In fact, in 2000 there was less than half the amount of harvestable cane as recorded on the same plants in 1996. The impact of intensive cane harvesting is readily apparent when comparing cane length distribution classes in 1996 with 2000 (Figure 1). For example, we recorded 37 canes greater than 20 m in length on marked *C. zollingeri* in 1996, but only three canes greater than 20 m on the same plants four years later.

Table 3: *Calamus zollingeri* Ramet Production and Growth over 4 Years on Marked Plants

	Marked Plants (n = 74)	
	1996	2000
Mean number ramets/genet	12.4**	15.6**
Mean number harvestable ramets/genet (cane > 10 m)	1.0*	0.7*
Mean ramet (cane) length	22.4 m**	11.4 m**
Mean length harvestable ramet (canes > 10 m)	26.0 m**	17.3 m**
Total length harvestable Cane (all plants)	1953 m	880 m

**   means significantly different at P=0.005 based on paired sample t test.
*   means significantly different at P=0.05 based on paired sample t test.

These data have several important implications for the management of wild rattan. First, repeated cane harvesting (i.e., cutting mature ramets at approximately 1 m height) appears to stimulate the production of new ramets. Indeed, we recorded an average of 3.5 new ramets produced per genet and 4.7 m of cane growth on marked plants one year after harvesting. The production of new ramets and rapid cane growth suggests that large numbers of canes could continue to be available in the future. Secondly, cane harvesting has significantly reduced mean cane length which, in turn, reduces returns to labour and requires collectors to travel further into the forest. Thirdly, the cutting of all mature ramets may preclude sexual reproduction. This is supported by the absence of flowering and fruiting on marked rattan during two years of bi-weekly monitoring. While *C. zollingeri* is clearly capable of vigorous vegetative growth, sexual reproduction is important for maintaining species vigour and diversity over the long-term.

4. Effects of Cutting Trees for Use as Floater Logs

Transporting rattan to market in the southern LLNP region entails dragging rattan to a river, cutting and bundling cane, cutting small trees for floatation, and then floating the bundles down the Lariang River for 2-14 days to a roadside access point. The use of rivers to transport rattan, timber and other forest products is common throughout Asia. In the Lariang River drainage small trees are cut and dried to float rattan and tree cutting is reported to be a serious threat to biological diversity and forest conservation in LLNP (BCN, 1996; Schweithelm *et al.*, 1992).

Eight tree species were regularly used to float rattan (Table 4). Not surprisingly, floater trees were light-weight, fast-growing, pioneer species. Floater logs averaged 25 cm diameter and 3 m in length, and typically floated a 50-60 kg bundle of cane. Canes were cut to 4 m length before bundling with the number of canes/bundle varying with cane diameter (i.e., weight). We recorded an average of 135 tons of rattan extracted each year from the case study watershed (from October 1996 to October 1998). If one conservatively assumes that each rattan bundle weighed 50 kg, a total of 2350 logs were required to float cane down river each year from the case study watershed.

Figure 1: Distribution of Calamus zollingeri Cane Lengths

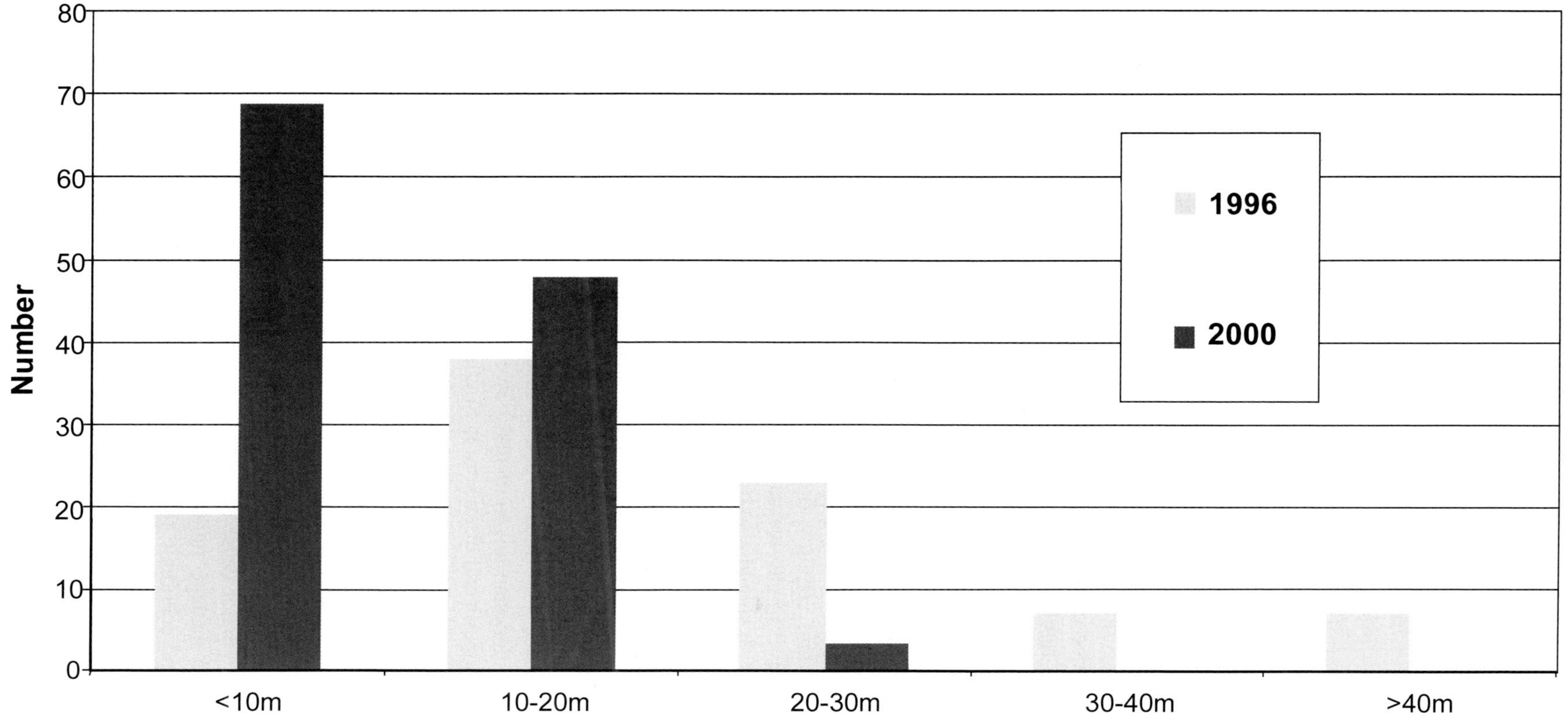

Table 4: Tree Species Used to Float Rattan

Species	Local Uma Name
Artocarpus teysmannii Miq.	tea uruh
Evodia latifolia D.C.	ki hio
Grewia multiflora Juss.	wokeh
Horsfieldia sp.	laru
Macaranga hispida Muell.Arg.	meapoh
Macaranga triloba (L.) Muell.Arg.	lengkoba
Pterospermum celebicum Miq.	entorodeh
Trema orientalis (L.) Bl.	wulajah

Floater logs were harvested almost exclusively from fallowed shifting cultivation fields and to a lesser extent from naturally disturbed riparian flood plains along the Lariang River. Over the four-year study period, we found no evidence of floater log cutting in primary forests within LLNP. According to rattan collectors, floater logs are rarely harvested from primary forests because suitable-sized, light woods are uncommon there and because primary forests are further from the river than fallowed swiddens (*pers. com.*). Thus, there is little evidence to support the claim that cutting small, early successional trees to float cane threatens primary forests or biodiversity in LLNP, at least within this case study watershed.

5. Research Limitations and Future Research and Development Needs

The monitoring methods employed in this study, while time consuming and reasonably rigorous, are not comprehensive and certainly do not assess all possible ecological impacts associated with wild rattan harvesting. For example, we gathered no information on invertebrate use of *C. zollingeri* or hunting by rattan collectors. It should also be emphasized that we monitored *C. zollingeri* for only four years. Repeated harvesting of all mature cane could, over a long period of time, adversely affect plant vigour or ramet production and growth. Thus, as conservation biologists argue, we can not prove that *C. zollingeri* cane harvesting is ecologically sustainable or that it is without adverse ecological effects. Nevertheless, the use of randomized and replicated sampling methods over a four year period is biometrically rigorous (Wong, 2000) and can provide the basis for assessing ecological impacts and initiating adaptive management practices.

Given the absence of effective and enforceable prohibitions against rattan harvesting, the continuing devolution of Indonesian governmental power to provincial and local authorities, and growing development pressures in Indonesia and elsewhere in Southeast Asia, one must ask – how will resident people secure their livelihoods if rattan harvesting is prohibited? In Central Sulawesi, as in much of the tropical world, domestic and international market forces encourage the cultivation of export cash crops (e.g., oil palm, cacao, coffee, etc.) in large plantation schemes (Collier *et al.,* 1994; Sunderlin, 1999). Under these conditions, if a forest has no value, it will likely be converted to something that does. When considered in this light, the potential ecological effects associated with *C. zollingeri* cane harvesting appear benign.

Comparing managed harvesting of wild rattan with agricultural plantations is also relevant because agriculture is the major cause of forest conversion in Southeast Asia and because the few large-scale efforts to cultivate and manage rattan have been on plantations (e.g., SAFODA, Malaysia). While long-term growth, yield and economic return data have not been published from these efforts, there is no agronomic reason why large-diameter, furniture-quality rattans could not be grown in large plantations. The long history of cultivating small and medium-sized rattan (*C. caesius* and *C. trachycoleus*) in swidden fallows in Kalimantan (Weinstock, 1983) suggests that large canes could be grown by small farmers as well.

This paper addresses ecological aspects of wild rattan gathering, a field in which I have expertise. However, even if harvesting wild rattan appears to be ecologically viable, socio-political and institutional aspects of cane extraction will require critical consideration. In addition, given declining wild rattan supplies it may prove necessary to cultivate rattan and that raises questions as to the type of production system one pursues. Key questions that policy makers must consider include: What is the most reliable and cost-efficient way of producing large-diameter rattan cane – management of wild populations or intensive cultivation by small farmers or in large plantations? What are the ecological effects associated with these different approaches? What economic costs and benefits are associated with managing wild populations vs. cultivating rattan in plantations? And what implications do these differing approaches have on the distribution of social and economic costs and benefits? In the following section I briefly review these issues in light of analyses by social scientists who have studied rattan and my own observations from Central Sulawesi.

6. Social, Economic and Institutional Challenges to Rattan Harvesting and Cultivation

Rattan is widely recognized as an important domestic and internationally-traded commodity. Indeed, rattan can be considered the "flagship" NWFP due to its unsurpassed importance in household, village, provincial and national economies. Rattan collection, trade, processing and manufacturing operate with a complex and dynamic socio-economic, political and ecological context. Crucial components include decentralized and dispersed cane collection; geographic centralization of manufacturing capacity; poor communications and infrastructure; ethnic, religious and social differences among collectors, traders and manufacturers; and the low priority of rattan among national governments.

Forests throughout Southeast Asia have been under formal state jurisdiction since colonial times, and both colonial and post-independence states have attempted to control and manage forest resources, albeit with great difficulty and limited success (Barr, 2000; Peluso, 1996; Sunderlin, 1999). In the case of rattan, efforts to control, regulate and manage cane harvesting have been largely absent or ineffective.

Since the 1970s, wild rattan supplies have drastically declined due to logging, forest conversion, over-harvesting, and forest fires. Some premier, large-diameter species such as *C. manan*, a solitary rattan that does not reproduce vegetatively, are nearly extinct (Dransfield and Manokaran, 1993). In some areas of Indonesia and elsewhere in Southeast Asia, rattan exploitation exemplifies what can occur under unregulated, open-access resource extraction conditions. However, in many regions the loss of rattan resources reflects political and economic choices by state and private industrial elites (backed by the military) to ignore traditional, customary resource tenure and forest management practices of resident ethnic minorities in the rush to exploit timber or convert forests to agricultural plantations. The "legal" destruction of wild and cultivated rattan for commercial logging and plantation agriculture is well documented among the Dayak of Kalimantan (Fried, 2000; Belsky, 1992) and has resulted in the destruction of traditional rattan production and management systems that have operated for generations. Studies by Mayer (1989), Dransfield (1988), Godoy and Feaw (1988), and Weinstock (1983) suggest that the social organization of small-holder rattan agroforestry systems were economically viable and reliably produced large quantities of cane in ways that were compatible with community economic and social well-being, and cultural identity (however, still not know if in a sustainable manner).

Given this history, is managed harvesting of wild rattan possible? If so, by whom and under what property rights arrangements? Given the success of smallholder rattan production systems, should they be protected, studied, used as models elsewhere? What are the comparative ecological, economic and socio-political benefits of smallholder vs. plantation rattan production systems? With regard to the latter, to what extent should concerns of ethnic minority cultural survival and economic justice enter into policy and resource development considerations?

Throughout Southeast Asia there is sufficient evidence to suggest that NWFPs, including rattan, have been successfully managed as a common property resource by traditional forest dwelling peoples for centuries (Lynch and Talbot, 1995; Peluso 1992; Peluso and Padoch 1996). In general, common property resource management succeeds where groups are relatively small and stable; where resource management perspectives, and issues of access and control are shared; and where enforcement is simple and inexpensive (Ostrom, 1990). However, historic common property resource management systems have been suppressed or usurped by colonial and post-colonial authorities in Indonesia and other Southeast Asian countries

(Peluso, 1996). Consequently, the social and institutional characteristics required for communal management of wild rattan face tremendous challenges in many regions.

In Central Sulawesi rattan collecting is an important livelihood strategy for young men and for households unable to secure food and income through other, preferred means (e.g., irrigated rice farming, shifting cultivation and perennial cash crop farming). Indeed, half of the households interviewed in a random household survey conducted in 1999 reported that selling rattan was their most important source of income (coffee and cacao was the second most important source, cited by 30% of respondents). Rattan gathering is widely acknowledged to be dangerous and demanding work that involves extended periods away from home. Not surprisingly, once men are married or have producing coffee and cacao farms, they engage in rattan collection less often than before. Coffee and cacao begin to yield four and three years after planting, respectively. In contrast, large-diameter rattan will not likely produce cane in less than 12 to 15 years. Nevertheless, residents in this study area expressed interest in rattan cultivation and preliminary results indicate that *C. zollingeri* seedlings thrive when intercropped in shade-grown coffee and cacao farms (Siebert, 2000).

A narrow consideration of economic costs and benefits suggests that returns from rattan gathering and cultivation compare poorly with perennial cash crop alternatives. However, it is important to remember that rattan is a primary or secondary source of cash income for tens of thousands of forest-dwelling people throughout Southeast Asia (DeBeer and McDermott, 1989) and is an irreplaceable source of emergency income for thousands more (Siebert and Belsky, 1985). Furthermore, rattan gathering and cultivation by small-holders in either swidden fallows or as an intercrop in traditional agroforestry systems provides important social and environmental benefits that tend to be ignored in narrow cost/benefit analyses. Foremost among these benefits are: (i) reduced economic risk due to less dependence upon volatile coffee and cacao markets; (ii) the potential to increase total returns; (iii) reduced insect and disease infestation rates that may result from greater species and structural diversity; and (iv) the maintenance of high levels of biological diversity and thus at least partial compatibility with biodiversity conservation objectives.

Given the relatively low financial returns from wild rattan harvesting and long period of yield deferral when cultivating large-diameter canes, whether by small farmers or large estates, significant private investment in rattan management or cultivation is unlikely without outside (i.e., non-governmental, state or international) subsidies. Tayor and Zabin (2000) argue that support of community resource management such as this should be viewed not as a subsidy, but rather as payment for goods and services (e.g., carbon sequestration, functional watersheds, and biodiversity conservation) provided by intact forests. This approach could be broadened to include compensating people who are denied access to historic resources due to the establishment of protected areas. Targeted external funding of this sort could provide sufficient financial incentive for small farmers and rattan collectors to cultivate and manage rattan in and around protected areas. Increased returns from rattan harvesting might also be realized through green certification as advocated by the Rainforest Alliance and has been granted to timber harvested in certifiably sustainable ways.

7. Conclusions

Declining supplies and strong market demand suggest that rattan resources will become increasingly scarce, particularly for large-diameter canes. Two general approaches could be pursued to increase rattan supplies: management of wild populations and/or smallholder or estate cultivation. Both strategies entail significant challenges, particularly regarding the unfavourable financial returns (narrowly calculated) of rattan in comparison to cash crop alternatives. The two approaches will also have profoundly different effects on different sectors of society, particularly in the case of smallholder vs. estate cultivation.

I argue that efforts to manage wild rattan or to cultivate rattan in small farms or plantations should focus on: (i) large-diameter, furniture-quality, clustering species (i.e., those that produce multiple canes and sprout new ramets when cut); (ii) the establishment of rigorous and standardized monitoring protocols; and (iii) careful consideration of social and economic costs, benefits and their distribution among different sectors of society. Promising large-diameter, clustering and coppicing rattan species include *C. zollingeri*, *C. merrillii* and *C. subinermis*. Widespread adoption of rigorous and standardized monitoring methods will help elucidate long-term ecological and social effects associated with rattan harvesting and cultivation, and provide information that can be used to modify and adapt management systems within the context of dynamic, unpredictable and chaotic ecological, social, economic and political environments. Long term monitoring and adaptive management practices could also be used to benefit different segments of society. Careful attention to social justice and equity issues could help insure that those most dependent upon rattan resources benefit from future investments, while simultaneously reducing their need to convert protected and production forests to farms.

Successful rattan cultivation and management will require that local people participate in all aspects of the enterprise, not merely work as wage labourers. When people benefit from natural forests (e.g., managing wild rattan harvesting) or agricultural development (e.g., cultivating rattan), incentives can be created to retain forests and traditional, complex agroforestry practices, and the associated public goods and services that these systems provide (Western and Wright, 1994).

Managed harvesting of wild rattan and rattan cultivation will likely require significant long-term financial assistance, as well as technical and marketing support. It is essential that the size and type of support compliment local resource managers (i.e., resident people) and their institutional capabilities. Policy makers should pay particular attention to providing small farmers and rattan collectors with adequate economic incentives, particularly vis-à-vis perennial cash crop alternatives, and to developing secure, stable and enforceable resource management institutions and property rights. Private, state and international support for rattan management and cultivation can be justified as compensation for the public benefits provided by natural forests and diverse agroecosystems and for the loss of historic access to resources by forest dwellers living in and around protected and production forests.

Acknowledgements:

I greatly appreciate the assistance, knowledge and friendship of Daud and Arnol from Moa, and the valuable comments provided by Jill Belsky on earlier drafts of this manuscript. This work was supported by USAID grant HRN-G-00-93-00047-00 and the University of Montana.

REFERENCES

Anderson, A. (ed)., 1990. Alternatives to Deforestation. Columbia Univ. Press, NY.

Barr, C., 2000. Will HPH reform lead to sustainable forest management?: Questioning the assumptions of the "sustainable logging" paradigm in Indonesia. CIFOR, Bogor, Indonesia.

BCN. 1996. Biodiversity Conservation Network Annual Report. Asia/Pacific Region, WWF, Washington, D.C.

Belsky, J., 1992. Balancing Forest and Marine Conservation with Local Livelihoods in Kalimantan and North Sulawesi. Final report to the Indonesia Natural Resource Management Project. USAID/ARD, Jakarta, Indonesia.

Boot, R. and R. Gullison. 1995. Approaches to developing sustainable extraction systems for tropical forest products. Ecological Applications 5: 896-903.

CIFOR. 1999. The World Heritage Convention as a Mechanism for Conserving Tropical Forest Biodiversity. CIFOR, Bogor, Indonesia.

Collier, G.; D., Mountjoy; R. Nigh, 1994. Peasant agriculture and global change. BioScience 44:398-407.

DeBeer, J. and M. McDermott. 1989. The Economic Value of Non-timber Forest Products in Southeast Asia. Netherlands Committee for IUCN, Amsterdam.

Dransfield, J., 1988. Prospects for rattan cultivation. Advances in Economic Botany 6:90-200.

Dransfield, J. and N. Manokaran (eds). 1993. Rattans. Plant Resources of South-East Asia, PROSEA # 6. Bogor, Indonesia.

Freese, C., 1997. The "use it or lose it" debate. In: Freese, C. (ed). Harvesting of Wild Species. The John Hopkins University Press, Baltimore. pp. 1-48..

Fried, S., 2000. Tropical forests forever? A contextual ecology of Bentian rattan agroforestry systems. In: C. Zerner (ed). People, Plants and Justice: The Politics of Nature Conservation. Columbia Univ. Press, NY. pp. 204-233.

Godoy, R. and K. Bawa. 1993. The economic value and sustainable harvest of plants and animals from the tropical forest: Assumptions, hypotheses, and methods. Economic Botany 47:215-219.

Godoy, R. and T. Feaw. 1988. Smallholder Rattan Cultivation in Southern Borneo, Indonesia. Harvard Institute for International Development. Cambridge, MA.

Hall, P. and K. Bawa. 1993. Methods to assess the impact of extraction of non-timber tropical forest products on plant populations. Economic Botany 47:234-247.

Hegarty, E. and G. Caballe. 1991. Distribution and abundance of vines in forest communities. In: F. Putz and H. Mooney (ed). The Biology of Vines. Cambridge Univ. Press, Cambridge, UK. pp. 313-335.

Kramer, R., C. vanSchaik; J. Johnson (eds). 1997. Last Stand: Protected Areas and the Defense of Tropical Biodiversity. Oxford Univ. Press, Oxford, UK.

Lynch, O. and K. Talbott. 1995. Balancing Acts: Community-based Forest Management and National Law in Asia and the Pacific. World Resources Institute, Washington, D.C.

Mayer, J., 1989. Rattan cultivation, family economy, and land use: a case from Pasir, East Kalimantan. In: Forestry and Forest Products. German Forestry Group (GFG) Report No. 13, Samarinda, Indonesia.

Ostrum, E., 1990. Governing the Commons: The Evolution of Institutions for Collective Action. Cambridge University Press, Cambridge, UK.

Peluso, N., 1996. Rich Forests, Poor People: Resource Control and Resistance in Java. University of California Press, Berkeley, CA.

Peluso, N., 1992. The ironwood problem: (mis)management and development of an extractive rainforest product. Conservation Biology 6:210-219.

Peluso, N. and C. Padoch. 1996. Changing resource rights in managed forests of West Kalimantan. In: C. Padoch and N. Padoch (eds). Borneo in Transition: People, forests, conservation and development. Oxford Univ. Press, Kuala Lumpur.

Peters, C., 1994. Sustainable Harvest of Non-timber Plant Resources in Tropical Moist Forest: An Ecological Primer. The Biodiversity Support Program, World Wildlife Fund, Washington, D.C.

Rice, R.; R. Gullison; J. Reid. 1997. Can sustainable management save tropical forests? Scientific American 276(4):44-49.

Redford, K., 1996. Getting to conservation. In: K. Redford and J. Mansour (eds). Traditional Peoples and Biodiversity Conservation in Large Tropical Landscapes. America Verde Publications, The Nature Conservancy, Arlington, VA. pp. 251-265.

Schweithelm J.; N. Wirawan; J. Elliott; J. Khan. 1992. Sulawesi Parks Program Land Use and Socio-Economic Survey: Lore Lindu National Park and Morowali Nature Reserve. The Nature Conservancy, Jakarta.

Siebert, S., 2001. Nutrient levels in rattan (*Calamus zollingeri*) foliage and cane and implications for harvesting. Biotropica (accepted).

Siebert, S., 2000. Survival and growth of rattan intercropped with coffee and cacao in the agroforests of Indonesia. Agroforestry Systems 50:95-102.

Siebert, S., 1993. The abundance and site preferences of rattan (*Calamus exilis* and *Calamus zollingeri*) in two Indonesian national parks. Forest Ecology and Management 59:105-113.

Siebert, S. and J. Belsky. 1985. Forest-product trade in a lowland Filipino village. Economic Botany 39:522-533.

Struhsaker, T., 1998. A biologist's perspective on the role of sustainable harvest in conservation. Conservation Biology 12:930-932.

Sunderlin, W., 1999. Between danger and opportunity: Indonesia and forests in an era of economic crisis and political change. Society & Natural Resources 12: 559-570.

Taylor, P. and C. Zabin. 2000. Neoliberal reform and sustainable forest management in Quintana Roo, Mexico: Rethinking the institutional framework of the Forestry Pilot Plan. Agriculture and Human Values 17:141-156.

Weinstock, J., 1983. Rattan: ecological balance in a Borneo rainforest swidden. Economic Botany 37:58-68.

Western, D. and R. Wright. 1994. The background to community-based conservation. In: D. Western and R. Wright (eds). Natural Connections. Island Press, Washington, D.C. pp. 1-12.

Wong, J., 2000. The biometrics of non-timber forest product resource assessment: a review of current methodology. Dept. for International Development (DFID), UK.

RATTAN IN THE TWENTY-FIRST CENTURY
– AN OUTLOOK

Cherla B. Sastry

1. Introduction

This gathering is well aware that rattans are the most important group of forest species after timber, especially in Asia. Humans have used rattan for livelihood and subsistence for many centuries throughout the documented history of mankind (Anon, 1983 and 1991; Abd. Latif, 2000; Johnson, 1997; Manokaran, 1990; Pabuayon, 2000; Sarma, 1989; Sastry, 2000; Tomba, Lapis *et al.*, 1993). Although rattan is confined mainly to Asia, the material found its way to many other parts of the world, such as ancient Egypt, Europe during the Renaissance period, and France during the reigns of Louis XIII and Louis XV of France (Anon, 1983).

This Expert Consultation on Rattan Development has already discussed the nature of the resource, the versatility, importance, multiple uses and the ongoing national and international R&D activities to develop the resource. This consultation has been organized at an appropriate time when rattan, a multi-billion dollar commodity, is in short supply and a general decline has affected the growth of the industry. Southeast Asia is by far the largest producer and exporter of rattan and rattan products globally. The purpose of this meeting is to assess the current situation, identify the major issues facing the rattan industry, and formulate a set of recommendations and an action plan towards economic and technical cooperation among the concerned countries for the development of rattan globally.

Rattan is by far the most important Non-Wood Forest Product (NWFP) in international trading. Yet, until recently, this group of plants received only benign attention from all but a small band of enthusiasts. Worldwide, over 700 million people trade in or use rattan for a variety of purposes, such as the beautiful furniture for which the material is universally known. There are, however, no quantitative estimates of the true economic/social value of rattan (Anon, 1998; Abd. Latif, 2000; Belcher, 1999; Manokaran, 1990; Pabuayon, 2000).

The global trade (domestic and export) and subsistence value of rattan and its products is estimated at over US$7,000 million per annum (Anon, 1991; Abd. Latif, 2000; Belcher; 1999; Manokaran, 1990; Pabuayon, 2000; Sastry, 2000; Soedarto, 1999). Undoubtedly, furniture is the most popular rattan product. Besides furniture, other products include carpet beaters, walking sticks, umbrella handles, sporting goods, hats, ropes, cordage, bird cages, matting, baskets, panelling, hoops, ammunition boxes and a host of other utility products.

Although rattan's ecological role is much less studied than that of bamboo, it is likely that rattan species with subterranean stems or those with widely radiating, horizontally growing roots could have a significant role in preventing soil displacement (Lakshmana, 1993; Nur Supardi, Hamzah and Wan Razali, 1999; Wan Razali, Fransfield and Manokaran, 1992).

Unlike bamboo, rattan is an integral part of the tropical forest ecosystem owing to its climbing habit. The innumerable pinnate leaves, which extend up to 7 m or more in length in many large species with their mosaic arrangement, play a major role in intercepting the splash effect of rain. The species also play a vital role in enriching the soil by their leaf litter, which adds to the organic content of the soil (Lakshmana, 1993). These ecological and other economic benefits of rattan as noted above, raise the value of standing forest (Anon, 1998; Lakshmana, 1993; Rawat and Khanduri, 1999).

The forests of the world provide an essential renewable resource, but they are nonetheless a finite resource. Predictions for this century are that the demand for the timber by the various wood-based industries in Asia and elsewhere will exceed existing supply. The problem of supplying forest products and other essentials to an ever-increasing population would become more acute with each passing year. What we do as we go on from here is more critical than ever before (Sastry, 2000).

2. Overview

Rattans, spiny climbing palms with some 600 species, are strictly Old World in origin. Its distribution is limited to tropical and subtropical Asia, where ten of the 13 known genera are endemic to equatorial Africa. True rattans are not known in Latin America. The greatest diversity is in the Malay Peninsula and Borneo. There is a secondary centre of diversity in New Guinea (Anon, 1991; Wan Razali, Dransfield and Manokaran, 1992).

In recent years, rattan and other NWFPs, e.g. bamboo and medicinal plants, have been accorded international priority in contrast to their previous categorization as "minor forest products". This heightened attention results from the recognition of their increasing economic, ecological and socio-cultural importance and the consequent need for sustainable use of the resource (Anon, 1991; Johnson, 1997).

The full socio-economic potential of rattan is yet to be realized and no quantitative estimates of the true value are available. Domestic trade and subsistence use of rattan create benefits estimated at US$3 billion per annum and another US$4 billion are generated through global exports. Additional benefits may accrue from the intervention in the sector to systematize resource use, management, marketing and processing (Anon, 1991; Manokaran, 1990; Wan Razali, Dransfield and Manokaran, 1992). Furthermore, today only a small portion of the approximately 600 species of rattan found is used for commercial purposes. If more of the presently under-utilized and lesser-known species are added to this list, the benefits could be wide ranging.

Rattan is almost entirely collected from natural forests and rattan gardens. In recent years, uncontrolled harvesting and deforestation have led to resource exhaustion of the desired species in many rattan-producing countries in Asia. The rattan sector cannot be discussed without first mentioning Indonesia and its dominance over world rattan trade. Indonesia has a clear advantage over other countries with its overwhelming supply of wild and cultivated rattan (80 percent of the world raw materials)[10]. The estimated production capacity for all species is about 12.4 million tonnes from the 11.5 million ha of the country's forested areas rich in rattan. The annual allowable cut is estimated at 700,000 tonnes. Indonesia's actions, therefore, will have a large impact on the global rattan market (Anon, 1998; Soedarto, 1999). The Philippines, Malaysia, China, Thailand and other countries of Indochina are also important contributors to global rattan trade, despite dependence of some of these countries on supplies of raw materials from Indonesia to augment domestic supplies.

Although Asia is the dominant player, it accounts for only about 58 percent of the world trade in rattan activity, the remaining 42 percent being held by industrialized countries, importing Southeast Asian rattan (Anon, 1999). Compared with world trade of all furniture (US$80–100 billion), rattan furniture trade represents less than 4 percent. However, in Asia the output of the rattan furniture industry represents well over 25 percent of all furniture industry output, and it is growing (Anon, 1983; Johnson, 1997).

The rattan industry is highly fragmented with over 90 percent of all factories employing less than 50 people, i.e. cottage and small-scale enterprises. In general, rattan furniture manufacturing is highly labour-intensive, employing well over one million people in Asia, of whom about 500,000 work in the manufacturing sector and another 700,000 are involved in the collection (and primary processing) and transportation of raw material (a majority on a seasonal basis). The average investment per worker in a modern rattan factory is about US$2,000, whereas it is ten times that much in a conventional furniture plant (Anon, 1983, 1991 and 1999).

In the 1970s and 1980s, the rattan industry in Southeast Asia and China grew at rates between 20 percent and 50 percent per year. The mid-1990s saw a significant downturn in most of Asia, especially in resource-poor

---

[10] Note from the editors: this figure is an estimation and varies from paper to paper.

countries, owing to shortage of raw material, restrictive government policies and the economic crisis (Anon, 1998 and 1999).

There are several forest research institutes, National Agricultural Research Systems (NARS), and regional and international organizations carrying out (or involved in) R&D on rattan. The latter category includes the International Union of Forestry Research Organizations (IUFRO), FAO and other United Nations agencies, CGIAR institutions, ITTO and the Asian Development Bank (ADB). A recent entrant into international forestry R&D, in particular NWFP, is the International Network for Bamboo and Rattan (INBAR). Regional Networks like the Forestry Research Support Programme for Asia and the Pacific (FORSPA) and the Asia-Pacific Association of Forestry Research Institutions (APAFRI) and the African Rattan Network are also actively involved in rattan development. A Rattan Information Centre (RIC), financed by IDRC Canada, was set up in Malaysia in 1982 for disseminating information on all aspects of rattan (Anon, 1991; Abd. Latif, 2000; Sastry, 2000; Rawat and Khanduri, 1999)

3. Major Issues

3.1 Resource Base

As a result of the rapid growth of the industry from the 1970s until the early 1990s, there was overexploitation and wasteful utilization of the resource and consequent depletion of the stock, especially of the desired species. Since the mid-1990s, the dwindling supplies of rattan owing to overexploitation and steady loss of forest habitat are posing a serious threat to the rattan industry. As a result, many Asian countries have experienced declining exports and closing of several operations. The hardest hit are the Philippines and China (Anon 1998 and 1999; Abd. Latif, 2000; Pabuayon, 2000; Soedarto, 1999).

To avoid further depletion of resources, governments of major rattan-producing countries in Asia embraced a ban on export of raw cane and/or heavy duties on export of semi-processed products. While there was an initial glut of raw material supplies in resource-rich countries such as Indonesia and Malaysia, the shift from the traditional practice of exporting the raw rattan to semi-processed and finished product has promoted local industries (Anon, 1999). Abd. Latif reports a significant increase (almost 200%) in the export value of finished products (mainly furniture) from Malaysia in the 1990s as a result of the export ban. The exports are expected to reach US$ 40 million over the next decade (Abd. Latif, 2000). Indonesia reported similar experiences although much slower growth, given their already leading role in world trade, i.e. US$200 million in 1987 (effective date) to an average of over US$300 in the 1990s (Soedarto, 1999).

To cope with the increasing global demand for rattan, there is an urgent need for sustainable management of the resources. A first step is to determine accurately the extent of the resource, both for establishing plantations of the desired species and for improved management of the existing stock in their natural habitats. Figures for most countries are approximate or absent (Anon, 1998). However, a step in the right direction has been taken recently through the joint efforts of FRIM, INBAR and DFID-UK in developing techniques for the inventory of rattan (Nur Supardi, Hamzah and Wan Razali, 1999).

Forest departments manage rattan stocks by:
1. limiting harvests to an allowable cut; and
2. managing resource flows by granting or selling harvesting rights.

Licensing rules vary between countries in eligibility and cost, but the aim of all is to limit overexploitation. In practice, license holders rarely adhere to harvesting guidelines, since there is neither uniformity nor strict enforcement by the field officers. This is a major reason for the depletion of stock, wasteful harvesting practices and the loss of royalties to governments. To promote sustainable resource management, some countries are exploring long-term tenure control through community-based forestry management institutions (Anon, 1998; Pabuayon, 2000).

3.1.1 Technology issues

Rattan passes through many hands that perform one or several levels of processing before it reaches its final state. In most developing countries, rattan processing is still at a craft level carried out in a great number of tiny "workshops" (Aguilar and Miralao, 1985; Anon, 1983, 1991 and 1999). Because of the high labour intensity of the work (i.e. scraping, drying, splitting, sizing, bending, cording and chemical treatments), even when mechanized, it is essential to ensure application of good designs and modern technologies to meet the standards for export. It is assumed that raw materials are not a limiting factor.

In recent years, many technical developments have taken place in the manufacture of rattan products. However, skilled workers and good supervisors are in short supply for higher-end processing. Also, for a majority of the small-scale processors, lack of credit availability and technical assistance limit adoption of modern/efficient technology (Anon, 1983 and 1999; Abd. Latif, 2000). Large manufacturers dominate the sector with better/sophisticated machinery, by adapting more contemporary designs and quality control measures to produce high-quality products that can fetch higher prices. In Indonesia and the Philippines, they have also found ways to reduce the cost of production by farming out to smaller firms specific tasks, mainly in the primary processing area, thus benefiting the sector as a whole (Anon, 1998 and 1999; Abd. Latif, 2000; Pabuayon, 2000).

Given rattan's potential as an industrial material, several countries in Southeast Asia have adapted low-cost automation and mechanization to improve the productivity of their factories. Some governments provided incentives by way of supportive policies, soft loans and tax breaks to the domestic industry. In addition, a few like Malaysia have set up service centres at the district level to provide training, technology transfer and other support. A Small-Scale Entrepreneurs Development Unit (SSEDV) was also created, with financial support from the World Bank and the Government, to provide technical backstopping and related training support to the industry. An Agro-forestry Unit established at FRIM, the Government's Forestry Research Institute, provided training and planting material for planting rattan in rubber plantations by smallholders. The results of all these efforts are increased foreign exchange earnings and employment opportunities in both the rural and urban sectors (Abd. Latif, 2000; Anon, 1998; Pabuayon, 2000;).

3.1.2. Plantation development

Plantations of rattan, either in logged-over forest areas or as an agroforestry crop in rubber or other tree plantations, are needed in order to relieve pressure on overexploited natural forests and to ensure stable supplies of desirable species for the industry. Although significant advances have been made in our understanding of rattan as a potential plantation crop, there is still much that is unknown. As noted earlier, even the basic data on the wild resources is lacking. Rattan plantation development is, with a few exceptions, well behind schedule owing to technical or financial problems (Anon, 1998; Abd. Latif, 2000; Pabuayon, 2000; Salleh, 2000; Soedarto, 1999; Wan Razali, Dransfied and Manokaran, 1992)

To date, more than 31,000 ha have been planted in Malaysia with the large diameter *Calamus manan.* Out of this, 7,000 ha have been planted in rubber plantations throughout the country. Other large plantations, amounting to 10 000 ha, have been established mainly with *C. caesius* and *C. trachycoleus.* Other rattan species that have been looked into include *C. scipionum, C. palustris* and C. subinermis (Abd. Latif, 2000).

Private sector cultivation of rattan, from both large and small-scale plantations, has fallen below expectations and failed to respond to local raw material scarcities. Although an estimated 37,000 ha of mostly high-value rattan species are grown in Indonesia, a paltry 6,000 ha is found in the Philippines, where scarcity is more pronounced (Anon, 1998). China has established over 20 000 ha of rattan plantations in public domain, employing both domestic and imported species.

Policy initiatives (incentives and regulations) to increase small-scale rattan cultivation have had limited impact as government policies and resulting economic conditions make investment in other resources more attractive. In addition to economic constraints, there are other factors that hinder small farmers and pose high risk for many large investors:

1. rattan has a long gestation period (at least 10 to 12 years);
2. no secure tenure over resources; and
3. difficult market conditions.

Though both large and small-scale plantations in Asia (Indonesian and Malaysian experience) are returning profits, other land uses are becoming more lucrative. Nevertheless, more appropriate government interventions in enhancing rattan cultivation are seen in several countries, justified by the economic benefits accrued to the rural households in Indonesia and smallholder rubber plantations in Malaysia (Anon, 1998).

Thus, domestic policies must support rattan plantation development by improving the incentive structure. This involves providing tenurial security to rattan gatherers and planters, credit and technical assistance for plantation development, and favourable harvesting and marketing arrangements. Basic infrastructure like transport and effective mechanism to link sellers with local and foreign buyers is also needed to improve profitability of rattan production, processing and manufacturing activities (Pabuayon, 2000).

In addition, incorporation of plantations into community-based forest management schemes, with or without vertical integration in processing, could be an important policy direction. Some lessons can be learned here from the success of the rattan plantations established in Kalimantan a century and a half ago (Anon, 1991 and 1998; Belcher, 1999).

While the true rattans are not known in Latin America, interest has been building for cultivating Asian rattans in Argentina, Belize, Bolivia, Colombia, Trinidad and Cuba. In recent years, Cuba has successfully introduced rattan from Vietnam, Malaysia and China in a 2,000 ha plantation, with the help of IDRC and INBAR (Anon, 1991).

It may be too early for Africa to be planning large-scale rattan plantations. Rattan is confined to the equatorial rainforests, and is of little economic importance at the present although it has gained recognition as an underexploited crop in West Africa. Kenya and Zambia have received financial and technical support from IDRC for research and introduction of rattan species from Asia. DFID-UK and IFAD, through INBAR, have recently initiated systematic taxonomic and socio-economic research, respectively, on rattan in Ghana, Nigeria, Cameroon, Ethiopia, Tanzania and Uganda (Anon, 1991).

3.1.3. R&D and information exchange

As mentioned earlier, there are several forest research institutes, NARS and regional and international organizations involved in the research or development of rattan. National research on rattan is at an advanced level in Asia, with several active projects funded by international agencies. Some progress has been achieved in the past two decades on rattan silviculture and ecology, plantation technology and development of innovative technologies for low-cost mechanization and automation, grading, and inventory methods. However, continued effort is needed from all concerned, i.e. government, industry and international agencies, to maintain the gains achieved. Particular emphasis should be given on resource assessment and conservation, socio-economics and marketing and design aspects for furniture. Networking between institutions is vital to share knowledge and to benefit the countries less privileged in R&D capabilities.

A Rattan Information Centre (RIC) of regional coverage was established in FRIM in 1982 with financial support from IDRC. It acts as a comprehensive depository of rattan literature, document and retrieval system, publishing regular news bulletins and disseminating information to interested parties. Continued support is strongly recommended to enhance its regional and international role.

There is a felt need for a Regional Rattan Research Centre in Asia to better serve the interests of rattan. Indonesia is a logical location, given its pre-eminent position in global rattan trade. This was also the recommendation by experts (hired by IFAD) who evaluated both the informal and the formal INBAR networks. There is also a need to establish a rattan seed bank to ensure good-quality seeds of good progenies in order to improve plantation quality and yield (Anon, 1991 and 1999; Salleh, 2000).

4. Outlook

It is difficult to gaze through the crystal ball and make predictions on the future of rattan in the twenty-first century, when the basic data needed for the forecast is utterly lacking. There also remain many problems about the trade and cultivation of cane, i.e. uncertainty about the future patterns of supply and demand, about the shifts in global rattan trade, about the economics of cultivation and yields and the lack of uniform policies that affect the sector. There are, thus, more questions rather than answers on the future of rattan. The following generalizations are based mostly on the Asian situation, since the region is the dominant force and a major contributor to global rattan trade.

By all counts, shortage of cane is definitely felt in the region and the problem of resource supply is more glaring than ever before (Manokaran, 2000; Salleh, 2000). There seems to be a decline in the planting of rattan by national forest departments despite the need to increase the supplies. Under these circumstances, do we look at a regional approach to planting, or is it not possible? In Peninsular Malaysia, manufacturers claim that rattan from 10-year old C. manan plantings is of poor quality. This did not appear to be the case for rattan from 15-year old plantings. We still need answers to these and other characteristics of the species. Some growers have also reported problems in harvesting of small-diameter plantation cane (Salleh, 2000).

Given the glut of supply in the country, Indonesia has since lifted its ban on the export of raw cane to increase its foreign exchange earnings (Manokaran, 2000; Pabuayon, 2000). The Philippines is sourcing rattan again from Indonesia to feed its starving industry and to revive sagging exports. Canes are being smuggled from some parts of Indo-China to China and Thailand to keep up their industry and exports. While these countries are trying to revive their operations to the pre-1990s level, the Malaysian industry seems to be suffering from a large number of closings of unprofitable mills either because of stiff competition from neighbouring countries and/or raw material supply problems. The market is thus experiencing some uncertainties.

Figure 11: Rattan bundles in the Philippines (Belcher)

The markets in the consumer countries in Europe, North America, Japan and other industrialized nations seem to be steadily growing for rattan. However, there is an urgent need for a marketing study and future prospects of rattan in those countries.

It is evident that, in view of the state of uncertainty, there is a need to review the ground situation and markets more thoroughly through field visits, to chart a course for rattans in the new millennium. Nevertheless, there is a future for rattan and it is closely tied to the health and wellbeing of the tropical forests.

5. Strategic Directions

Rattans were once abundant in the tropical forests of Asia but have become a scarce produce today. This is primarily because of two reasons – overexploitation and a shrinking forest area where they grow. Natural regeneration seems to be inadequate, and there is a general decline in the planting of rattan for various reasons – economic, technical and policy-related. There is an urgent need to rectify this situation and ensure the future for rattan, given its economic, ecological and socio-cultural importance to nearly a billion people in the developing world. Some strategic directions towards that end include:

- Sustainable management of the resource through actions that include the development of rattan plantations (both in agroforestry situations and in degraded forest areas, including shifting cultivation) and home gardens. Serious efforts are needed to address the problems of reckless harvesting, loss of productivity and poor management.
- Suitable conservation measures must be implemented urgently (De Zoysa and Vivekanandan, 1994; Rao and Rao, 1996). A number of Asian rattans are under serious threat from loss of habitat and over-exploitation. This will require an initial survey of available stock, hot spot areas, distribution of populations and current levels of exploitation.
- Continuous product and market development and formation of effective business partnerships.
- Strengthening of institutional support structures and government-private sector coordination, including financing schemes for small and medium enterprises. There is also an urgent need for policy interventions/support for macro economic and sectoral policies affecting the rattan sector.
- International tie-ups with furniture makers in the consumer countries to promote rattan products in the green market.
- Enhancing technology adoption and commercialization, and strengthening institutional networking to share knowledge.
- Continuing R&D in response to dynamic changes in markets and to address medium and long- term objectives.

REFERENCES

Abd. Latif, M., 2000. Production and utilization of bamboo, rattan and related species: management and research considerations. *XXI IUFRO World Congress 2000. Sub-Plenary Papers and Abstracts.* Vol. 1. pp. 393-406.

Aguilar, F.M. Jr. & Miralao, V.A., 1985. *Rattan furniture manufacturing in Metro Cebu. A case study of an export industry.* Ramón Magsaysay Award Foundation. Philippines. Handicraft Project Paper Series No. 6. 227 pp.

Anon. 1983. *Manual on the production of rattan furniture.* UNIDO. 108 pp.

Anon. 1991. *Research needs for bamboo and rattan to the year 2000.* Tropical Tree Crops Programme, International Fund for Agricultural Research. 81 pp.

Anon. 1998. *Assessment of socio-economic issues and constraints in the bamboo and rattan sectors.* INBAR Unpublished report. 22 pp.

Anon. 1999. A report of the workshop on the expansion of trade in rattan and rubber wood furniture. Unpublished Report. ESCAP. 14 pp.

Belcher, B., 1999. *The bamboo and rattan sectors in Asia: An analysis of production-to-consumption systems.* INBAR Working Paper No. 22. 84 pp.

De Zoysa, N. & Vivekanandan, K., 1994. *Rattans of Sri Lanka: An illustrated field guide.* Sri Lanka Forest Department and International Development Research Centre. 82 pp.

FAO. 1997. *Non-wood forest products: Tropical palms,* by D.V. Johnson. Bangkok, RAP Publication No.10. 166 pp.

Lakshmana, A.C., 1993. *Rattans of south India.* Bangalore, Evergreen Publishers. 180 pp.

Manokaran, N., 2000. Personal communication with the author.

Manokaran, N., 1990. *The state of the rattan and bamboo trade.* RIC Occasional Paper. No. 7. FRIM, Malaysia.

Nur Supardi, M.N., Hamzah, KA. & Wan Razali, W.M., 1999. *Considerations in rattan inventory practices in the tropics.* INBAR Technical Report No. 14. 57 pp.

Pabuayon, I., 2000. *Addressing rattan technology needs for Asia.* Paper presented at the XXI IUFRO World Congress 2000, Kuala Lumpur, Malaysia.

Rao, R.& Rao, A.N., eds. 1996. Bamboo and rattan genetic resources and use. *Proceedings of the Second INBAR-IPGRI Biodiversity, Genetic Resources, and Conservation. Working Group Meeting and Report on the Workshop on Rattan Resources and their Development in Indonesia held in Indonesia in 1995.* 77 pp.

Rawat, J.K. & Khanduri, D.C., 1999. *National report on the state of bamboo and rattan in India.* Unpublished report. INBAR. 31 pp.

Salleh, M.N., 2000. Personal Communication with the author.

Sarma, S.S., 1989. *Plants in Yajurveda..* Tirupathi (India), K.S. Vidya Peetha. 286 pp.

Sastry, C., 2000. *Bamboo in the new millennium: Opportunities and challenges.* Paper presented at the XXI IUFRO World Congress 2000. Kuala Lumpur, Malaysia.

Soedarto, K., 1999. *The state of bamboo and rattan development in Indonesia.* Unpublished report. INBAR. 33 pp.

Tombac, C.C., Lapis, A.B. *et al.,* 1993. *Indigenous people and rattan.* FORSPA Publication No. 5. 12 pp.

Wan Razali, W.M., Dransfield, J. & Manokaran, N., 1992. *A guide to the cultivation of rattan.* Malayan Forest Record No. 35. 293 pp.

POLICY AND INSTITUTIONAL FRAMEWORK FOR SUSTAINABLE DEVELOPMENT OF RATTAN

M.N. Salleh

1. Introduction

Rattan has been recognized as one of the most important non-timber products from the forest. Statistics abound to the fact that it is a unique plant with many excellent qualities. However, literature also abounds with statements that rattan is a depleting resource, that the natural forests are being depleted of their rattan populations. Natural regeneration appears limited, and rehabilitation by planting appears sporadic. This is reflected in the market by less rattan furniture, prices becoming more expensive and the quality declining. The author submits that this is a global problem and not a local or regional problem. As a result of the declining resource, the whole rattan industry is threatened. The result is the appearance in the market of substitutes, particularly in the form of plastics, and rattan furniture becoming beyond the reach of the common man. The danger is that, if rattan really disappears from the market, it is not only the furniture and related industries that would have lost a unique resource. Man would have lost a tradition that had developed over generations. The question that waits for an answer is, why are the rattan resources not managed on a sustainable basis?

2. Resource Depletion

The causes of depletion are straightforward, well known and well documented:

2.1 Loss of habitat

With the conversion of much of the tropical forests to other land uses, the habitat of rattan disappears as well, and so does the resource. The classic example is the disappearance of the lowland forests of Malaysia, mostly for conversion to industrial plantation crops of rubber and oil palm. While the author is unable to obtain statistics on actual loss of species, there would be definitely a loss of genetic diversity within the species that were endemic in these ecosystems.

2.2 Overexploitation

There has been in the past an overexploitation of this resource from the wild. The author has not been able to obtain recent statistics, but literature does indicate that this is happening in all the regions of the world, with no exception.

2.3 Inadequate replenishment

The rate of regeneration and replenishment by artificial means is inadequate. Here again, the author is unable to obtain figures, but he is confident that other papers presented in this consultation will provide statistics on the status of the three above issues.

2.4 Inadequate knowledge

While there has been increasing attention to research on rattan in the past decades, there are still many technical issues and challenges that need to be overcome, if the rattan industry were to truly develop. Much of the research of the past has been in the documentation and observation arena. While the information and knowledge generated are useful, there is little "developmental" research to solve the critical problems that need to be overcome to develop a sustainable and viable industry.

The result of the loss of habitats, overexploitation and inadequate replenishment is of course a depleting resource and its unsustainable use.

3. Institutional Issues

3.1 Policy

The author has never come across a policy in any country that states that the country's policy is to deplete its rattan resources. Policies are always positive and very encouraging. While rattan is sometimes not referred to as a specific resource, it is usually acknowledged as an important non-timber resource that must be conserved and utilized on a sustainable basis.

However, as with other non-timber forest products, the policy on rattan is subsumed within a larger national or state forest policy. The author has not come across any specific national or state rattan policy. Is such a policy necessary, or have forest policies adequately addressed the issues of sustainable rattan development? The author submits that a policy is only a statement of intent or broad desire. As long as the forest policy adequately addresses the issues of sustainable development and use of rattan, then it is adequate. However, if rattan is an important non-timber resource in the southeast Asian countries, the author suggests that the forest policies be reviewed to ensure that rattan is mentioned specifically, and that subpolicies be developed specifically for rattan. These policies should not only state broad desires but also commit governments to undertake adequate and appropriate measures to protect and develop the resource on a sustainable basis.

3.2 Strategies and plans for action

What is important is the commitment for action through the presence of definite plans and strategies. States should develop such plans and strategies for the sustainable harvesting and development of rattan resources. This should include the allocation of adequate resources, including human and financial, on an extended period of at least five years. The commitment over a longer period is desirable and important due to the need for planning and for maintenance of the crop after planting. Unless there is that commitment for action, there is no way that sustainable use and development of the resource will take place.

This is similar to what has happened and will continue to happen in forestry in the exploitation of timber resources. The fact that the tropical forests are declining and being "managed" in an unsustainable manner is basically due to the fact that there is no long-term plan and commitment at regeneration and rehabilitation. We all know the importance of planning in forestry. Yet, in practice forest managers do not practise what they preach. Of course, beautiful plans are of little use if they are not adhered to and put into practice. In this regard, political support is critical.

3.3 Research

Research on rattan needs to be changed from information generation to problem solving and new innovative initiatives, addressing the technical problems faced by past and potential investors in the rattan-growing business. Various authors have written on the need for research and fields suggested include species-site matching, light requirements, selection and breeding, development of genetic markers, genetic variation studies and manufacturing, processing and engineering.

While these areas are relevant and important, the author submits that one of the most important aspects of research is to reduce the period of the rosette stage of rattan. It has been reported that the rosette stage can last for more than two years, and yet even the *Guide to the cultivation of rattan* (FRIM MFR No. 35, 1991) has only one mention of the "rosette" in the whole book. Yet, the extended period of the rosette stage is very discouraging to the rattan grower as the rattan just "sits" there during that period without appearing to "grow". If the period of the rosette stage can be reduced, then the growth can be enhanced, and the frustration of the "stagnant" stage can be overcome.

Figure 12: Rotan factory in Jambi, Sumatra (Dransfield)

Problems arising from the climbing nature of rattan have been reported, such as breaking of the branches of the "host" trees and difficulty in harvesting. Can rattan be grown like sugar cane, growing straight up to a limited height and harvested before it starts climbing? If this can be done, it would revolutionize the growing of rattan. Problems of maturity of the cane, reducing taper and lengthening the inter-node length are issues that have to be addressed first. Research must be mobilized to maximize the use of modern biotechnology to solve these problems and others that may be related.

3.4 Certification

It is only a matter of time that all forests will have to be certified to get access to the international markets. The writing is on the wall. Non-wood products that reach the international market will soon feel the pressure of certification. Thus, it augurs well for the industry to look ahead and prepare itself for the day when certification is not an option but a necessity. Research must address the development of an appropriate certification scheme. The industry must be prepared, but in view of the current unsustainable situation, it would not be easy.

3.5 Financial incentives

If governments are serious about developing a sustainable rattan industry, they must provide adequate and appropriate financial incentives. What the form that these incentives should take would depend upon the resource to be developed, the characteristics of the industry and the location. Incentives for forestry in general have not been very effective, and a complete new look is necessary, if such incentives were to truly make an impact.

At the same time, there are international funding mechanisms that should be mobilized to address this issue of sustainable rattan management. The use of the Global Environment Facility (GEF) is a possibility, as biodiversity is one of the areas eligible for funding support under the GEF.

There are now discussions on trading biodiversity similar to efforts at carbon trading under the Convention on Climate Change. This will not be as straightforward as carbon trading, but can rattan biodiversity be traded on the international market under the Convention on Biodiversity? If so, the funds raised can be utilized to promote the sustainable management and development of the rattan resource and industry.

3.6 A champion is needed

In order for all these ideas and many more that will be presented during this consultation to succeed, they need a champion, one that believes in the cause and will commit to pursue the vision to its successful end. Where can the champion be found? Can the International Network on Bamboo and Rattan play that role? If so, how can it play that role? If not, why not and what are the alternatives? The author admits that he does not have the answer, but this consultation should seriously discuss this. In any case, the champion should not be a government agency. Government agencies just do not have the institutional ability to undertake such "developmental" activities. Rattan planting and development must be viewed as a business, and the involvement of the private sector is critical. At worst, it should be a joint venture with the private sector or an independent international agency with the capacity and tenacity to pursue this effort to a successful end.

One of the main reasons for lack of success in rattan planting and development in the past is the fact that institutions undertaking such activities have other primary responsibilities, such as the forestry departments or rubber plantation owners. Rattan is a minor crop within the total agenda of the institution and never gets the priority that is necessary. A dedicated institution at the national or state level, which is formed solely for the promotion of rattan, may be the only answer. A Rattan Development Corporation with seed money from government and with the involvement of the private sector may be able to play the "champion's" role.

4. Conclusions

The problem of sustainable development of rattan globally is an area of concern. This was one of the reasons why INBAR was established. The fact that this consultation is being held augurs well for rattan and INBAR, as there is a group of concerned citizens of the world who are willing to contribute ideas to the global effort. We need to mobilize this team to assist in championing the cause.

The problems are many but not insurmountable. A "champion" is needed to champion the cause and a dedicated institution such as a Rattan Development Corporation with the involvement of the private sector may be the answer. We need a friendly country to try these ideas and the financial resources to pursue the idea to a successful completion. INBAR and the Forestry Department of FAO should join forces to take this idea forward.

**IDENTIFICATION AND DISCUSSION
ON REQUIRED ACTIONS TO ENHANCE THE SUSTAINABLE
DEVELOPMENT OF THE RATTAN SECTOR**

KEY ISSUES FOR ACTION

Wulf Killmann

Introduction

A number of issues, connected with the rattan resource and its harvesting, processing and trade, require attention and action.

The resource

Ninety percent of the rattan stems processed today still come from natural forests. However, out of about 650 known rattan species, only 20 of them are used. Inventories of natural rattan resources are insufficient and, in most cases, non-existent. There are no standards for the measurement and assessment of the resource. With some small exceptions in Central Kalimantan, natural rattan resources not only are not managed, but they are even over-harvested. This leads to a partial shortage.

Meanwhile about 120,000 hectares of rattan plantations have been established. For a successful rattan plantation, important factors that must be considered are the selection of site, species and tree crops to be under-planted. So far experience has been gathered with under-planting tree species such as *Pinus* spp., *Acacia mangium*, *Hevea brasiliensis*, and with planting in logged-over forests. The envisaged double use of the plantations (host trees and rattans) has shown only limited success in practice due to harvesting problems (rattan spines) and weight of rattan (breaking of host tree branches). In some cases the conversion of rattan into oil-palm plantations was considered a more economic land-use than rattan growing.

Issues to be addressed:
- Shortage of rattan resources in some countries;
- Management of natural resources;
- Rattan inventories;
- Introduction of resource assessment and measuring standards;
- Introduction of sustainable harvesting methods;
- Regulation of harvesting;
- Selection of host vegetation for rattan plantations;
- Plantation establishment and management;
- Plantation economics.

Processing and trade of rattan

A considerable amount of the resource is lost due to wastage in harvesting, transport and processing. Some of this wastage is on account of mechanical damage to the stem, other to losses due to fungi and insects. Protective methods and processing techniques are well known and established. However, lack of training and lack of standards lead to additional losses during processing and to quality deficiencies of the end products.

Issues to be addressed:
- Reduction of wastage in harvesting, transport and processing;
- Training at all levels;
- Quality standards to be agreed upon;
- Standards to be introduced and monitored;
- Investment for training and equipment.

Enabling policy

Resource collection, processing and trade of rattan generate employment and income in both rural and urban areas in the producer countries. Rattan exports also provide foreign exchange earnings and thus contribute positively to the trade balance of these countries. With the dwindling supply of the natural resource and an impending decline of the rattan industry, policy measures are required to enable the sector's continuing contribution to local, regional and national economies.

Issues to be addressed:
- Regulation of ownership of, and access to, the resource;
- Introduction of licensing rules;
- Verification of beneficiaries;
- Introduction of incentives for resource management (in natural forests or in plantations);
- Promotion of downstream processing.

Conclusion

Natural rattan resources are declining in quantity and quality. Reduced quality of raw material and end products lead, on the one hand, to increased costs and prices for high-quality products and, on the other, to substitution of rattan by other natural and industrial materials. Export reduction results in a reduction of foreign exchange earned by this market segment and to a decrease in employment and income in the rural and suburban areas of producer countries. The health of the rattan sector thus has an impact on the socio-economic fabric of rural and urban society. Management, collection, processing and trade of the rattan resource cannot be seen in isolation, but rather the strengthening of this sector has to be integrated with other agricultural, forestry and small-scale industry concepts in rural and suburban areas.

CONCLUSIONS AND RECOMMENDATIONS

Based on the above presentations and ensuing discussions, the meeting identified the lack of management of rattan as the key problem. In addition, it was underlined by the participants that rattan is the flagship of NWFP, and the success or failure of the management and development of rattan would have a far-reaching impact on the management and development of other NWFP. The present state of failure in the management of rattan had resulted in the following: depleting resource, diminishing supply, increased threat to the rattan industry, disappearing genetic resource, reduced rural income and loss of opportunity for socio-economic and forestry development.

In order to address the above problems and related issues, three working groups were set up:
- Rattan supply issues from natural forests and plantations (*Chairman: Mr. Tesoro*);
- Socio-economic issues: rural incomes, industrial development and economic incentives *(Chairman: Mr. Belcher)*; and
- Environment and conservation of rattan (*Chairman: Mr. Vongkaluang*).

The task of the working groups was to revisit the issues raised during the meeting with regard to the respective topics and to prepare a report identifying key problems, possible solutions and strategies for action at the local, regional and international levels, including institutional and policy issues; and formulating recommendations for all relevant stakeholders (Appendix 3).

The results of the three working groups and their review in the plenary led to the following major conclusions and recommendations, the content of which was approved in principle in the closing plenary session. After the session, however, further editing was provided by the FAO Secretariat, and the edited version was circulated after the meeting so as to have full endorsement of the final text by all participants.

The participants agreed that this meeting was the first step of a process in the formulation and application of global concerted collaboration among key stakeholders in the sustainable development of the rattan sector. The meeting recommended that FAO and INBAR should bring the recommendations and conclusions of this meeting to the attention of their regional intergovernmental bodies for their review and possible endorsement.

A.   Key observations

Based on the papers presented and ensuing discussions, the meeting emphasized the economic, socio-cultural and ecological importance of rattan to a large number of people in the world and noted that rattan resources in their natural range of tropical forests in Asia and Africa were being depleted through overexploitation, inadequate replenishment, poor forest management and loss of forest habitats. There was a need to ensure a sustainable supply of rattan through improved and equitable management.

The meeting recalled that:
- there were approximately 600 species of rattans, of which some 10 percent were commercial species. Many species, including some of commercial importance, had very restricted natural ranges. The majority of the world rattan resources (by volumes and by number of species) were in one country – Indonesia.
- rattan was an important commodity in international trade. At the local level, it was of critical relevance for rural livelihood strategies as a primary, supplementary and/or emergency source of income. Rattan collection complemented agriculture in terms of seasonal labour and as a source of capital for agricultural inputs.
- the rattan sector was characterized by a variety of stakeholders with different needs and interests, such as rattan growers, raw material collectors, manufacturers and traders, and it functioned within a complex and dynamic socio-economic, political and ecological context. Rattan was gathered by unorganized or organized collectors, the latter either under contract or in debt relationships with traders and

farmers/cultivators. In addition, there was a loss of traditional rattan management practices and, at the same time, increasing competition for resources. Linkages between industry, traders, collectors/cultivators and research and development efforts were weak. Rattan manufacturing and trade were fragmented and diverse in size and markets, with a focus on export.

The meeting highlighted that taxonomic knowledge of species was patchy and available information conflicting. Likewise patchy was the knowledge of biological aspects, e.g. pollination and gene flow. In spite of the IUCN Red List review of 1998, the conservation status of rattans was not well known and it was difficult to assess and monitor. In addition, rattan species were assumed not to be "safe" in protected areas or in national parks, as harvesting in such areas was usually permitted or tolerated. It was also assumed that the genetic basis of rattan species was narrowing. Some species were likely to be at risk of extinction.

The meeting underlined that there could be no sustainable supply of rattan, if the forests in which they grew were not managed sustainably. In its natural habitat, rattan was not as yet managed, and rattan received low priority in national forest and conservation policies. There was no dedicated rattan development institution in any country as rattan was usually subsumed within the forestry services, and the few existing national rattan programmes were weak and with limited research and development capacity. With a few exceptions, national forest inventories did not include rattan, and information on the resource base was scarce. However, in large tracts of degraded and logged-over forests, (re-)introduction and management of rattan had the potential to complement significantly the value of these forests.

The meeting was informed that significant advances had been made in the understanding of growing rattan as a plantation crop. Although community-based or smallholder rattan gardens could be profitable in some situations, the profitability of industrial-scale plantations in Asia was currently uncertain, as other land uses were more lucrative. As a result of this, private-sector investment in industrial-scale rattan plantations had declined. The meeting took note that existing rattan plantations had been converted into more profitable crops like oil palm.

The meeting was further informed that rattan production was also affected by the low return to gatherers, resulting in weak incentives for sustainable rattan harvesting and management. A number of factors contributed to the low returns. Foremost among these were uncertain property rights, the dispersed nature of production and inconsistent cane quality. In several countries, prices and competition had been affected by the remoteness of collecting areas and poor transportation; "illegal" harvesting; poor market information; lack of organization among collectors; large post-harvest losses due to insect and fungal infestation; prohibitive tax policies; export barriers; and informal taxes that depressed raw material prices.

The meeting noted that international agencies such as INBAR, CIFOR, IPGRI, FAO and ITTO addressed rattan management, either directly or indirectly, within their programmes. National focal points for member countries of INBAR on rattan information had been established with the primary function to identify key stakeholders and their increasing involvement, to collect statistical data and exchange information in the local languages.

B.   Conclusions

In the light of the above, the meeting concluded that there was a wide variety of potential interventions that could assist the different stakeholder groups. Raw material producers and smallholders could be encouraged to, and assisted in, managing local resources on a more sustainable and productive basis, through the establishment of community forest management practices, long-term concessions, local land-use planning and the provision of resource and/or land tenure rights, in conjunction with approved management plans.

At the processing level, needs were particularly great at the artisanal level. Potential interventions that might assist industry included improving entrepreneurship and competitiveness; training of advisers; improving post-harvest treatment and quality control; market deregulation and improved market information; establishment of design centres; and trade fairs. Also, given the nature of the resource users and the industry being generally cottage and small scale, employing socially disadvantaged groups, rattan products could become ideal commodities for promotion as rainforest conservation products.

The meeting identified the following key actions to be initiated immediately for enhancing a more sustainable supply of rattan:

Resources:
- intensifying *ex situ* and *in situ* conservation efforts in a more coordinated and organized manner among countries in the regions;
- developing suitable methods for resource assessments, including studies on growth, yield, basic biology and taxonomy of rattan species;
- improving techniques of enrichment planting and management of rattan in degraded forests, and a wide dissemination of the available guidelines for rattan planting.

Products:
- increasing the knowledge of the properties of commercial species and of the potential of underutilized/ lesser known species;
- improving and disseminating technologies for reducing post-harvesting losses, biological deterioration, improved storage and processing;
- introduction of quality grading.

Policies and institutional support:
- awareness raising on the importance of the rattan sector to decision-makers at all levels;
- institutional strengthening and coordination regarding rattan conservation, management and processing issues, including the promotion of increased government and private sector cooperation /coordination to enhance the contribution of rattan to poverty alleviation and economic prosperity;
- providing tenure security to rattan gatherers and planters by incorporating them into community-based forest management schemes;
- introducing incentive schemes for rattan cultivation to increase the economic benefits for rural households and smallholder plantations in Asia, such as providing credit and technical assistance for small-scale plantation development and favourable harvesting and marketing arrangements;
- introducing market deregulation to benefit rattan collectors and traders (i.e. removing transport barriers; support for improved collection and dissemination of market information; extension in processing techniques);
- providing comprehensive training and support to local specialists in rattan-producing countries in taxonomy, management and processing, complemented with "twinning arrangements" among relevant institutions in the regions.

C. Recommendations

The expert consultation recommended for immediate follow-up:

- to FAO to:
  - develop consistent terms and definitions on rattan and its products;
  - harmonize existing measurement concepts and methodologies for rattan resources inventories and for accurate collection of statistics on rattan products.

- to INBAR to:
  - establish a list server on rattan;
  - conduct, with the assistance of CIFOR, a study on the economics of large- and small-scale rattan plantations;
  - conduct, with the assistance of FAO, a study on improving forest policies and relevant regulations with regard to rattan;

- commission a study on potential alternative market mechanisms to provide greater transparency and competition in rattan trade (e.g. auction mechanisms);
- update and publish the existing rattan bibliography on its web site.

The expert consultation recommended that the <u>governments</u> of countries with rattan resources be encouraged:

- At the <u>national</u> level to:
  - develop and implement a national rattan strategy involving all stakeholders in a participatory process;
  - include rattan as an integral component of national forest and conservation policies and forest management plans, giving due attention to rattan, where appropriate, in the national and regional processes on Criteria & Indicators for Sustainable Forest Management;
  - establish specific pilot projects focused on critical issues such as property rights and management institutions, opportunities and constraints to community-based resource management, and post-harvest treatment;
  - strengthen national research programmes/activities through enhancing the network of rattan research and development activities, including the establishment of "rattan scholarships".

- In support to actions at the <u>international</u> level to:
  - commission the development of a five-year international rattan development programme with the primary objectives of promoting and undertaking rattan development activities with partner institutions in the various regions and strengthening global networking in rattan research and development. This international programme would enhance national institutional capabilities and examine the possibilities and merits of INBAR establishing/strengthening nodal point(s) in national institutions as permanent focal point(s) to continue long-term programmes on rattan research and development;
  - revive the Rattan Information Centre (RIC) established in 1982 in Malaysia;
  - support awareness-raising campaigns on conservation, management and processing of rattan (e.g. impacts of insufficient taxonomic/biological knowledge of rattan conservation issues) with senior policy and decision-makers at international development, conservation, research and funding agencies, as well as senior government officials in rattan producing/consuming countries.

The expert consultation emphasized the potential of enhancing regional cooperation through information exchange; collaborative research and development; training; and material exchange to promote rattan as a vehicle for achieving social, economic and environmental sustainability in rattan-producing countries. To this end, the expert consultation called for a concerted effort of governments, the private sector, NGOs and relevant international agencies.

APPENDICES

APPENDIX 1

LIST OF CONTRIBUTORS

Aboitiz Booth, Josephine. President, Mehitabel Furniture Co.Inc., P.O. Box 331, Cebu City, Philippines – e-mail: Jbooth@Mehitabel.com.ph

Altarelli, Vanda. FAO, Rural Sociologist, Investment Centre Division, Viale delle Terme di Caracalla, 00100 Rome, Italy – e-mail: Vanda.Altarelli@fao.org

Amaral, Weber. IPGRI, Forest Genetic Resources Coordinator, Via delle Sette Chiese, Rome, Italy – e-mail: w.amaral@cgiar.org

Belcher, Brian. CIFOR, Programme Leader, FPP, PO Box 6596, Jakarta, Indonesia - e-mail: b.belcher@cgiar.org

Dransfield, John. Herbarium, Royal Botanic Gardens, Kew, Richmond, Surrey TW9 3AB, United Kingdom – e-mail: J.Dransfield@rbgkew.org.uk

Evans, Tom. Oxford Forestry Institute, South Park Road, Exford OX1 3RB, United Kingdom – e-mail: Tom.evans@plantsciences.oxford.ac.uk

Hunter, Ian. INBAR, Director-General, PO Box 9799, Beijing, People's Republic of China – e-mail: ihunter@inbar.org.cn

Hwan-Ok Ma. Project Manager, Forest Industry, ITTO, Pacifico-Yokohama, 1-1-1 Minato-Mirai, Nishi-ku, Yokohama, Japan – e-mail: itto@or.jp

Killmann, Wulf. FAO, Director, Forest Products Division, Viale delle Terme di Caracalla, 00100 Rome, Italy – e-mail: Wulf.Killmann@fao.org

Kollert, Walter. GTZ, c/o National University, Faculty of Forestry, P.O. Box 5653, Vientiane, Lao People's Democratic Republic – e-mail: profep.wk@laonet.net

Liese, Walter. Chief Technical Adviser, Institute for Wood Biology, University of Hamburg, Leuschnerstr.91, Hamburg, Germany – e-mail: Wliese@aol.com

Lou Yiping, INBAR, Acting Program Manager, Ecological Security, P.O. Box 9799, Beijing, People's Republic of China – e-mail: yplou@inbar.org.cn

Løyche Wilkie, Mette. FAO, Forestry Officer, Forest Management, Forest Products Division, Viale delle Terme di Caracalla, 00100 Rome, Italy – e-mail: Mette.LoycheWilkie@fao.org

Nielsen, Kaare Riis. IFAD, Asia Division, Via del Serafico, Rome, Italy – e-mail: k.nielsen@ifad.org

Razak, Abdul. Director-General, Forest Research Institute Malaysia, Kepong 52109, Kuala Lumpur, Malaysia – e-mail: razak@frim.gov.my

Renuka, C. Kerala Forest Research Institute, Peechi 680 653, Thrissur District, Kerala, India – e-mail: Renuka@kfri.org

Russo, Laura. FAO, Forestry Officer, Non-Wood Forest Products, Forest Products Division, Viale delle Terme di Caracalla, 00100 Rome, Italy – e-mail: Laura.Russo@fao.org

Salleh, Mohammed. Director-General, TropBio Research Sdn Bhd, 338 Jalan Tuanky Abdul Rahman, Kuala Lumpur, Malaysia – e-mail: Mnsalleh@pd.jaring.my or tropfor@po.jaring.my

Sastry, Cherla. Scientist, Kerala Forest Research Institute, University of Toronto, 1601, 1 Reidmont Avenue, Scarborough, ONT M1S 4V3, Canada. e-mail: Csastry@home.com

Siebert, Stephen. Professor, School of Forestry, University of Montana, Missoula, Montana 59812, USA – e-mail: Siebert@forestry.umt.edu

Silitonga, Toga. Senior Researcher, Forest Products Research Institute, Jl. Gunung Batu No. 5, Bogor 16610, Indonesia – e-mail: FAO-IDN@field.fao.org

Skagerfält, Joacim. Desk Officer, Department for Natural Resources and the Environment, SIDA, S-105 25 Stockholm, Sweden – e-mail: joacim.skagerfalt@Sida.se

Sunderland, Terry. African Rattan Research Programme, Programme Manager, c/o Limbe Botanic Garden, BP 437, Limbe, Cameroon - e-mail: Afrirattan@aol.com

Tesoro, Florentino. Director, Forest Products Research and Development Institute, UPBL College. 4031, Laguna, Philippines - e-mail: Ftesoro@laguna.net

Tong, Hong Lay. IPGRI, Specialist, Rattan, Bamboo and FGR, Via delle Sette Chiese, Rome, Italy – e-mail: l.hong@cgiar.org

Vantomme, Paul. FAO, Forestry Officer, Non-Wood Forest Products, Forest Products Division, Viale delle Terme di Caracalla, 00100 Rome, Italy – e-mail: Paul.Vantomme@fao.org

van Valkenburg, Johan. Tropenbos Foundation, Specialist, NTFP, c/o PROSEA Office, P.O. Box 341, 6700 AE Wageningen, Netherlands – e-mail: johan.vanvalkenburg@pros.agro.wau.nl

Vongkaluang, I. Associate Professor, Department of Forest Biology, Kasetsart University, 50 Phahonyothin Road, Chatuchak, Bangkok 10900, Thailand – e-mail: Fforisv@nontri.ku.ac.th

Walter, Sven. FAO, Forestry Officer, Non-Wood Forest Products, Forest Products Division, Viale delle Terme di Caracalla, 00100 Rome, Italy – e-mail: Sven.Walter@fao.org

APPENDIX 2

SUMMARY NOTES PREPARED BY THE SECRETARIAT

With the purpose to facilitate the discussions and deliberations of the meeting, the following draft notes, with a summary of key problems as identified in the background papers for the meeting, may help participants to identify and prioritize main issues to be highlighted in the outcome of the meeting:

<u>**Resources**</u>:

> **cane** and shoots (focus of this meeting is on cane, while recognizing the potential of shoots as food supplement at local level)

> what: 13 Genera; 600 species; Table 1: Rattan species table
>   Need for thorough taxonomic knowledge and rigour (at regional level)
>   some species close to extinction (at national- regional level)

> where: Tropical Asia: in order of importance: Indonesia, Malaysia, Phillipines, Papua, Indochina; South Asia; Australia; Central and West Africa (Table 2)

Inventory of rattan resources? resource surveys? Only few countries have inventory of rattan resources (with different methodologies and results are in different measuring units; lengths, tonnes, m^3)

Rattan cane is collected from: not even these figures (or estimates ?) are sure !!!

> 90% from forest (with rattan quality and sizes in decline, and some species becoming endangered (forests are overharvested or unmanaged for rattan)
> 10 % from plantations
> > What type of plantation schemes (indigenous, small growers, industrial....?+ what type of property right arrangements; management rules; licences
> > (potential of Latin America for rattan introduction) ???

Resource management:
Silviculture of natural forests for rattan: not applied (some enrichment planting in logged-over forests, but rattan management is not included into silvicultural systems)

Domestication of species: ongoing and key element is site selection and species/site match- propagation techniques (wildlings, seeds, +tissue culture); in *situ; ex situ* resource conservation ("rattan reserves")

"Lessons-learned" on rattan plantation establishment and maintenance are maturing
Industrial scale rattan plantation presently less economical attractive (low cane price)
not perceived as a problem in case of rattan cultivation for shoots (locally) or at the small holder level

Mixed rattan tree crop plantations: ongoing (but technical and economic conflicts with other crop: rubber; oilpalms, foodcrops...or damaging to other crop)

Plantation of rattan in degraded forests: "starting" (but here also the risks that the forest/rattan plantation can be converted to oilpalm or other financially more rewarding agriculture crops)

Importance also of Environmental role of rattan (control of soil erosion with clumps species); but also negative environmental impact (residual cane left in crows; cutting of floater logs...)

Harvesting

Rules and legislation governing the harvesting of rattan

Do they exist? are they applied? controlled by whom? (by forest service? or through community systems, private sector....?)

Limiting to allowable cut if known; ascertaining sustainable extraction levels?

Governance of prices paid to harvesters for raw cane (avoiding monopoly by cane buyers)

In natural forests and/or open access based resources: restricting/managing harvesting rights/ (long term rattan cutting licences; <10years?; payment of royalties for harvesting to who and for what?

Need for a "rattan fund (deposit)" for funding rattan plantations/enrichment planting

Role of national forest service for assigning harvesting tenders or assigned rattan cutting rights to local communities; preference being given to assigning harvesting rights to local indigenous communities/forest dependent people

Technical constraints of harvesting

Improve harvesting technology (increase pulling power devices) or others to avoid high wastage (from unharvested cane left in tree crowns till rejected/unsuited cut cane sticks left in the forest)

Improve transport facilities for raw cane from harvesting site till sales point (+ uncertainties of "check points" (including unofficial payments) along transport route from forest till road site/first processing for transporter/harvester

There is an increasing road access to more and more remote forest areas for rattan harvesting

Environmental impact of cane harvesting and transport: (f.ex. cutting of floater logs for river transport of cane sticks)

Processing:

High wastage during cutting to size, losses to insect/fungi attack (30%);
Dipping with preservative or diesel oil (pollution risks);
Proper seasoning (harvesting in dry season only for better drying);
Curing;
Grading (based on dimensions and surface). "Rattan stick grading rules" exist (Bhat 1996) but not widely used.

Technological constraints: - many rattan species but only few are commercial species (approx. 20). Anatomical properties can be a basis for identifying commercial uses

Wide range of processing techniques from artisanal/village level till industrial plant level; but at all levels availability of skilled labour for processing into furniture is very difficult to recruit.
Rattan industry is a highly fragmented - cottage level industry
Rattan Uses: (at the industrial level) mainly (or only) furniture?
Can rattan furniture design cope with "changing" raw material (more smaller diameter cane).
For non-timber furniture, rattan is technically superior to other plant products

Trade:

Poor knowledge of importance of rattan for subsistence use; and local rattan trade

Lack of (reliable) statistic's on rattan production and trade (f. ex. "total value of rattan" estimated at 7,000 M$ annually, of which 3,000 M$ at subsistence/local level and 4,000 M $ for international trade) (Table 3). However this is a much disputed estimate.

Available figures mix and add raw material, semi-processed and values of furniture, raw cane, semi-processed.

No socio-economic quantification of the rattan sector by country exist?

Competition in cane sticks for local trade and processing or for export (importance of imposing/lifting export bans for raw/semi-processed); Supply of raw cane limited and uncertain for importing – intermediate/processing countries; illegal and/or unrecorded trade in rattan cane (smuggling)

Rattan trade and production <u>trends</u> increased till mid '90 and then decreased or remain equal

Industry: rattan industry growth is declining since 1990? (shortage of quality raw material; highly fragmented industry with low "lobbying" marketing and financial power; single market as use is mainly furniture? Prefabricated rattan furniture pieces are bulky products with relatively low value: declining interest by large furniture stores (Japan); competition from substitutes: plastic/wicker; …)

One country dominates world supply of cane: Indonesia

Certification of rattan? (guidelines in preparation)

Emerging market: international marketing of rattan shoots as an oriental food delicatesse

Policy issues:

Lack of specific references to rattan (or rattan rules) in the existing forest policy and legislation

All countries have promotion of added-value locally - but these measures are not accompanied with resource development policies
-        export bans (rattan sticks- semi-processed) (illegal trade – smuggling)
-        elimination of export taxes for finish rattan products
-        provision of economic incentives for resource cultivation/management if exist (but lower then those for rubber/oil palm plantations)
foreign investments (from resources till processing) allowed?
protectionism affecting free movement of rattan seeds

Need for the set-up of national rattan development boards in major producers

Social aspects:

Rattan is a key product for supporting livelihoods of rural/forest dwellers (often with no substitute available to these people once the cane is gone), including for food (shoots)

Not well know yet: the impact of "devolution" of forest lands to be managed by rural communities according to their "rules"; if this will be beneficial to the sustainability of rattan resources or if it will lead to a " boom and burst" cycle.

The impact of government decisions/incentives/bans on rattan resources access and rattan processing rights on livelihoods of forest dependent local people. In many cases even the more powerful wood industry has failed to raise incomes of forest dependent peoples based on sustainable forest operations (Indonesia)

Preference be given to forest dependent local communities first? or "open" the forests to any "investor"?

Younger people no more interested in rattan gathering/processing (as time consuming/poor rewarding and better job opportunities exist; mostly of interest to older people (Thailand)

High competition of rattan growing with other "cash crops" (like oil palm, coffee, etc.) and between plantation grown rattan versus wild gathered ? as rattan gathering is often a secondary job!

Institutional :
Weak representation of the " rattan sector" within the existing institutional set-up of countries
Lack of or conflicting data on resources, production value export, no uniform measuring units or conversion factors making regional aggregation of data on rattan impossible. No systematic and reliable figures available at regional level on rattan resources, production and trade (different measuring units, levels of processing from raw material till semi finished; or very frequently even conflicting data).

Rattan research more focused on "information gathering" rather then "problem solving" (ex.: How to shorten the rosetta stage of rattan?)

International agencies and stakeholders on rattan: INBAR
also: Rattan Information Centre (Malaysia); African Rattan Research Network, IDRC, IUFRO, FAO, IFAD, ITTO, ADB.

Table 1. Distribution of rattan genera and species by country and region

Countries/Region	Genera														
Region	Ca**	Co.	Ce.	Da.	Er.	Ko.	La.	My.	On.	Pc.	Po.	Pg.	Re.	Ng.	Ns.
China	+			+						+				3	40
Indochina*	+			+		+				+				4	33
Thailand	+		+	+		+		+		+	+			7	50
Myanmar	+			+		+		+		+				5	30
India	+			+		+				+				4	46
Philippines	+			+		+				+				4	54
Malaysia	+	+	+	+		+		+		+	+	+		9	104
Java	+		+	+		+				+				5	25
Kalimantan	+		+	+		+				+	+	+	+	8	105
Sumatra	+		+	+		+		+						5	75
Celebes	+			+		+								3	28
New Guinea	+			+		+								3	50
Sri Lanka	+													1	10
Solomon Isl.	+													1	3
Australia	+													1	8
Western Africa	+				+		+		+					4	24
Ns per genus	400	1	6	115	7	26	7	1	5	16	5	3	1		
Total														13	600

Notes: * The data cited from references 1, 2, 3, 9, 10 and 11 .
   ** Vietnam, Laos and Kampuchea are included.
   *** Abbreviations:
Ca. = Calamus, Co. = Calospatha, Ce. = Ceratolobus, Da. = Daemonorops, Er. = Eremospatha,
Ko. = Korthalsía, La. = Laccosperma, My. = Myríalepis, On. = Oncocatamus, Pc. = Plectocomía,
Po.= Plectocomíopsis, Pp. = Pogonotium, Re. = Retispatha,

Ng = Number of genera
Ns = Number of species

Source: Research on Rattans in China, IPGRI INBAR2000.

Table 2. Rattan resources in Africa and Asia-Pacific

Africa:

All locations in Africa where rattans have been recorder to date. The area within the circles indicate important areas where rattan still remains to be collected.

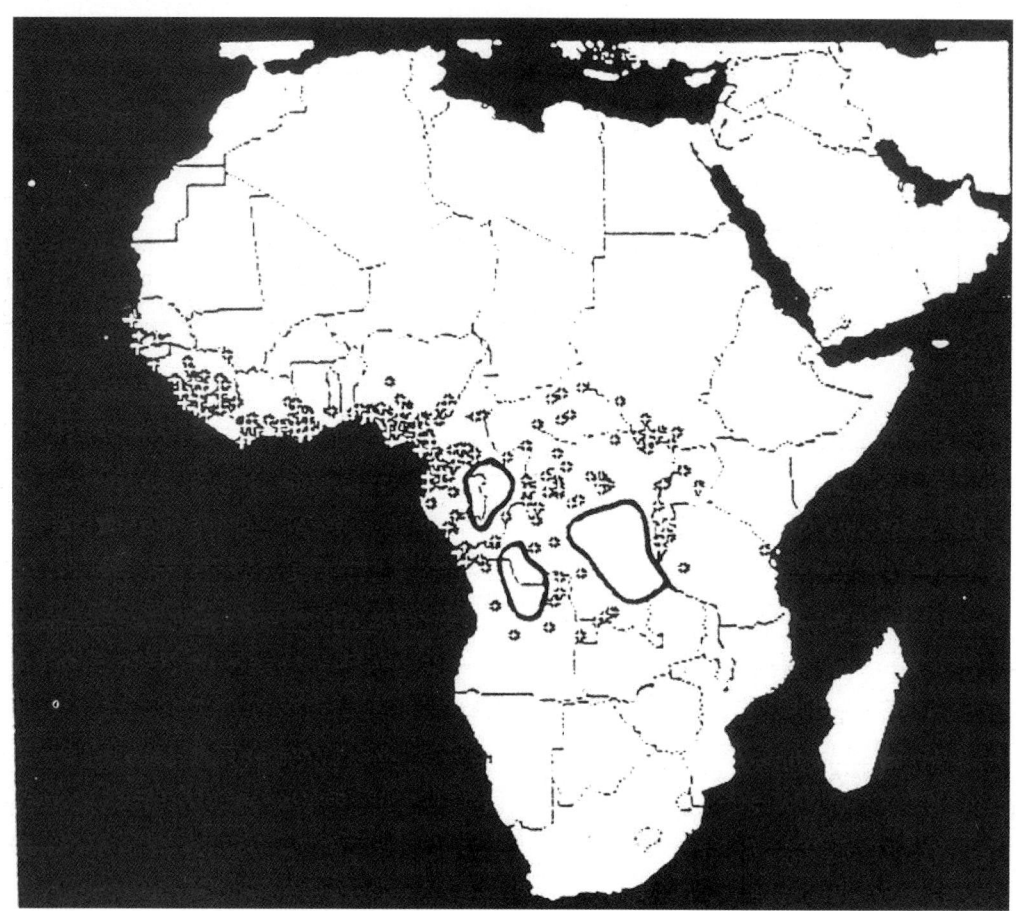

Source: Morakinyo, 1995, The Rattans of Nigeria

Asia:

in order of importance of resources

Indonesia, Malaysia, Papua New Guinea
Philippines, Laos, Cambodia
Myanmar, Vietnam, India, Australia
Bangladesh, Nepal, Sri Lanka, China, Thailand,
Pacific Islands (Solomon)

Table 3. Country table on rattan resources, production and trade

Country	Forest area (Mill. ha)*1	Forest Rattan areas (Mill. ha)*3	Plantations (1000 ha)*3	Rattan production (1000 Tons/year)*3	Rattan trade Volumes/ Values *3
Indonesia	103.6	11.5 – 3 30 Degr.forest area	37	**570** (80-85-90% World mark) 10% from plantations prod. Capa: 12,400 T (700 AAC	Furniture: (1999) 590.021T/ 1.147 B$
Malaysia	15.3		31 Sabah: 23 ('96) 15 in natural forests	500 mills 21.7 M $	Furniture: 24 M$/y (1999)
PNG	36.9 - 36				
Philippines	6.5 - 5.4	3 – 1.7	6 - 11	Stock: 5.5Bm (97) Pr. 108mlm + imp: 112T (98)	Furniturexp: (1999) 112.89 M$
Laos	12.4 - 9.5	2.2		**0.1**	Exp: 1.6Mm (1998) LD 117.503$
Vietnam	7.6 - 8.3		25pl + 60 forest	**144 (86-92) – 25**	25 t/y demand 35M$ exp/y
Cambodia	9.8	18.2% of forest area			
Myanmar	26.8				
India	50.3 - 51				
Bangladesh	0.7			3.5 Mill. Ft (81-87)	
Nepal	4.7				
Sri Lanka	1.6		0.248+0.146		80.000$ Exp 1986)
China	99.5*2		20	4-6 (=90% prod)	Demand: 30.000T M$ 100 –exp: 60%
Thailand	11.1		.5		Imp: 79Mbath Exp: 120 Mbath
Cameroon	19.5				
Gabon	17.8				
Congo	19.5				
Nigeria	13.6			180.000m/month (Lagos)	1.1 M$/y val
Eq. Guinea	1.7				
Cuba	1.5		2		
Total				7.000 M $ /y Value	

*1: First Figure: Area of Natural Forest (1995) – FRA 2000
Second Figure: "Forest" area as given in background papers
*2: Only a minor fraction is tropical forest
*3: As given in background papers (eventually different figures by country)

APPENDIX 3

OUTPUT OF THE THREE WORKING GROUPS

WORKING GROUP 1: Rattan Resource

Rattan was once abundant in the tropical forests (of Asia) but has become a scarce resource today. The causes of depletion are straightforward, well known and well documented – loss of habitat, overexploitation, inadequate replenishment and poor management of the resource. There is an urgent need to rectify the situation and ensure the future for rattan, given its economic, ecological and socio-cultural importance to nearly a billion people in the developing world.

Outlined below are some specific problems facing the resource supply and suggested actions/ solutions:

From natural forests

Forest inventories in most countries do not include rattan. This needs to be rectified and suitable methods developed for rapid appraisal, including estimates of growth and yield. Taxonomic inventories should be an essential component of this important activity. Forest management plans should include rattan as an integral component where appropriate.

There is very little involvement of forest communities and indigenous people in the management and development of rattan resource. Inclusion of these people and giving them long-term user rights would go a long way in the sustainable management of the resource.

Large tracts of degraded and logged-over forests remain under-utilized. These are potential areas for regeneration of rattan in their natural habitat. Improved techniques for enrichment planting are needed, including development of better planting materials. *In-situ* conservation of genetically rich areas would go a long way in enhancing the availability of high-quality planting materials.

Enormous wastes occur during harvesting of rattan and improved harvesting techniques and regulations would contribute to the reduction of these wastes. In addition, adequate treatments against biological deterioration should be developed and applied close to the collection points as soon as possible after harvest to reduce wastage, improve quality and increase revenues to gatherers.

Present supply comes from about 25 species out of the 650 species found in the rattan-producing areas. The potential of underutilized/lesser known species must be studied on a priority basis to enhance the supply of the resource.

From plantations/rattan gardens

Although significant advances have been made in the understanding of rattan as a potential plantation crop, there is still much that is unknown. Though both large- and small-scale plantations in Asia are returning profits other land uses are becoming more lucrative. As a result of this and other reasons such as lack of technical and policy-related support, private-sector cultivation of rattan, from both large- and small-scale plantations, has fallen below expectations and failed to respond to local raw material scarcities. There is a need for more appropriate government interventions in enhancing rattan cultivation given the economic benefits that accrue to the rural households in Indonesia and smallholder rubber plantations in Malaysia. Thus domestic policies must support rattan plantation development by improving incentives structure. This involves providing tenurial security to rattan gatherers and planters, credit and technical assistance for plantation development, and favourable harvesting and marketing arrangements. In addition, incorporation of plantations into community-based forest management schemes, with or without vertical integration in processing, could be an important policy direction.

An initial step in this direction is to review and document existing successes/failures of rattan plantations.

Strategies for action

A first priority is to create, where necessary, and/or strengthen the national institutional support system to address the above issues.

A second priority is to ensure that forest policies address sustainable management of rattan. In addition, more government and private sector cooperation/coordination should be promoted to enhance the contribution of rattan to poverty alleviation and economic prosperity.

Regional cooperation should be enhanced to promote rattan as a vehicle for achieving social, economic and environmental sustainability through information exchange, collaborative research and development, training and material exchange. This could be achieved through the establishment of an internationally supported Regional Rattan Research and Development Centre located in the resource-rich area. In addition to providing rattan management and development technologies, the centre could also provide the needed manpower training for other regions. In time this could evolve into an international expert centre for rattan. A first step in this direction could be a five-year regional rattan development programme to initiate the activities.

Given its long and successful forestry experience and stature in both the donor community and the developing countries, FAO should champion this initiative, with support from other international agencies (ITTO, UN, IFAD), relevant CGIAR centres (CIFOR, IPGRI, ICRAF) and INBAR.

WORKING GROUP 2: Socio-Economic Issues

Rattan is widely recognized as an important commodity, indeed the "flagship" NTFP for income and employment generation at many levels. This includes raw material production, transport, trade, processing, manufacturing and export. Rattan is critically important in rural livelihood strategies as a primary, supplementary and emergency source of income and as a source of capital for agricultural inputs. Rattan collection complements agriculture in terms of seasonal labour and is especially important for young households as a bridge to other livelihood activities.

Rattan collection, trade and manufacturing function within a complex and dynamic socio-economic, political and ecological context. Crucial components include: geographic centralization of manufacturing capacity (e.g. Java and Cebu City); poor communications and infrastructure; ethnic, religious and social differences (particularly in Indonesia); and the low priority of rattan among national governments. Indeed national governments have often functioned as a barrier to effective market functioning and resource management. As a consequence, since the colonial era traditional management practices and social institutions crucial for sustainable resource management have broken down or been usurped by the state. This problem has been exacerbated in recent decades by increased market penetration; increasing competition for resources from outsiders; and weak linkages between industry, traders, collectors/cultivators and research and development efforts. Nevertheless the potential exists to increase the profitability of the rattan, particularly in the international furniture market by improving production quality and quantity.

The rattan sector includes a diverse array of stakeholders with variable needs and interests. Key stakeholders include raw material produces, traders and processors/manufacturers. Raw material producers include unorganized collectors of wild rattan, organized collectors under contract or in debt relationships with traders and farmers/cultivators. Trade includes small- and large-scale operations and those focussed on raw material exports. In the processing and manufacturing sector there are small and large producers and marketing foci on domestic and international markets.

Key issues

The primary problem confronting rattan production is the low return to producers, especially among "contract" harvesters in patron-client relationships. This situation results in weak incentives for sustainable rattan harvesting and management. A number of factors contribute to the low returns. Foremost among these are weak property rights regimes. Other important contributing factors include: the dispersed nature of production and inconsistent cane quality, which contributes to weak bargaining power by collectors. Government policies such as prohibitive export taxes and bans have depressed prices and competition in several countries. Prices are also suppressed by the remoteness of collecting areas and poor transportation; illegal harvesting (i.e. harvesting in historic or traditional areas which are now designated concessions or protected areas and thus "need" to pay bribes); poor market information; market failure; lack of organization among collectors and large post-harvest losses due to insect and fungal infestation.

Rattan production is characterized by overharvesting and destructive harvesting practices, the reduction in resource availability due to forest conversion and the fact that collecting is typically an activity of young men who have limited long-term commitment or interest in collection. Rattan collection and cultivation is also characterized by increasing competition from other livelihood opportunities and the failure of the market to acknowledge or incorporate its role and importance.

Given this situation, there are a wide variety of potential interventions that could assist the different stakeholder groups. Raw material producers could be assisted through the establishment of community forest management practices, long-term concessions, local land-use planning and the provision of resource and/or land tenure rights, in conjunction with approved management plans. These and other reforms to tenure and institutional conditions could provide resource users with an incentive to manage local resources on a more sustainable and productive basis. Raw material producers could also benefit by deregulation (i.e. removing transport and export restrictions; support for improved collection and dissemination of market information; extension of methods to reduce post-harvest losses; improve storage; and support for local collector organizations).

Key issues faced by traders include high risk, high costs (due to checkpoints) and poor market information. These might be remedied through improving post-harvest treatment, market deregulation to reduce corruption and improved market information.

At the industry level, needs are particularly great at the low end where the industry is characterized by lack of entrepreneurship, poor design, inefficiency, low quality, increasing competition and substitution, lack of support and lack of market formation. Potential interventions that might assist industry include improving competitiveness via the establishment of design centres, the training of advisers, trade fairs and greater market research.

Conclusions and recommendations

There is a need to promote national rattan strategies in producer countries on a participatory basis, involving all stakeholders. These efforts could lead to the establishment of pilot projects focussed on critical issues such as property rights and management institutions, opportunities and constraints to community-based resource management and post-harvest treatment. It is recommended to focus on some countries, including Indonesia, the Philippines and Cambodia, with others such as Malaysia and Cameroon to follow. This effort will need the assistance of a strong coordinator and leader, which it is suggested should be undertaken by strengthening the personnel capacity in INBAR with assistance and support from FAO and ITTO, as required.

WORKING GROUP 3: Environment and Conservation
Focus on conservation

Preamble – There are some 600 species of rattans, of which approximately 10 percent are commercial species representing less than 10 percent of the total growing stock of rattans in the forest. There is a high level of endemism and the majority of the world rattan resources (by volumes and by number of species) are in one country – Indonesia.

Problems (more important at the global than at the national level)
At the **global level** (Africa, Asia and Pacific)
1. Patchy taxonomic knowledge; and unresolved or conflicting taxonomic information about species delimitation, as was presented in the background papers.
2. Conservation status of rattans is not well known despite the IUCN Red List Book review of 1998.
3. Rattan areas are being seriously affected by habitat loss and fragmentation.
4. It is assumed that there is a Narrowing Genetic basis of rattan species (but with no scientific evidence yet available to proof it, neither is it sufficiently well known).
5. Little is known about the basic biology of pollination and gene flows.
6. Negative impact of forest disturbance factors (from outside the "rattan" sector); for example, overhunting (Africa) or lack of hunting (Asia) of seed dispersals, such as wild pigs, causes lack of regeneration.
7. Logistical (and methodological) difficulties to assess/monitor the conservation status of rattan species, including lack of an agreed-upon "baseline".
8. Overharvesting of commercial species is causing resource depletion.
9. Rattans are not "safe" in protected forest areas (or in national parks), particularly in view of the fact that rattan harvesting (as well as the gathering of other NWFP) in such areas by indigenous/local communities is permitted/tolerated (perceived to be more the case in Asia than in Africa).

Solutions (strategy) - Recommendations (in the short term)
1. Regional-based strategy for institutional strengthening and capacity building in all aspects of improving knowledge of rattan taxonomy in the relevant countries. There is also a need for basic biological knowledge and information on yield, growth and inventory; comprehensive training and support to "targeted" key agencies (and local experts) in rattan-producing countries, complemented with (interregional and intraregional) "twinning arrangements" among rattan taxonomic programmes; academic and expert exchange programmes; reference collection centres; coordination of further needed taxonomic field work; curriculum development; extension and elaboration/wide dissemination of user-friendly rattan taxonomic aids, etc.
2. Awareness-raising campaigns on the impact of insufficient taxonomic/biological knowledge of rattan conservation issues in general (and of the conservation status of commercial rattan species in particular) regarding the above-mentioned problems, particularly No. 8 and 9), targeted:
    a) at the **global level** to senior policy and decision makers at international development, conservation, research and funding agencies (who are not aware of the critical resource depletion level of commercial rattan species as compared to total available stock of all rattan species), including the need for taxonomic knowledge and rigour as an essential basic requirement in such fields as:
        i.   rattan resource inventory for assessing available commercial stock and AAC (forest rattan inventories to be specific on commercial species stock and yield) ,
        ii.   rattan certification schemes (for those focussing on the sustainability of the resources of the species commercially used)
        iii. biological and technological research, etc.
    b) at the level of **national** governments of rattan producing/consuming countries, to senior government officials in relevant ministries such as forestry, environment, industry and trade in:
        i.   the same topics as above, including the need for institutional strengthening and ministry coordination regarding rattan conservation issues;
        ii.   the need to intensify *ex-situ* and *in-situ* conservation efforts in a more coordinated and organized manner.
3. Standardization of assessment methodologies for quantifying the resource base.